Introduction to
Real Analysis

Introduction to Real Analysis

John DePree

New Mexico State University

Charles Swartz

New Mexico State University

WILEY

John Wiley & Sons

New York • Chichester • Brisbane • Toronto • Singapore

Library of Congress Cataloging in Publication Data:

Depree, John D., 1933–
    Introduction to real analysis.

    Bibliography: p. **347**
    Includes index.
    1. Functions of real variables. 2. Mathematical
analysis. I. Swartz, Charles, 1938–   . II. Title.
QA331.5.D46   1988   515.8   87-34622
ISBN 0-471-85391-7

Printed in the United States of America

10 9 8 7 6 5 4 3 2 1

Preface

Traditionally there has been a separation of advanced undergraduate and beginning graduate real analysis into two courses, "Advanced Calculus" and "Introduction to Real Analysis," according to the presumed needs of the students; the former was taken by students in engineering and other applied areas and the latter was taken by students whose primary interest lies in the mathematical sciences. At our institution, these courses have coalesced into a course called "Introduction to Real Analysis," and it contains substantial portions of each of its predecessors. The reasons for this merger are partly to effect an economy of scale, partly because of the increased sophistication of the needs of those interested in the application and use of mathematics in their own primary areas, and partly because of the insufficient mathematical preparation of many students. This text evolved from a set of lecture notes that was written for this new course.

We try to begin gently by reviewing and extending some familiar ideas such as the set of real numbers and the convergence of sequences and series of real numbers. The student should have gained familiarity with this introductory material during his or her presumed work in the calculus. The material is very

v

computational in nature and offers an opportunity to become acquainted with
the level of rigor that will be demanded. Euclidean space is introduced early,
and analysis in this n-dimensional space is stressed throughout the first third
of the text. The study of continuity and differentiation in $\mathbb{R}^n$ is carried out
with minimal frills by using the sequential methods introduced earlier; the
emphasis is on describing the main results quickly, thoroughly, and efficiently.
This approach should benefit students in disciplines such as statistics, operations research, and engineering, for such students often take only one
semester of analysis in the hope that it will adequately provide them with the
background necessary to understand the mathematical analysis encountered in
these applications. A particular aid to these students should be the section on
optimization in $\mathbb{R}^n$, which includes the Kuhn–Tucker theorem on Lagrange
multipliers.

The middle third of the text gives a unified treatment of integration theory
through the introduction of the gauge integral. Because of its simplicity and
ease of exposition, the Riemann integral is the vehicle by which most students
of analysis are introduced to integration theory. As deeper applications are
encountered, more sophisticated theories of integration are required, and soon
there is a variety of integrals, for example, Riemann, Stieltjes, Riemann-
Stieltjes, Daniell, and Lebesgue, each requiring its own foundation and structure. Perhaps the most important of these are the integrals of Riemann and
Lebesgue. It is recognized that the Lebesgue integral is much superior to the
Riemann integral, but because of the technical difficulties encountered in
defining and developing the Lebesgue integral, it is often considered inap-
propriate for a beginning real analysis course. The gauge integral, introduced
in Chapter 13, seems to be an almost perfect combination of the desired
properties of the Riemann and Lebesgue integrals. It is conceptually almost as
easy to describe as the Riemann integral, and it has all the powerful convergence properties of the Lebesgue integral. The gauge integral is studied in
detail; applications and examples are given to illustrate the utility and power
of the integral. There has been no attempt at achieving maximum generality.
The exposition presents the most important results in a form that should be
sufficient for most applications encountered by the student.

The last third of the text introduces and studies metric spaces in detail. In
this context, a more detailed study of topological notions is made. By this
time, the students should have reached a level of sophistication at which these
abstract results can be understood and appreciated. The previous work on
convergence, continuity, differentiation, and integration in $\mathbb{R}^n$ provides a
wealth of examples of metric spaces. Applications of the abstract ideas that
are being studied can be given to integral equations, differential equations, and
function spaces. The development is in sharp contrast to many introductory
texts in which metric spaces are defined early and then examples are given as
the exposition develops.

For students emerging from a beginning calculus course sequence, the first
18 chapters, with the exception of Chapter 16, when they are covered at a

moderate pace should constitute a good two-semester course in elementary real analysis. The chapters marked by an asterisk are independent, with the exception that the contraction mapping theorem **23.1** is used in the proof of the inverse function theorem (**30.1**). These marked chapters can be used to fill out a course according to the needs, interests, and sophistication of the students.

We are grateful to those who have helped and encouraged us in this endeavor and particularly to our students who graciously tolerated this work through its many revisions from embryonic class notes to this final version. Special thanks are due to Richard Carmichael, Joe Howard, Joe Kist, Douglas Kurtz, and David Ruch, each of whom worked through the text and offered many valuable suggestions.

JOHN DEPREE
CHARLES SWARTZ

Contents

CHAPTER 1

Preliminaries

In this text, we introduce real analysis. It is assumed that the reader is familiar, at least on an informal or intuitive level, with the basic elements of the calculus. We will reexamine some of the basic concepts of the calculus on a more formal level and establish in a rigorous fashion some of the basic ideas encountered in the calculus. Such a rigorous study is necessary if we are to understand some of the more advanced topics encountered in the richest of all mathematical fields, analysis, and its wealth of applications in mathematics, physics, chemistry, economics, the computer sciences, and other fields. For example, Fourier series is a subject that occupies a central position in mathematical analysis and has numerous applications to such diverse subjects as boundary value problems in mathematical physics, signal detection and image processing in engineering, and many areas in probability and statistics. The study of Fourier series has motivated the development of some of the most important topics in mathematical analysis, for example, the modern definition of function, set theory, the study of convergence of series of functions, and the development of the Lebesgue integral. An investigation of

Fourier analysis requires an understanding of the basic principles of mathematical analysis; it is these basic principles, and others, that we will study in this book. We begin with some preliminaries.

We'll be talking about mathematics and it will be expedient to have, along with English, a common mathematical vocabulary and language through which we can communicate. Some of this will be introduced immediately; the rest as it's needed. Much of the language is surely already known to the reader, and so, its reintroduction should be little more than an opportunity to go gently into our subject.

By a *set A*, we mean a well-defined collection of objects such that it can be determined whether or not any particular object is a member of A. For example, biologists might be interested in the set of all Orca whales in Puget Sound on September first, and students in this particular course will eventually develop a deep appreciation for the properties of and concepts associated with the set of real numbers. Generally, capital letters will be used to denote sets of objects and lowercase letters to denote the objects themselves. If a is an object in the set A we say that a is a member of A and write $a \in A$ (see the notation list below). The specially designed letters $\mathbb{N}$, $\mathbb{Z}$, $\mathbb{Q}$, and $\mathbb{R}$ will stand for the sets of natural numbers (positive integers), integers, rational numbers, and real numbers, respectively; these sets will be discussed extensively in Chapter 2. In the meantime, the knowledge of these sets gained by the reader in previous studies should be adequate. Often a set A can be described by specifying a property P that is uniquely satisfied by each of its elements; in this case, we adopt the notation

$$A = \{ a : a \text{ satisfies property } P \}.$$

For example, if n is an integer, the set of *divisors* of n is $\{ d : d$ is an integer and $n = cd$ for some integer $c \}$.

The following shorthand will be used extensively:

" $\in$ " means "is an element of,"	" $\forall$ " means "for every"
" $\notin$ " means "is not an element of"	" $\exists$ " means "there exists"
"iff" means "if and only if."	

In Chapter 2, and indeed throughout the text, we shall be examining the structure of functions and other abstract mathematical concepts defined on or related to the set of real numbers. As is often the case in mathematics, much of our work will consist of making conjectures, rephrasing these as theorems, and then "proving" these theorems, that is, providing convincing arguments that the proposed theorems are correct. A problem sometimes encountered in this process is recognition of the point at which one should have been convinced, that is, the "end" of the proof. We'll try to help this recognition by placing a

"□" at the end of each proof. We have also placed a "□" at the ends of examples.

Before proceeding much further, it may be useful to review some of the language used in stating theorems, lemmas, propositions, and their corollaries. We are talking about relationships between statements, say Statement P and Statement Q. A theorem that states that "Q is true if P is true" might take any one of the following equivalent forms:

P implies Q;

If P is true, then Q is true;

The truth of P is sufficient for the truth of Q;

The truth of Q is necessary for the truth of P;

P being true is a sufficient condition for Q to be true;

Q being true is a necessary condition for P to be true;

If Q is false, then P is false. (This last statement is called the *contrapositive* of the second statement above, that is, of "if P is true, then Q is true." In trying to prove an implication, it is sometimes easier to establish its contrapositive.)

Sometimes a situation in which "P implies Q" and "Q implies P" occurs. The following statements are equivalent to this:

P is true if, and only if, Q is true,

The truth of P is a necessary and sufficient condition for the truth of Q.

If every element of a set A is also an element of a set B, then A is called a *subset* of B and we write $A \subseteq B$, or $B \supseteq A$. If there is also $b \in B$ such that $b \notin A$, then A is a *proper subset* of B, and we may write $A \subset B$ or $B \supset A$; this distinction in notation is not religiously adhered to by the users of set theory and so we too, when it's convenient, will take liberties with "$\subset$," "$\subseteq$," "$\supset$," and "$\supseteq$." It is a great convenience to have available the notation of a set with no elements. We call this set the *empty set*, denote it by $\varnothing$, and agree that it will be a subset of every set. If A and B are subsets of another set U, then set inclusion of A in B can be illustrated using a Venn diagram in which U is represented as the interior of a rectangle and A and B are represented by the regions indicated (Figure 1.1).

Again, if A and B are sets, the *union* of A and B is the set of all objects that belong to either A or B; we denote the union of A and B by $A \cup B$. Thus $A \cup B = \{a: a \in A \text{ or } a \in B\}$. The *intersection* of A and B, denoted by $A \cap B$, is the set of all objects that belong to both A and B; thus $A \cap B = \{a: a \in A \text{ and } a \in B\}$. Venn diagrams illustrating union and intersection are shown in Figure 1.2.

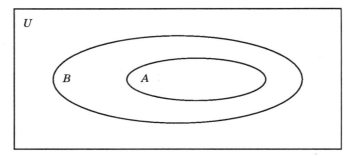

Figure 1.1

If for each $a \in A$, there is defined a set E_a, we then have a family of sets $\mathscr{E} = \{E_a: a \in A\}$. The set A is called an *index set* for the family $\mathscr{E}$ and $\mathscr{E}$ is said to be *indexed* by A. The family of intervals $\mathscr{I} = \{I_n: I_n = \{x: x \in \mathbb{R}$ and $1/n \le x \le 1 + 1/n\}, n \in \mathbb{N}\}$ is indexed by the set of natural numbers $\mathbb{N}$.

The *union* of the family of sets $\{E_a: a \in A\}$ is the set of all objects that belong to at least one E_a and is denoted by $\bigcup_{a \in A} E_a$; if $A = \{1, \ldots, n\}$, we write $\bigcup_{i \in A} E_i = E_1 \cup E_2 \cup \cdots \cup E_n$; if $A = \mathbb{N}$, we write $\bigcup_{i \in \mathbb{N}} E_i = \bigcup_{i=1}^{\infty} E_i$. The *intersection* of the family of sets $\{E_a: a \in A\}$ is the set of all objects that belong to all of the E_a and is denoted by $\bigcap_{a \in A} E_a$. If $A = \{1, \ldots, n\}$, we write $\bigcap_{i \in A} E_i = E_1 \cap E_2 \cap \cdots \cap E_n$; if $A = \mathbb{N}$, we write $\bigcap_{i \in \mathbb{N}} E_i = \bigcap_{i=1}^{\infty} E_i$. In the example above, $\bigcup_{n \in \mathbb{N}} I_n = \bigcup_{n=1}^{\infty} I_n = \{x: 0 < x \le 2\}$ and $\bigcap_{n \in \mathbb{N}} I_n = \bigcap_{n=1}^{\infty} I_n = \{1\}$. This last set, which consists of only one element, is called a *singleton*. The *difference*, $F \setminus E$, of two sets F and E is the set of all objects that belong to F but do not belong to E. If E is a subset of F, $F \setminus E$ is called the *complement* of E in F. The shaded areas in the Venn diagrams in Figure 1.3 illustrate difference and complement, respectively.

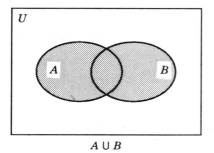

$A \cup B$

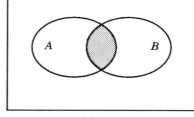

$A \cap B$

Figure 1.2

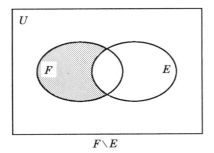 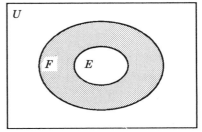

$F\setminus E$ Complement of E in F

Figure 1.3

Two sets E and F are *disjoint* if $E \cap F = \varnothing$; the family of sets $\{E_a: a \in A\}$ is said to be *pairwise disjoint* if $E_a \cap E_b = \varnothing$ for $a, b \in A$, $a \neq b$.

The set operations of taking differences, unions, and intersections were studied by the English mathematician Augustus DeMorgan (1806–1873); his results are given in Proposition 1. The reader should check the validity of these results by using the Venn diagram in Figure 1.4.

Proposition 1 (DeMorgan's Laws). For each $a \in A$ let E_a be a set and let F be a set. Then

(i) $F\setminus \bigcup_{a \in A} E_a = \bigcap_{a \in A}(F\setminus E_a)$.
(ii) $F\setminus \bigcap_{a \in A} E_a = \bigcup_{a \in A}(F\setminus E_a)$.

Proof. To show that two sets are the same, it suffices to show that each is a subset of the other.

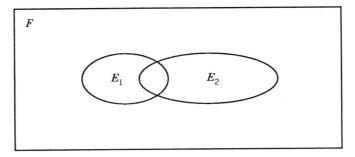

Figure 1.4

For this, let $x \in F \setminus \bigcup_{a \in A} E_a$. Then $x \in F$ and $x \notin \bigcup_{a \in A} E_a$. Hence, $x \notin E_a \ \forall a \in A$. That is, $x \in F \setminus E_a \ \forall a \in A$ or $x \in \bigcap_{a \in A} (F \setminus E_a)$. Thus, we have $F \setminus \bigcup_{a \in A} E_a \subseteq \bigcap_{a \in A} (F \setminus E_a)$.

Next, let $x \in \bigcap_{a \in A} (F \setminus E_a)$. Then $x \in F \setminus E_a \ \forall a \in A$ or $x \in F$ and $x \notin E_a \ \forall a \in A$. Thus, $x \in F$ and $x \notin \bigcup_{a \in A} E_a$, that is, $x \in F \setminus \bigcup_{a \in A} E_a$. Hence, $\bigcap_{a \in A} (F \setminus E_a) \subseteq F \setminus \bigcup_{a \in A} E_a$, and we have (i).

(ii) is similar and is left to Exercise 1. $\qquad\qquad\qquad\qquad\qquad\qquad$ □

An *ordered pair* (a, b) is an object associated with the individual objects a and b; the order in which a and b are written is considered essential to the identity of (a, b). Therefore, unless $a = b$, (a, b) and (b, a) are distinct ordered pairs. (This is, of course, in contrast to the sets $\{a, b\}$ and $\{b, a\}$, which are equal.) Thus two ordered pairs, (a, b) and (c, d), are equal if and only if $a = c$ and $b = d$.

If A and B are sets, the *cartesian product*, named after the French geometer René Descartes (1596–1650), of A and B is the set of all ordered pairs (a, b), such that $a \in A$, $b \in B$, and is denoted by $A \times B$. The reader, through long standing use, is undoubtedly familiar with the representation of the real numbers as points on a line resulting in a model called the *real line*. In this model, two distinct points are arbitrarily chosen and associated with the real numbers 0 and 1. If the line is drawn horizontally, the left chosen point is usually associated with 0 and the right-hand point with 1. The measured distance between these two points defines a unit of distance for the line, and the rest of the real numbers can be associated with points on the line using this distance; if x is a positive real number, it is associated with the point on the line that is x units of distance to the right of 0. If y is negative, it is associated with the point that is $|y|$ (read "absolute value") units to the left of 0. The point corresponding to 0 is called the origin on the real line. See Figure 1.5. Many of the terms just used ("distance," "positive," "negative," "absolute value") will be defined later. For now, the reader's intuitive grasp of these concepts will be adequate. If the cartesian product of $\mathbb{R}$ with itself is formed, the result is the *cartesian plane*. A nice geometric representation of the cartesian plane can be constructed using two real lines intersecting at a right angle at their origins as shown in Figure 1.6. The ordered pair $(x, y) \in \mathbb{R} \times \mathbb{R}$ is associated with the point which is $|x|$ from the vertical axis, to the right if x is positive, on the vertical axis if $x = 0$, to the left if x is negative, and $|y|$ from the horizontal axis, above if y is positive, on the horizontal axis if $y = 0$, and below if y is negative.

Figure 1.5

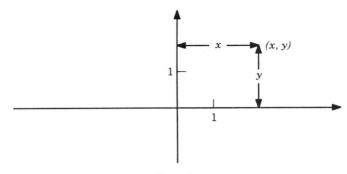

Figure 1.6

A *function (map)* f from A into B, often written $f\colon A \to B$, is a rule that associates with each $a \in A$ a unique element $b \in B$; we write $f(a) = b$ and call b the *value* of f at a. The set A is called the *domain* of f, and the set $f(A) = \{b\colon b = f(a) \text{ for some } a \in A\}$ is called the *range* of f. See Figure 1.7.

The function f is *onto* B if $f(A) = B$ and is *one-one* if $f(a) = f(a')$ implies $a = a'$. If $f\colon A \to B$ and $E \subseteq A$, then $f(E) = \{y\colon f(x) = y \text{ for some } x \in E\}$ is called the *direct image* or *image* of E under f; if $F \subseteq B$, the set $f^{-1}(F) = \{x \in A\colon f(x) \in F\}$ is called the *inverse image* of F under f. The *graph* of f is the set

$$G(f) = \{(x, f(x))\colon x \in A\} \subseteq A \times B.$$

If the domain and range of f are both subsets of $\mathbb{R}$, then the graph of f can sometimes be displayed as a set of points in the plane with coordinates $(a, f(a))$, a in the domain of f. For example, in Figure 1.8, the graph of the function f defined by the equation $f(x) = \sqrt{x}$ is shown. The domain and range are each the set of positive real numbers. The graph is the set of points $G(f) = \{(x, \sqrt{x})\colon x \geq 0\}$ in the cartesian plane.

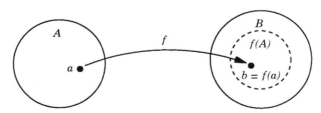

Figure 1.7

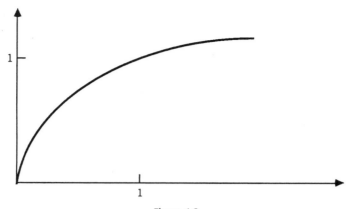

Figure 1.8

This idea can be extended to functions from $\mathbb{R}^2$ into $\mathbb{R}$ or from $\mathbb{R}$ into $\mathbb{R}^2$. Because of our inability to create the physical models, this process breaks down for higher dimensions.

The *identity function* $i = i_A$ on the set A is the function defined by $i(a) = a$ for $a \in A$.

If f and g are functions such that f is defined on a domain A, g is defined on a domain A', and $f(A) \subseteq A'$, then the *composition* of g and f, denoted by $g \circ f$, is the function that assigns to each $a \in A$ the value $g(f(a)) = g \circ f(a)$. See Figure 1.9.

A very useful kind of function is that of a sequence in a set A: a *sequence* in A is a function whose domain is $\mathbb{N}$, the set of positive integers, and whose range is contained in A, that is, $f \colon \mathbb{N} \to A$. The set $\mathbb{N}$ specifies an ordering of the elements of A that are used in the sequence. The values of f are usually denoted by $f_n = f(n)$, $n \in \mathbb{N}$, and the sequence f is often denoted by $\{f_n\}$, or

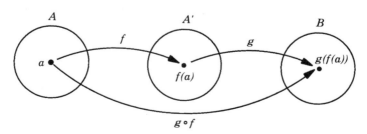

Figure 1.9

$\{f_n\}_{n=1}^{\infty}$, or $f_1, f_2, \ldots$. If $f: \mathbb{N} \to A$ is onto A, we say that the elements of A are "arranged in a sequence," $f_1, f_2, \ldots$, or $f_1, f_2, \ldots$ is an "enumeration of the elements of A." The family $\mathscr{I}$ introduced above is a sequence of intervals on the real line. Other examples of sequences are $\{1/n\}$, $\{(-1)^{n-1}\}$, and $\{\sin \pi x/n\}$. Writing out the arrangements in these examples, we obtain $1, \frac{1}{2}, \frac{1}{3}, \ldots, 1, -1, 1, -1, 1, \ldots$, and $\sin \pi x, \sin \pi x/2, \sin \pi x/3, \ldots$.

If f is a one-one function from A onto B, then the function $f^{-1}: B \to A$, defined by $f^{-1}(b) = a$ iff $f(a) = b$, is called the *inverse* of f. Note that f^{-1} is well-defined since f is one-one. In the example, $f(x) = \sqrt{x}$, given above, f is a one-one function and so, it has an inverse. f^{-1} that can be found by solving the equation $y = f(x) = \sqrt{x}$. Thus $x = y^2$, $y \geq 0$ yields $f^{-1}(y) = y^2$, $y \geq 0$. The graph of f^{-1} is the set

$$G(f^{-1}) = \left\{ (x, f^{-1}(x)): x \in \text{domain of } f^{-1} \right\}$$

$$= \left\{ (x, y): y = f^{-1}(x) \right\}$$

$$= \left\{ (x, y): f(y) = x \right\}$$

$$= \left\{ (x, y): \sqrt{y} = x \right\}$$

$$= \left\{ (x, y): y = x^2, x, y \geq 0 \right\}.$$

The reader should make a sketch of the graph of f^{-1} in the cartesian plane and observe that the graph of f^{-1} is the reflection of the graph of f in the line $y = x$. Exercise 16 shows that this is generally the case.

If there is a one-one function f from A onto B, we say that A and B can be put in a *one-one correspondence* or that A and B are *equivalent*. Intuitively, A and B have the "same number of elements." We write $A \sim B$. It is routine to show that (Exercise 7)

$$A \sim A,$$

$$A \sim B \text{ iff } B \sim A, \tag{1}$$

$$A \sim B, B \sim C \text{ implies } A \sim C.$$

A relation " $\sim$ " satisfying the conditions in (1) is called an *equivalence relation*.

EXAMPLE 2. The set of positive integers, $\mathbb{N}$, and the set of even positive integers, E, are equivalent. In fact, the function $f: \mathbb{N} \to E$, defined by $f(n) = 2n$, defines a one-one correspondence between $\mathbb{N}$ and E.

The set $\mathbb{N}$ and the set of all integers, $\mathbb{Z}$, are equivalent since the function

$$f(n) = \begin{cases} n/2, & n \text{ even,} \\ -(n-1)/2, & n \text{ odd,} \end{cases}$$

gives a one-one correspondence between $\mathbb{N}$ and $\mathbb{Z}$.

If $(0,1)$ is the set of all real numbers x satisfying $0 < x < 1$ and if $(0,1]$ is the set of all real numbers x satisfying $0 < x \leq 1$, then $(0,1)$ and $(0,1]$ are equivalent. For let f be defined by

$$f(x) = \begin{bmatrix} 3/2 - x, & 1/2 < x \leq 1, \\ 3/4 - x, & 1/4 < x \leq 1/2, \\ 3/8 - x, & 1/8 < x \leq 1/4; \\ \vdots \end{bmatrix}$$

then f defines a one-one correspondence between $(0,1]$ and $(0,1)$. (Make a sketch!)   □

Let $\mathbb{N}_n$ denote the set consisting of the first n positive integers. A set A is *finite* if $A = \varnothing$ or $A \sim \mathbb{N}_n$ for some n; a set is *infinite* if it is not finite. A set A is *countably infinite* if $A \sim \mathbb{N}$ and is *countable* if it is either countably infinite or finite; A is *uncountable* if it is infinite but not countable.

Thus, from Example 2, the set of even positive integers and the set of all integers are countably infinite. We show later that the set of rational numbers, $\mathbb{Q}$, is also countably infinite.

The following result implies that countably infinite sets are, in a certain sense, the "smallest" infinite sets; that is, a countably infinite set cannot contain an uncountable subset. For the proof of this result, we use the *principle of induction*, which was introduced by Giuseppe Peano (1858–1932) as one of his axioms for the positive integers:

If S is a nonempty subset of the set $\mathbb{N}$ of positive integers such that

$$\text{(i) } 1 \in S \quad \text{and} \quad \text{(ii) if } k \in S, \quad \text{then } k+1 \in S, \tag{2}$$

then $S = \mathbb{N}$.

To see how this is used in a "proof by induction," consider the statement: every integer $n > 1$ is either prime or a product of primes. (Recall that a *prime* is a positive integer $n > 1$ whose only positive integer divisors are 1 and n.) Let $k = n - 1$ and let S be the set of all positive integers $k + 1$ for which the statement is true. Then the statement is trivially true for $k = 1$ and so $1 \in S$. Assume the statement is true for every positive integer m with $1 < m < k + 1$. If $k + 1$ is not prime, it has a positive divisor d with

$1 < d < k + 1$. Then we can write $k + 1 = cd$, where $1 < c < k + 1$. Since both c and d are less than $k + 1$, they are both products of primes. Hence $k + 1$ is a product of primes and $k + 1 \in S$. We have shown that $1 \in S$ and that if $2, 3, \ldots, k \in S$, then $k + 1 \in S$. It follows from the principle of induction that $S = \mathbb{N}$, that is, the statement is true for every positive integer n.

We also use another property, called the *well-ordering property*, of the natural numbers:

$$\text{If } B \subseteq \mathbb{N} \text{ and } B \text{ is nonempty, then } B \text{ has the smallest element.} \qquad (3)$$

Proposition 3. Every infinite subset of a countably infinite set is countably infinite.

Proof. Let A be a countably infinite set and let E be an infinite subset of A. Let $f \colon \mathbb{N} \to A$ be a one-one correspondence between $\mathbb{N}$ and A.

To construct a function $g \colon \mathbb{N} \to E$ which is onto, we first construct a sequence $\{n_k\}$ of positive integers as follows: The well-ordering property assures us that there is a smallest positive integer n_1 such that $f(n_1) \in E$. Suppose that we have chosen positive integers $n_1, \ldots, n_k$ such that $n_i < n_{n+1}$ for $i = 1, \ldots, k - 1$ and $f(n_i) \in E$. Using the well-ordering property again, let n_{k+1} be the smallest positive integer such that $n_{k+1} > n_k$ and $f(n_{k+1}) \in E$. If $S = \{k \colon n_k$ has been chosen as above$\}$, it follows from the principle of induction that $S = \mathbb{N}$ and we have constructed a sequence $\{n_k\}_{k=1}^{\infty}$.

If $g \colon \mathbb{N} \to E$ is defined by $g(k) = f(n_k)$, then g defines a one-one function which, by the construction, is onto E. $\qquad \square$

The following is a useful criterion for countability.

Proposition 4. A nonempty set A is countable if, and only if, there is a function $f \colon \mathbb{N} \to A$ that is onto A.

Proof. If the set A is countable, the existence of such a function f is clear. Suppose next that there is a function $f \colon \mathbb{N} \to A$ that is onto. For $a \in A$, $f^{-1}(a) \neq \varnothing$, so pick $a' \in f^{-1}(a)$. Then $g \colon a \to a'$ defines a one-one function from A into $\mathbb{N}$. Since $A \sim g(A)$, it follows from Proposition 3 that A is countable. $\qquad \square$

From Proposition 4 we can establish the countability of a large number of sets.

Proposition 5. Let $\{E_n\}_{n=1}^{\infty}$ be a sequence of countable sets. Then $E = \bigcup_{n=1}^{\infty} E_n$ is countable.

Proof. First assume that each E_n is countably infinite. For each n, let

$\{x_{nk}\}_{k=1}^{\infty}$ be an enumeration of the points of E_n. From Proposition 4, it suffices to give an enumeration of the points of E. For this consider the infinite array:

$$\begin{bmatrix} x_{11} & x_{12} & x_{13} & \cdots \\ x_{21} & x_{22} & x_{23} & \cdots \\ x_{31} & x_{32} & x_{33} & \cdots \\ \vdots & & & \end{bmatrix} \tag{4}$$

The arrows indicated in this "infinite matrix" yield an enumeration of E, that is, $x_{11}, x_{21}, x_{12}, x_{31}, x_{22}, x_{13}, \ldots$.

If any E_n is finite, let $x_{n1}, \ldots, x_{nk_n}$ be an enumeration of the points in E_n. By setting $x_{nj} = x_{nk_n}$ for $j \geq k_n$, we can apply the method of the above proof to produce a function from $\mathbb{N}$ onto E. $\square$

Corollary 6. Let A be countable and let $B_n = \{(a_1, \ldots, a_n): a_i \in A\}$ (i.e., B_n is the cartesian product of A with itself n times). Then B_n is countable.

Proof. The proof is by induction on n. B_1 is countable by hypothesis. Assume that B_{n-1} is countable. The elements of B_n are of the form (a, b) where $a \in B_{n-1}$ and $b \in A$. Thus, $B_n = \bigcup_{b \in A} \{(a, b): a \in B_{n-1}\}$ is countable by Proposition 5. $\square$

Corollary 7. The set of rational numbers $\mathbb{Q}$ is countably infinite.

Proof. Let $\mathbb{Z}_0$ be the set of all nonzero integers. Then $\mathbb{Z} \times \mathbb{Z}_0$ is countable, by Corollary 6, and $f(a, b) = a/b$ defines a mapping from $\mathbb{Z} \times \mathbb{Z}_0$ onto $\mathbb{Q}$. Proposition 4 implies that $\mathbb{Q}$ is countable. $\square$

It is actually the case that the set of algebraic numbers is countable (a real number x is an algebraic number if x is the root of a polynomial with integer coefficients) (Exercise 14).

Finally, we give an example of an uncountable set.

Proposition 8. Let E be the set that consists of all sequences whose elements are 0 or 1. Then E is uncountable.

Proof. Let $C \subseteq E$ be countably infinite, $\{x_n\}$ an enumeration of the points of C, and $x_n = \{s_{nk}\}_{k=1}^{\infty}$. Define a sequence x in E as follows: $x_k = 0$ if $s_{kk} = 1$ and $x_k = 1$ if $s_{kk} = 0$. Thus, $x \neq x_n$ for any n since x and x_n differ in the nth coordinate. That is, $x \notin C$. It follows that E is uncountable since every countable subset of E is a proper subset of E. $\square$

Analogous to the decimal expansion, which can be viewed as a representation of a positive real number as a sequence whose terms are integers between 0 and 9, every real number can also be uniquely represented in a "binary expansion" as a sequence of 0s and 1s (see Exercise 3.46). The reader familiar with the binary expansion will recall that each real number x, $0 \le x \le 1$, has a binary expansion of the form $.a_1 a_2 \ldots$, where $a_i = 0$ or 1. Some such x's have two binary expansions (just as some real numbers have two decimal expansions); for example, $\frac{1}{2} = .1000 \ldots = .0111 \ldots$. However each such x is a rational number, and so, the set of all such x's is countable. It follows from Proposition 8 that $[0, 1] = \{x \in \mathbb{R}: 0 \le x \le 1\}$ is uncountable and Proposition 3 implies that $\mathbb{R}$ itself is uncountable.

EXERCISES 1

1. Prove Proposition 1 (ii).

2. Show that $E \cap (\bigcup_{a \in A} E_a) = \bigcup_{a \in A}(E \cap E_a)$, $E \cup (\bigcap_{a \in A} E_a) = \bigcap_{a \in A}(E \cup E_a)$.

3. Let $f: A \to B$, $g: B \to C$ be one-one and onto. Show that $(g \circ f)^{-1} = f^{-1} \circ g^{-1}$.

4. Show that $f^{-1}(\bigcap_{a \in A} E_a) = \bigcap_{a \in A} f^{-1}(E_a)$, $f^{-1}(\bigcup_{a \in A} E_a) = \bigcup_{a \in A} f^{-1}(E_a)$, $f^{-1}(E \setminus F) = f^{-1}(E) \setminus f^{-1}(F)$.

5. Show that $f^{-1}(f(E)) \supseteq E$ and $f(f^{-1}(E)) \subseteq E$. Does equality hold in general?

6. If $f: A \to B$ is one-one and onto, show that $f \circ f^{-1} = i_B$ and $f^{-1} \circ f = i_A$.

7. Establish (1).

8. If O is the set of positive odd integers, exhibit a one-one correspondence between O and $\mathbb{N}$.

9. If $[0, 1)$ ($[0, 1]$) denotes the set of real numbers x such that $0 \le x < 1$ ($0 \le x \le 1$), show that $(0, 1) \sim [0, 1] \sim [0, 1) \sim \mathbb{R}$.

10. Use the principle of induction to establish each of the following:

(a) $\sum_{k=1}^{n} k = n(n + 1)/2 \ \forall n \in \mathbb{N}$.

(b) $\sum_{k=1}^{n} k^2 = n(n + 1)(2n + 1)/6 \ \forall n \in \mathbb{N}$.

(c) $n < 2^n \ \forall n \in \mathbb{N}$.

(d) If $a, r \in \mathbb{R}$ and $s_n = a + ar + ar^2 + \cdots + ar^{n-1}$, then $s_n = a(1 - r^n)/(1 - r) \ \forall n \in \mathbb{N}$ provided $r \neq 1$.

11. Show that every subset of a finite set is finite.
    Hint: Use induction.

12. Show that every infinite set has a countably infinite subset.
    Hint: Use induction.

13. Show that every infinite set is equivalent to one of its proper subsets.

14. Show that the set of algebraic numbers is countable.
    Hint: Consider the roots of the polynomials of degree $\leq n$ for each n.

15. Let $f: E \to F$. Show that $f(f^{-1}(B)) = B \ \forall B \subseteq F$ iff f is onto. Show that $f^{-1}(f(A)) = A \ \forall A \subseteq E$ iff f is one-one.

16. If A and B are subsets of $\mathbb{R}$ and $f: A \to B$ is one-one, show that the graph of f^{-1} in the cartesian plane is the reflection of the graph of f in the line $y = x$.

CHAPTER 2

The Real Numbers

Real analysis can be fairly well described as the study of concepts that have their roots in the set of real numbers. We assume that the reader is familiar with the algebraic properties of the real numbers $\mathbb{R}$ and the subsets $\mathbb{Q}$ and $\mathbb{N}$ of $\mathbb{R}$ consisting of the rational numbers and the positive integers, respectively. These are listed in Axioms A below. Familiarity with the basic order properties of $\mathbb{R}$ is also assumed; these are listed in Axiom B. The order completeness property of $\mathbb{R}$ is probably less familiar; however, it is fundamental to our study and we illustrate its importance by giving several applications in Theorems 1 to 5.

A. The Field Axioms. For all x, y, $z \in \mathbb{R}$,

A1. $x + y = y + x$.

A2. $(x + y) + z = x + (y + z)$.

A3. $\exists\, 0 \in \mathbb{R}$ such that $x + 0 = x$ for all $x \in \mathbb{R}$.

A4. For each $x \in \mathbb{R}\ \exists -x \in \mathbb{R}$ such that $x + (-x) = 0$.

A5. $xy = yx$.

A6. $(xy)z = x(yz)$.

A7. $\exists\, 1 \in \mathbb{R}$ such that $1 \neq 0$ and $x \cdot 1 = x$ for all $x \in \mathbb{R}$.

A8. if $x \neq 0$, $\exists\, x^{-1} \in \mathbb{R}$ such that $x(x^{-1}) = 1$.

A9. $x(y + z) = xy + xz$.

B. Order Axioms. The subset P of positive real numbers satisfies

B1. $x, y \in P$ implies $x + y \in P$.

B2. $x, y \in P$ implies $xy \in P$.

B3. $x \in P$ implies $-x \notin P$.

B4. $x \in \mathbb{R}$ implies $x = 0$, $x \in P$, or $-x \in P$.

We write $x < y$ iff $y - x \in P$ and $x \leq y$ iff $x < y$ or $x = y$; $x > y$ $(x \geq y)$ iff $y < x$ $(y \leq x)$. If $x \in \mathbb{R}$, the *absolute value* of x, denoted by $|x|$, is defined by $|x| = x$ if $0 \leq x$ and $|x| = -x$ if $x < 0$.

Clearly $\pm x \leq |x|$. It follows that $x + y \leq |x| + |y|$ and $-(x + y) \leq |x| + |y|$. These last two inequalities imply the important *triangle inequality*.

$$|x + y| \leq |x| + |y| \qquad \forall\, x, y \in \mathbb{R}. \tag{1}$$

The name of the inequality (1) derives from the fact that it also holds in more general settings, in particular in the cartesian plane $\mathbb{R}^2 = \mathbb{R} \times \mathbb{R}$. In this setting, x, y, and $x + y$ represent vectors, and the inequality can be interpreted as saying that the length of any side of a triangle does not exceed the sum of the lengths of the other two sides.

Another useful inequality that follows from the triangle inequality is

$$|x - y| \geq |\,|x| - |y|\,| \qquad \forall\, x, y \in \mathbb{R}. \tag{2}$$

To prove this, note that $|x| = |x - y + y| \leq |x - y| + |y|$ or $|x| - |y| \leq |x - y|$. By symmetry $|y| - |x| \leq |y - x| = |x - y|$ (Exercise 2a), so that $|\,|x| - |y|\,| \leq |x - y|$.

A nonempty set that satisfies axioms A is called a *field*; one that satisfies A and B is called an *ordered field*. When we have finally finished specifying $\mathbb{R}$, we will, of course, have required that $\mathbb{R}$ be an ordered field. It's not hard to show that the set $\mathbb{Q}$ of rational numbers is also an ordered field (see Exercise 3). Thus it is essential to be able to uniquely describe the set $\mathbb{R}$ and in particular to know that we have distinguished it from all other ordered fields.

Let $S \subseteq \mathbb{R}$. An element $b \in \mathbb{R}$ is an *upper bound* for S if $x \leq b$ $\forall\, x \in S$; c is a *least upper bound* or *supremum* for S if c is an upper bound for S and $c \leq b$ for any upper bound b for S. If c is a supremum for S, we write $c = \text{lub}\,S = \sup S$ (c is unique by Exercise 4). We say that S is *bounded above* if S has an upper bound.

Similarly, $a \in \mathbb{R}$ is a *lower bound* for S if $a \leq x \; \forall x \in S$; c is a *greatest lower bound* or *infimum* for S if c is a lower bound for S and $a \leq c$ for any lower bound, a, of S. If c is an infimum for S, we write $c = \text{glb } S = \inf S$ (c is unique by Exercise 4). S is said to be *bounded below* if S has a lower bound.

C. Completeness Axiom. Every nonempty subset of $\mathbb{R}$ that is bounded above has a supremum.

For the existence of infima, see Exercise 5. A field that satisfies A, B, and C is called a *complete ordered field*. We have not considered the problem of the existence of a complete ordered field. It can be shown that such a field can be constructed from the set of natural numbers. (For the construction see E. Landau, 1957.) It can also be shown that such a field is unique. (E. J. McShane and T. A. Botts, 1959.) The set of real numbers $\mathbb{R}$ enjoys many desirable properties as a consequence of its being a complete ordered field. We examine some of these in the following results.

Theorem 1 (Archimedean Principle). If $x \in \mathbb{R}$, then there is a positive integer n such that $x < n$.

Proof. Let $A = \{k \in N : k \leq x\}$. If $A = \varnothing$, we are finished. Assume $A \neq \varnothing$. Then A is bounded from above and $a = \sup A$ exists. Since $a - 1 < a$, it follows that there is $k \in A$ such that $a - 1 < k$. But then $a < k + 1$; so $k + 1 \notin A$, that is, $k + 1 > x$. Set $n = k + 1$. $\square$

Corollary 2 is an application of the Archimedean principle.

Corollary 2. Between any two real numbers there is a rational number.

Proof. Let $x, y \in \mathbb{R}$, $x < y$. Suppose first that $x \geq 0$. Then Theorem 1 implies that $\exists$ a positive integer $q > (y - x)^{-1}$, or equivalently, $1/q < y - x$. The set of positive integers n such that $y \leq n/q$ is nonempty (Theorem 1) and, therefore, has a smallest element p. Then $(p - 1)/q < y \leq p/q$ and $x = y - (y - x) < p/q - 1/q = (p - 1)/q$. So $x < r = (p - 1)/q < y$.

If $x < 0$, $\exists$ a positive integer n such that $n > -x$. Then $n + x > 0$ and $\exists$ a rational r with $n + x < r < n + y$ and $r - n$ is a rational between x and y. $\square$

A similar result holds for irrationals.

In Exercise 11 it is asked to show that $\sqrt{2}$ is irrational, that is, $\sqrt{2} \notin \mathbb{Q}$. Thus $A = \{x \in \mathbb{Q} : x^2 \leq 2\}$ is a bounded subset of $\mathbb{Q}$ that has no least upper bound in $\mathbb{Q}$, that is, $\mathbb{Q}$ is not complete.

Corollary 3. Between any two real numbers there is an irrational number.

Proof. Let $x, y \in \mathbb{R}$ with $x < y$. By Corollary 2, there is a rational r between $x/\sqrt{2}$ and $y/\sqrt{2}$. Since $\sqrt{2}$ is irrational (Exercise 11), $r\sqrt{2}$ (Exercise 12) is an irrational between x and y. □

As a final application of the completeness axiom, we show the existence of "nth roots" for positive real numbers. Recall that the binomial theorem states that for $n \in \mathbb{N}$

$$(a + b)^n = \sum_{k=0}^{n} \binom{n}{k} a^{n-k} b^k. \tag{3}$$

The *binomial coefficient* $\binom{n}{k}$ is given by

$$\binom{n}{k} = \frac{n(n-1) \cdots (n-k+1)}{k!} = \frac{n!}{k!(n-k)!} \tag{4}$$

where $k! = 1 \cdot 2 \cdot \cdots \cdot k$ for $k \in \mathbb{N}$ and $0! = 1$.

Theorem 4. Let $a > 0$ and $n \in \mathbb{N}$. Then there is a unique positive x such that $x^n = a$.

Proof. The uniqueness is clear, since $0 < y < x$ implies $0 < y^n < x^n$.

Set $E = \{t > 0 : t^n < a\}$. If $t = a/(a+1)$, then $0 < t < 1$ implies $t^n \leq t < a$, so $t \in E$. Hence, $E \neq \varnothing$. Also, if $t_0 = a + 1$, then $t > t_0$ implies $t^n \geq t > a$; so $t \notin E$ and t_0 is an upper bound for E.

Set $x = \sup E$. We show that $x^n = a$.

Suppose $x^n < a$. Choose h such that $0 < h < 1$ and

$$h < (a - x^n)/((1 + x)^n - x^n).$$

Then

$$(x + h)^n = \sum_{k=0}^{n} \binom{n}{k} x^{n-k} h^k \leq x^n + h\left[\binom{n}{1} x^{n-1} + \binom{n}{2} x^{n-2} + \cdots + \binom{n}{n}\right]$$

$$= x^n + h[(1 + x)^n - x^n] < x^n + (a - x^n) = a$$

which implies that $x + h \in E$ and $x + h > x$. This contradicts the definition of x.

Suppose $x^n > a$. Choose k such that $0 < k < 1$ and

$$k < (x^n - a)/((1 + x)^n - x^n)$$

For $t \geq x - k$ we have

$$t^n \geq (x - k)^n = x^n - \binom{n}{1}x^{n-1}k + \binom{n}{2}x^{n-2}k^2 - \cdots + (-1)^n\binom{n}{n}k^n$$

$$= x^n - k\left[\binom{n}{1}x^{n-1} - \cdots + (-1)^{n-1}\binom{n}{n}k^{n-1}\right]$$

$$\geq x^n - k\left[\binom{n}{1}x^{n-1} + \binom{n}{2}x^{n-2} + \cdots + 1\right]$$

$$= x^n - k\left[(1 + x)^n - x^n\right] > x^n - (x^n - a) = a$$

The second inequality above follows from (2) and Exercise 2d.

Thus, $x - k$ is an upper bound for E. This contradicts the definition of x. Hence, $x^n = a$. $\qquad\qquad\qquad\qquad\qquad\qquad\qquad\qquad\qquad\qquad\qquad\qquad\qquad\square$

It is convenient to adjoin to $\mathbb{R}$ two additional elements, ∞ and $-\infty$ (not belonging to $\mathbb{R}$) with the order properties $-\infty < x < \infty \;\forall\, x \in \mathbb{R}$ and call the resulting set the *extended real numbers*. The set $\mathbb{R} \cup \{\infty\} \cup \{-\infty\}$ will be denoted by $\overline{\mathbb{R}}$. If $E \subseteq \mathbb{R}$ is nonempty and is not bounded above (below) we set $\sup E = \infty$ ($\inf E = -\infty$). With these conventions, every nonempty subset of $\mathbb{R}$ has a supremum and infimum (in $\overline{\mathbb{R}}$).

We adopt the following conventions for algebraic operations on ∞ and $-\infty$:

$\infty + x = x + \infty = \infty$ unless $x = -\infty$;

$(-\infty) + x = x + (-\infty) = -\infty$ unless $x = \infty$;

if $a > 0$, then $a \cdot \infty = \infty \cdot a = \infty$ and $a \cdot (-\infty) = (-\infty) \cdot a = -\infty$;

if $a < 0$, then $a \cdot \infty = \infty \cdot a = -\infty$ and $a \cdot (-\infty) = (-\infty) \cdot a = \infty$.

The motivation for these definitions should be clear after the study of limits in Chapter 3.

If $a, b \in \mathbb{R}$ and $a < b$, we define the following intervals generated by a, b:

$$[a, b] = \{t : a \leq t \leq b\},$$

$$(a, b) = \{t : a < t < b\},$$

$$[a, b) = \{t : a \leq t < b\},$$

$$(a, b] = \{t : a < t \leq b\};$$

$[a, b]$ is called the *closed interval* with end points a and b, and (a, b) is called the *open interval* with end points a and b. Similarly, we write $[a, \infty) =$

$\{t: \ a \le t < \infty\}$, $(a, \infty) = \{t: \ a < t < \infty\}$, $(-\infty, a) = \{t: \ -\infty < t < a\}$, $(-\infty, a] = \{t: \ -\infty < t \le a\}$, and $(-\infty, \infty) = \mathbb{R}$. The *length* of an interval I with endpoints a and b, $a < b$, is defined as $b - a$ and denoted by $\ell(I)$. It's sometimes convenient to formally extend this terminology to a single real number a and refer to the "interval" $[a, a]$ as having length zero.

EXERCISES 2

1. Show that
   (a) $x \le y$ implies $x + z \le y + z \ \forall z \in \mathbb{R}$.
   (b) $x \le y$ and $y \le z$ implies $x \le z$.
   (c) $x \le y$ and $y \le x$ implies $x = y$.

2. Show that
   (a) $|x| = |-x|$.
   (b) $|x| \le a$ iff $-a \le x \le a$.
   (c) $|xy| = |x| \, |y|$.
   (d) $|x_1 + \cdots + x_n| \le |x_1| + \cdots + |x_n|$. (Use equation 1 and the principle of mathematical induction.)

3. Show that $\mathbb{Q}$ is an ordered field.

4. If b is a supremum (infimum) for E, show that b is unique.

5. Show that every nonempty set that is bounded from below has an infimum.

6. If $a \in \mathbb{R}$ and $b > 0$, show that $\exists \, n \in \mathbb{N}$ such that $a < nb$.

7. Let $f: \ I \times J \to \mathbb{R}$. Show that $\inf \{ f(x, y): \ (x, y) \in I \times J \} = \inf_{x \in I} \inf_{y \in J} f(x, y) = \inf_{y \in J} \inf_{x \in I} f(x, y)$. State and prove an analogous statement for suprema. What about $\sup_{x \in I} \inf_{y \in J} f(x, y)$ and $\inf_{y \in J} \sup_{x \in I} f(x, y)$?
   Hint: $f(n, m) = (-1)^{n+m}$.

8. If $A, B \subseteq \mathbb{R}$, $A \ne \emptyset$, $B \ne \emptyset$, let $A + B = \{a + b: a \in A, b \in B\}$. Show that $\sup (A + B) \le \sup A + \sup B$ and $\inf (A + B) \ge \inf A + \inf B$. Are they equal?

9. If $a > 0$ and $b \in \mathbb{R}$, give a reasonable definition of a^b. (Assume that a^q has been defined as usual for $q \in \mathbb{Q}$.)

10. If $A \subseteq \mathbb{R}$, let $-A = \{-a: a \in A\}$. Show that $\sup A = -\inf(-A)$, $\inf A = -\sup(-A)$.

11. Show that $\sqrt{2}$ is irrational.

12. If $a \ne 0$ is rational and b is irrational, show that ab is irrational.

13. If $\{a_k\}$ and $\{b_k\}$ are real sequences, show that

$$\sup\{a_k + b_k : k \in \mathbb{N}\} \leq \sup\{a_k : k \in \mathbb{N}\} + \sup\{b_k : k \in \mathbb{N}\},$$

and

$$\inf\{a_k + b_k : k \in \mathbb{N}\} \geq \inf\{a_k : k \in \mathbb{N}\} + \inf\{b_k : k \in \mathbb{N}\}.$$

When does equality hold?

14. Use the principle of mathematical induction to prove the binomial theorem.

15. Let $\mathbb{C} = \{(a, b) : a, b \in \mathbb{R}\}$ and define addition and multiplication in $\mathbb{C}$ by $(a_1, b_1) + (a_2, b_2) = (a_1 + a_2, b_1 + b_2)$ and $(a_1, b_1) \cdot (a_2, b_2) = (a_1 a_2 - b_1 b_2, a_1 b_2 + a_2 b_1)$, respectively. $\mathbb{C}$ is called the set of *complex numbers*.

(a) Show that $\mathbb{C}$ is a field.

(b) Can $\mathbb{C}$ be ordered?

(c) What is the additive identity in $\mathbb{C}$?

(d) Describe the additive inverse of (a, b) as an ordered pair.

(e) What is the multiplicative identity in $\mathbb{C}$?

(f) Describe the multiplicative inverse of (a, b) as an ordered pair.

(g) Show that there is a one-one correspondence between real numbers a and complex numbers of the form $(a, 0)$. Thus the set of real numbers is embedded in the field $\mathbb{C}$. Since there is only one set of real numbers, we may write $a = (a, 0)$ for $a \in \mathbb{R}$.

(h) Show that there is $i \in \mathbb{C}$ such that $i^2 = -1$.

(i) Show that each $(a, b) \in \mathbb{C}$ can be written as $a + ib$.

CHAPTER 3

Sequences

The concept of limit is central to many of the most important problems of analysis; for example, the two basic problems of the calculus, that of finding the tangent line to a suitably nice curve and of computing the area of the region under a curve are both usually solved using one of several kinds of limits. We initiate the study of limits by considering one of the simplest kind, the limit of a sequence. After establishing the basic properties of limits of sequences, it is a straightforward matter to extend the ideas to sequences in euclidean spaces, limits of functions, sequences of functions, and so on.

Let S be a nonempty set. Recall that a *sequence* in S is a function f whose domain is the set of positive integers, $\mathbb{N}$, and whose range is a subset of S. The anatomy of a sequence is essentially its terms and the ordering induced on those terms by the positive integers. Each of the sequences $\{1 + (-1)^n/n\}$, $\{n^2 + 1\}$ and $\{\sin n\pi/2\}$ exhibits a different behavior as we progress through its terms. As n increases, the terms of the first sequence tend toward one, those of the second increase without bound, and those of the third cycle through the values 1, 0, -1, and 0. It is the behavior of the first sequence that interests us at the moment. We would like to make precise the idea of "nearness to one as

n becomes large." We want to show that $1 + (-1)^n/n$ will be as close to one as desired, provided only that we are sufficiently "far out" in the sequence, or, that the distance $|1 - [1 + (-1)^n/n]|$ between the terms of the sequence and 1 can be made arbitrarily small for n sufficiently large. A way to do this is to specify a small tolerance, say $\varepsilon > 0$, for this distance, and then demonstrate an $N > 0$ such that if we examine any terms in the sequence beyond the Nth one, the distance of that term from 1 will be within the specified tolerance. We can do this: for $|1 + (-1)^n/n - 1| = |(-1)^n/n| = 1/n < \varepsilon$ if $n > 1/\varepsilon$. So, given a tolerance $\varepsilon > 0$, if N is taken as the first integer bigger than $1/\varepsilon$, then $|1 + (-1)^n/n - 1| < \varepsilon$ for all $n \geq N$, that is, for all terms of the sequence beyond the Nth one. In this case, the sequence is said to converge to 1. These ideas are made explicit for the general case in Definition 1. Occasionally, it's more convenient to use the nonnegative integers, or even some infinite subset of the nonnegative integers, for the domain of $\{f_k\}$. This should cause no difficulty for the reader.

Definition 1. Let $\{x_k\}$ be a sequence in $\mathbb{R}$. The sequence $\{x_k\}$ *converges* to $x \in \mathbb{R}$ if $\forall \varepsilon > 0 \; \exists N \in \mathbb{N}$ (N depends only on ε) such that $k \geq N$ implies $|x_k - x| < \varepsilon$. If $\{x_k\}$ converges to x, we write

$$\lim_k x_k = \lim x_k = \lim_{k \to \infty} x_k = x, \qquad \text{or } x_k \to x.$$

Proposition 8 (i) below shows that the number x is unique and so the notation used above is unambiguous.

Intuitively, a sequence $\{x_k\}$ converges to x if the elements x_k become arbitrarily close to x for all sufficiently large values of k. Since $|x_k - x| < \varepsilon$ is equivalent to $-\varepsilon < x_k - x < \varepsilon$ or $x - \varepsilon < x_k < x + \varepsilon$, (see Exercise 2.2b) the statement $x_k \to x$ means that for each $\varepsilon > 0$, there is $N > 0$ such that all terms of the sequence after the Nth one lie in the open interval $(x - \varepsilon, x + \varepsilon)$. See Figure 3.1.

The terms of the sequence whose kth term is given by $x_k = (-1)^k$ alternate between $+1$ and -1 and do not appear to approach any fixed number for large values of k. In fact, if $\varepsilon < \frac{1}{2}$ is given and x is any proposed limit, then $|x_k - x| < \varepsilon$ and the triangle inequality (2.1) imply that

$$|x_k - x_{k+1}| = 2 \leq |x_k - x| + |x - x_{k+1}| < 1.$$

Consequently, there is no real number x that is the limit of $\{x_k\}$.

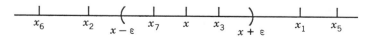

Figure 3.1

EXAMPLE 2. (i) $\lim_k c/k = 0 \ \forall c \in \mathbb{R}$. Suppose $\varepsilon > 0$, is given. The Archimedean principle (Theorem 2.1) guarantees that there is $N \in \mathbb{N}$ such that $N > |c|/\varepsilon$. Thus, for $k \geq N$, we have $|c/k| \leq |c/N| < \varepsilon$. Definition 1 is satisfied and so $\{c/k\}$ converges to 0.

(ii) Consider $\{k/(k + 3)\}$. Dividing the numerator and denominator of the kth term by k yields $k/(k + 3) = 1/(1 + 3/k)$ which apparently tends to 1 as k becomes large. To show that 1 is in fact the limit of the sequence, we must show that for each $\varepsilon > 0$, we can find $N \in \mathbb{N}$, which will generally depend on ε, such that $|k/(k + 3) - 1| < \varepsilon$ whenever $k > N$. We have $|k/(k + 3) - 1| = |-3/(k + 3)| = 3/(k + 3)$. Thus, we want $3/(k + 3) < \varepsilon$ or $k > (3 - 3\varepsilon)/\varepsilon$ for k sufficiently large. This suggests that N be chosen as the first integer larger than $(3 - 3\varepsilon)/\varepsilon$. Then for $k \geq N$, we have

$$\left| \frac{k}{k + 3} - 1 \right| = \frac{3}{k + 3} \leq \frac{3}{N + 3} < \varepsilon,$$

as required. $\square$

It's sometimes possible to establish convergence of a sequence by "squeezing" it between two convergent sequences with the same limit. The essence of such a situation is covered by the following lemma.

Lemma 3. If $\lim a_k = 0$ and $0 \leq b_k \leq a_k \ \forall k \in \mathbb{N}$, then $\lim b_k = 0$.

Proof. Let $\varepsilon > 0$. According to Definition 1, there is $N \in \mathbb{N}$ such that $|a_k| < \varepsilon$ for $k \geq N$. But then $|b_k| < \varepsilon$ for $k \geq N$. $\square$

Lemma 4. (Bernoulli's Inequality). Let $h > -1$. Then

$$(1 + h)^n \geq 1 + nh \ \forall n \in \mathbb{N}.$$

Proof. The proof is by induction on n. For $n = 1$ we have equality. Suppose the inequality holds for $n = k$. Then

$$(1 + h)^{k+1} = (1 + h)^k(1 + h) \geq (1 + kh)(1 + h)$$

$$= kh^2 + 1 + (k + 1)h \geq 1 + (k + 1)h,$$

and it follows that the inequality holds for $n = k + 1$. By the Principle of Induction, the inequality holds for all $n \in \mathbb{N}$. $\square$

EXAMPLE 5.

(i) If $|x| < 1$, then $x^k \to 0$.

(ii) If $p > 0$, then $\sqrt[k]{p} \to 1$.

(iii) $\sqrt[k]{k} \to 1$.

For (i), if $|x| < 1$, then $|x| = 1/(1 + h)$ where $h > 0$. Then $|x^k| = 1/(1 + h)^k \le 1/(1 + kh) < 1/kh$, by Bernoulli's inequality. Now apply Example 2 and Lemma 3.

For $p = 1$, (ii) is trivial; so suppose $p > 1$. Then $\sqrt[k]{p} > 1$ and we can write $\sqrt[k]{p} = 1 + h_k$ where $h_k > 0$. By Bernoulli's inequality, $p = (1 + h_k)^k \ge 1 + kh_k$ so that $0 < h_k \le (p - 1)/k$. Applying Example 2 and Lemma 3 gives $h_k = \sqrt[k]{p} - 1 \to 0$ or, equivalently, $\sqrt[k]{p} \to 1$. If $0 < p < 1$, then $1/p > 1$, which implies $(1/p)^{1/k} = 1/p^{1/k} \to 1$. By the Quotient Rule in Proposition 8(v) below, it follows that $p^{1/k} \to 1$.

For (iii), note that for $k > 1$, $\sqrt[k]{k} > 1$ so that $\sqrt[k]{k} - 1 = h_k > 0$. Thus

$$k = (1 + h_k)^k = \sum_{j=0}^{k} \binom{k}{j} h_k^j > \binom{k}{2} h_k^2 = k(k - 1)h_k^2/2;$$

this implies that $h_k^2 \le 2/(k - 1)$. Now Lemma 3, Example 2, and Exercise 1 imply that $h_k \to 0$, that is, $\sqrt[k]{k} \to 1$. ☐

As one might surmise, the behavior of an arbitrary sequence of real numbers can be quite bizarre; consider $\{(-n)^n\}$, whose terms alternate between increasingly large positive and negative numbers. In the event that a sequence converges, it must eventually settle down and become and remain close to its limit. In particular, a convergent sequence neither blows up nor does it wildly oscillate. To demonstrate this, we first describe what we mean by a bounded set.

Definition 6. A subset $E \subseteq \mathbb{R}$ is *bounded* if there is $B > 0$ such that $|x| \le B$ $\forall\, x \in E$.

Lemma 7. If $x_k \to x$, then $\{x_k : k \in \mathbb{N}\}$ is bounded. That is, a convergent sequence is bounded.

Proof. There is N such that $|x_k - x| < 1$ for $k \ge N$. Let $B = \max\{|x| + 1, |x_1|, \ldots, |x_{N-1}|\}$. Clearly $|x_k| \le B$ for $1 \le k < N$. For $k \ge N$, we have

$$|x_k| = |x_k - x + x| \le |x_k - x| + |x| \le 1 + |x| \le B.$$

So $\{x_k\}$ is bounded. ☐

Proposition 8. Let $x_k \to x$, $y_k \to y$. Then

    (i) (Uniqueness) if $x_k \to x'$, then $x = x'$;
    (ii) $x_k + y_k \to x + y$;
    (iii) if $t \in \mathbb{R}$, $tx_k \to tx$;
    (iv) $x_k y_k \to xy$;
    (v) if $y_k \ne 0 \;\forall\, k$ and $y \ne 0$, then $x_k/y_k \to x/y$.

Proof. Let $\varepsilon > 0$.

For (i), there is N_1 such that $k \geq N_1$ implies $|x_k - x| < \varepsilon/2$ and there is N_2 such that $k \geq N_2$ implies $|x_k - x'| < \varepsilon/2$. Let $N = \max\{N_1, N_2\}$. Then $k \geq N$ implies that

$$|x - x'| \leq |x - x_k| + |x_k - x'| < \varepsilon/2 + \varepsilon/2 = \varepsilon.$$

Since $\varepsilon > 0$ is arbitrary, this means that $|x - x'| = 0$ or $x = x'$.

For (ii) there is N_1 such that $k \geq N_1$ implies $|x_k - x| < \varepsilon/2$ and there is N_2 such that $|y_k - y| < \varepsilon/2$. Set $N = \max\{N_1, N_2\}$. If $k \geq N$,

$$\left|(x_k + y_k) - (x + y)\right| \leq |x_k - x| + |y_k - y| < \varepsilon/2 + \varepsilon/2 = \varepsilon.$$

(iii) is left to Exercise 5.

For (iv), Lemma 7 implies that there is $B > 0$ such that $|y_k| \leq B \ \forall k$. There are N_1 and N_2 such that $|x_k - x| < \varepsilon/2B$ for $k \geq N_1$ and $|y_k - y| < \varepsilon/2(|x| + 1)$ for $k \geq N_2$. Let $N = \max\{N_1, N_2\}$. If $k \geq N$, then

$$|x_k y_k - xy| \leq |x_k - x|\,|y_k| + |x|\,|y_k - y|$$

$$< (\varepsilon/2B)B + |x|\varepsilon/2(|x| + 1) \leq \varepsilon.$$

For (v), we first show that $1/y_k \to 1/y$. Note that there is $B > 0$ such that $|y_k| \geq B \ \forall k$ [$\exists N$ such that $k \geq N$ implies $|y_k - y| < |y|/2$ so $|y_k| > |y|/2$ for $k \geq N$. Now set $B = \min\{|y|/2, |y_1|, \ldots, |y_{N-1}|\}$]. There is N_1 such that $k \geq N_1$ implies $|y_k - y| < B|y|\varepsilon$. Thus, $k \geq N$ implies

$$|1/y_k - 1/y| = |y_k - y|\big/|yy_k| \leq |y_k - y|\big/|y|B < \varepsilon.$$

Consequently $1/y_k \to 1/y$. Combining this with (iv) yields (v). □

Proposition 9. If $x_k \to x$, $y_k \to y$ and $x_k \leq y_k \ \forall k$, then $x \leq y$.

Proof. If $x_k \geq 0$ and $x_k \to x$, then $x \geq 0$, for suppose $x < 0$. Then there is N such that $k \geq N$ implies $|x_k - x| < |x|/2$. Thus $x_k < x + |x|/2 = x/2 < 0$, which contradicts our hypothesis $x_k \geq 0$.

Now $y_k - x_k \geq 0$ and $y_k - x_k \to y - x$; so the above paragraph implies that $y - x \geq 0$. □

Proposition 8 is a useful tool that allows us to determine convergence of sequences whose terms are fairly complex algebraic combinations of simpler sequences. The drawback is that we usually need a priori information on the limits of these simpler sequences. There are, however, conditions that allow us to determine convergence without having to actually exhibit a limit. The easiest results along these lines are obtained when the terms of the sequence are eventually steadily increasing or steadily decreasing as in $\{1 - 1/n\}$ or $\{1/n\}$.

Definition 10. A real sequence $\{x_k\}$ is *increasing* (*decreasing*) if $x_k \leq x_{k+1}$ $\forall k (x_k \geq x_{k+1} \ \forall k)$. We write $x_k \uparrow (x_k \downarrow)$. A sequence $\{x_k\}$ is *monotone* if it is either increasing or decreasing.

If $x_k \uparrow$ and $x_k \to x$, we'll write $x_k \uparrow x$; a similar notation is used for convergent decreasing sequences.

The following easily obtained result characterizes convergence of monotone sequences.

Theorem 11. A monotone sequence $\{x_k\}$ converges if, and only if, it is bounded. If $x_k \uparrow$ and $\{x_k\}$ is bounded, $x_k \to \sup\{x_k : k \in \mathbb{N}\}$ $(x_k \downarrow, x_k \to \inf\{x_k : k \in \mathbb{N}\})$.

Proof. If $\{x_k\}$ converges, Lemma 7 states that it is bounded.

If $\{x_k\}$ is bounded, let $x = \sup\{x_k : k \in \mathbb{N}\}$. Let $\varepsilon > 0$. By the definition of supremum, there is N such that $x - \varepsilon < x_N \leq x$. If $k \geq N$, then $x - \varepsilon < x_N \leq x_k \leq x < x + \varepsilon$. $\qquad\square$

Note that the order completeness of $\mathbb{R}$ was utilized in this proof.

EXAMPLE 12. Let $a > 0$. We'll inductively construct a sequence that converges to $\sqrt{a}$. Choose $x_1 > 0$ arbitrarily and set

$$x_{k+1} = \frac{x_k + a/x_k}{2} \qquad \text{for} \qquad k \in \mathbb{N}.$$

We first show that $\{x_k\}$ is bounded below. Since x_k is a real root of the quadratic equation $x_k^2 - 2x_{k+1}x_k + a = 0$, the discriminant $4x_{k+1}^2 - 4a$ must be nonnegative, that is, $x_{k+1}^2 \geq a$.

Using the definition of x_{k+1}, we obtain

$$x_k - x_{k+1} = x_k - \frac{x_k + a/x_k}{2} = \left(x_k^2 - a\right)/2x_k \geq 0,$$

so that $\{x_k\}$ is decreasing for $k \geq 2$. Consequently, Theorem 11 implies that $x = \lim x_k$ exists. Since x must satisfy $x = (x + a/x)/2$, and $x \geq 0$, we obtain $x = \sqrt{a}$. $\qquad\square$

One of the basic limits discussed in a standard calculus course is

$$\lim_{k \to \infty} (1 + 1/k)^k = e,$$

which is the base for the natural logarithm. There are a number of ways to handle this limit. We present as an example a method using the binomial theorem to obtain a bounded sequence that increases to this limit.

EXAMPLE 13. Let $x_k = (1 + 1/k)^k$. An application of the binomial theorem yields

$$x_k = 1 + k\left(\frac{1}{k}\right) + \frac{k(k-1)}{2!k^2} + \frac{k(k-1)(k-2)}{3!k^3} + \cdots$$

$$+ k(k-1) \cdots \frac{1}{k!k^k}$$

$$= 1 + 1 + \left(1 - \frac{1}{k}\right)\Big/2! + \left(1 - \frac{1}{k}\right)\left(1 - \frac{2}{k}\right)\Big/3! + \cdots$$

$$+ \left(1 - \frac{1}{k}\right)\left(1 - \frac{2}{k}\right) \cdots \left(1 - \frac{k-1}{k}\right)\Big/k!.$$

As we pass from x_k to x_{k+1}, the terms after "1 + 1" increase and another term is added; so $x_k \uparrow$. Also

$$x_k \leq 1 + 1 + \frac{1}{2} + \frac{1}{2^2} + \cdots + \frac{1}{2^{k-1}} = 1 + 2\left(1 - \frac{1}{2^k}\right) < 3$$

(Exercise 2). Hence, $x = \lim x_k$ exists and $x \leq 3$. Moreover, $x_k \geq 2$, and so, $x \geq 2$.   □

Examples 12 and 13 illustrate how Theorem 11 can be used to establish the existence of a limit of a monotone sequence without actually finding that limit. However, most sequences are not monotone and Theorem 11 is not then applicable. One of the founders of modern mathematical analysis, Augustin–Louis Cauchy (1789–1857), gave a general necessary and sufficient condition for the convergence of sequences in $\mathbb{R}$ that does not require exhibition of the value of the limit.

Definition 14. A real sequence $\{x_k\}$ is a *Cauchy sequence* (or *Cauchy*) if for each $\varepsilon > 0$, there is N such that $j, k \geq N$ implies $|x_k - x_j| < \varepsilon$.

Intuitively, a sequence is Cauchy if its terms become arbitrarily close for large values of k. One must be very careful here and note that this must be true for all terms sufficiently far out in the sequence. It's possible to construct a sequence in which consecutive terms become arbitrarily close, but the sequence still diverges. See Exercise 41.

Proposition 15. If $\{x_k\}$ converges, then $\{x_k\}$ is Cauchy.

Proof. Let $\varepsilon > 0$. Since $\{x_k\}$ converges, say to x, there is N such that $k \geq N$ implies $|x_k - x| < \varepsilon/2$. Then, $k, j \geq N$ implies

$$|x_k - x_j| \leq |x_k - x| + |x - x_j| < \varepsilon. \qquad \square$$

As indicated above, the converse of Proposition 15 is also true, that is, every Cauchy sequence is convergent. We establish this result using the following lemma about "nested intervals."

Lemma 16.

(i) Let $I_k = [a_k, b_k]$, $a_k < b_k$, be such that $I_k \supseteq I_{k+1}$, $\forall\, k$. Then $\bigcap_{k=1}^{\infty} I_k \neq \varnothing$.

(ii) if in addition, $\ell(I_k) = b_k - a_k \to 0$, then the set $\bigcap_{k=1}^{\infty} I_k$ is a singleton, that is, contains exactly one element.

Proof. For (i) we have $a_1 \leq a_k \leq a_{k+1} < b_{k+1} \leq b_1$; so $\{a_k\}$ is increasing and bounded from above. Since $\mathbb{R}$ is order complete, $x = \sup\{a_k : k \in \mathbb{N}\}$ exists. Moreover, $a_k \leq x \leq b_k$ $\forall\, k$ and so, $x \in \bigcap_{k=1}^{\infty} I_k$.

For (ii), if $x, y \in \bigcap_{k=1}^{\infty} I_k$, then $|x - y| \leq b_k - a_k$ $\forall\, k$, so $|x - y| = 0$ and $x = y$. $\qquad \square$

It is important that the intervals be closed; see Exercise 11.

Theorem 17 (Cauchy Criterion). Let $\{x_k\}$ be a sequence in $\mathbb{R}$. Then $\{x_k\}$ converges iff $\{x_k\}$ is Cauchy.

Proof. Proposition 15 states that if $\{x_k\}$ converges, then it is Cauchy.

Suppose next that $\{x_k\}$ is Cauchy. We define inductively a sequence of positive integers $\{n_k\}$ as follows: let $n_1 = 1$ and choose n_{k+1} to be the smallest positive integer such that $n_{k+1} > n_k$ and $i, j \geq n_{k+1}$ implies $|x_j - x_i| < 1/2^{k+2}$. Set

$$I_k = \left[x_{n_k} - \frac{1}{2^k}, x_{n_k} + \frac{1}{2^k} \right].$$

We claim that $I_k \supseteq I_{k+1}$. For let $a \in I_{k+1}$. Then

$$|x_{n_{k+1}} - x_{n_k}| < \frac{1}{2^{k+1}}$$

and

$$|a - x_{n_{k+1}}| < \frac{1}{2^{k+1}};$$

so

$$|a - x_{n_k}| < \frac{1}{2^{k+1}} + \frac{1}{2^{k+1}} = \frac{1}{2^k}$$

and $a \in I_k$. Since $\ell(I_k) \to 0$, Lemma 16 implies that $\bigcap_{k=1}^{\infty} I_k = \{x\}$, for some x.

We now claim that $x_k \to x$. Let $\varepsilon > 0$. Choose k such that $1/2^k < \varepsilon$. If $j \geq n_{k+1}$, then

$$|x_j - x_{n_{k+1}}| < \frac{1}{2^{k+2}};$$

but $x \in I_{k+1}$, so

$$|x - x_{n_{k+1}}| \leq \frac{1}{2^{k+1}}.$$

Hence

$$|x_j - x| \leq \frac{1}{2^{k+2}} + \frac{1}{2^{k+1}} < \frac{1}{2^k} < \varepsilon \qquad \text{for } j \geq n_{k+1}. \qquad \square$$

If $\{x_k\}$ is a sequence and if $\{n_k\}$ is a sequence in $\mathbb{N}$ such that $n_k < n_{k+1}$, then the sequence $\{x_{n_k}\}$ is called a *subsequence* of $\{x_k\}$. For example, $\{x_{2k}\}$, $\{x_{2k+1}\}$ and $\{x_{2^k}\}$ are subsequences of $\{x_k\}$. In the proof of Theorem 17 we constructed a subsequence $\{x_{n_k}\}$ of the Cauchy sequence $\{x_k\}$ that converged to an element $x \in \mathbb{R}$. Then we showed that $x_k \to x$. This holds in general (Exercise 12).

It's clear that each subsequence of a convergent sequence also converges. This is formally established in the next proposition.

Proposition 18. If $x_k \to x$, then every subsequence $\{x_{n_k}\}$ of $\{x_k\}$ also converges to x.

Proof. Let $\varepsilon > 0$. $\exists N$ such that $k \geq N$ implies $|x_k - x| < \varepsilon$. Then $k \geq N$ implies $n_k \geq k \geq N$; so $|x_{n_k} - x| < \varepsilon$. $\qquad \square$

A more general statement, along with its converse, also holds and is given below in Proposition 19. The conclusion of this proposition is called the *Urysohn property* for sequences.

Proposition 19. The sequence $\{x_k\}$ converges to x if, and only if, every subsequence of $\{x_k\}$ has a subsequence that converges to x.

Proof. Suppose that $\{x_k\}$ does not converge to x. Then $\{x_k\}$ does not satisfy the definition for convergence. So, there must be at least one $\varepsilon > 0$ such that $|x_k - x| \geq \varepsilon$ for infinitely many k, that is, for each positive integer j, there is n_j with $|x_{n_j} - x| \geq \varepsilon$. Moreover, the n_j can be chosen so that $n_{j+1} \geq n_j$. Thus $\{x_{n_j}\}$ is a subsequence of $\{x_k\}$, but no subsequence of $\{x_{n_j}\}$ can converge to x.

We have shown that if $\{x_k\}$ does not converge to x, then there is a subsequence $\{x_{n_k}\}$ of $\{x_k\}$ such that no subsequence of $\{x_{n_k}\}$ converges to x. Recall that this is the *contrapositive* of the "if" part of the statement of the proposition and is equivalent to it. Thus we have established the sufficiency of the conclusion of Proposition 19.

The necessity follows from Proposition 18.   □

Convergence can also be established by considering two special subsequences, the subsequence of "even terms" and the one of "odd terms."

Proposition 20. Let $\{x_k\}$ be a real sequence. If both subsequences $\{x_{2k}\}$ and $\{x_{2k-1}\}$ converge to x, then $x_k \to x$.

Proof. Let $\varepsilon > 0$. There are N_1 and N_2 such that $k \geq N_1$ implies $|x_{2k} - x| < \varepsilon$ and $k \geq N_2$ implies $|x_{2k-1} - x| < \varepsilon$. If $N = \max\{2N_1, 2N_2 - 1\}$, then $k \geq N$ implies $|x_k - x| < \varepsilon$.   □

The example $\{(-1)^k\}$ shows that it's important that the subsequence of "even terms," $\{x_{2k}\}$, and the subsequence of "odd terms" converge to the same limit.

Clearly the sequence $\{k^2\}$ diverges, but it does so in a fairly orderly fashion by having its terms steadily increase. It will be convenient to assign an "infinite limit" to such a sequence. Intuitively, a sequence $\{x_k\}$ should have an infinite limit if the terms x_k become arbitrarily large for large values of k. This leads to the following definition.

Definition 21. The real sequence $\{x_k\}$ converges to ∞ if for each $M \in \mathbb{R}$, there is $N \in \mathbb{N}$ such that $k \geq N$ implies $x_k \geq M$. We write $x_k \to \infty$ or $\lim x_k = \infty$.

Sequences with $-\infty$ as limit are defined similarly (Exercise 29). Theorem 11 has the following extension.

Theorem 22. If $x_k \uparrow$, then $\{x_k\}$ converges to $\sup\{x_k : k \in \mathbb{N}\}$.

Proof. If $\{x_k\}$ is bounded, then Theorem 11 gives the result.

Assume that $\{x_k\}$ is not bounded. Then since $\{x_k\}$ is bounded below, for each $M > 0$, there is $N \in \mathbb{N}$ such that $x_N \geq M$. If $k \geq N$, then $x_k \geq x_N \geq M$; so $x_k \to \infty$.   □

The following proposition gives some of the algebra for infinite limits.

Proposition 23. Let $x_k \to \infty$.

   (i) If $\{y_k\}$ is bounded below, then $\lim(x_k + y_k) = \infty$.

   (ii) If $t > 0$, then $tx_k \to \infty$.

   (iii) $1/x_k \to 0$.

Proof. For (i), suppose $y_k \geq b$ for all k. Let $M > 0$. There is N such that $k \geq N$ implies $x_k \geq M - b$. Then $k \geq N$ implies $x_k + y_k \geq M$.

For (ii), if $M > 0$, $\exists N$ such that $x_k > M/t$ for $k \geq N$ or $tx_k \geq M$.

For (iii), let $\varepsilon > 0$. $\exists N$ such that $x_k \geq 1/\varepsilon$ for $k \geq N$ or $1/x_k \leq \varepsilon$.   □

The converse of Proposition 23 (iii) is also true.

Proposition 24. If $x_k > 0$ and $x_k \to 0$, then $1/x_k \to \infty$.

Proof. Let $M > 0$. There is N such that $x_k < 1/M$ for $k \geq N$, or equivalently $1/x_k > M$.   □

EXERCISES 3

1. If $p > 0$, show that $\lim_{k \to \infty} 1/k^p = 0$. What about $p \leq 0$? Assume that k^p has been defined and that the usual laws of exponents hold.

2. Use the principle of induction to show that

   (a) $(k + 1)! \geq 2^k \ \forall \, k \in \mathbb{N}$.

   (b) $1 + \dfrac{1}{2} + \dfrac{1}{2^2} + \cdots + \dfrac{1}{2^{k-1}} = 2\left(1 - \dfrac{1}{2^k}\right) \ \forall \, k \in \mathbb{N}$.

3. Show that $E \subseteq \mathbb{R}$ is bounded iff E is bounded from above and from below.

4. Is the converse of Lemma 7 valid?

5. Prove Proposition 8 (iii).

6. If $x_k \to x$ and $x_k \geq 0 \ \forall \, k \in \mathbb{N}$, show that $\sqrt{x_k} \to \sqrt{x}$.

7. If $x_k \to 0$ and $\{y_k\}$ is bounded, show that $x_k y_k \to 0$. Can the boundedness condition be dropped?

8. Is Theorem 11 valid without monotonicity?

9. Let $x_1 = 1$, $x_{k+1} = (2x_k + 3)/4$ for $k \geq 1$. Show that $\{x_k\}$ converges and find its limit.

10. Show that a Cauchy sequence is bounded without using Theorem 17 and Lemma 7.

11. Is Lemma 16 valid if $[a_k, b_k]$ is replaced by (a_k, b_k)?

12. If $\{x_k\}$ is Cauchy and has a subsequence $\{x_{n_k}\}$ which converges to x, show that $x_k \to x$.

13. If $x_k \to x$, show that $|x_k| \to |x|$. Does the converse hold?

14. If $\lim (x_k + y_k)$ exists, do $\{x_k\}$ and $\{y_k\}$ have to converge?

15. If $x_k \leq y_k \leq z_k$ and $\lim x_k = \lim z_k = a$, show that $y_k \to a$.

16. Does $\{\sqrt{k + 1} - \sqrt{k}\}$ converge?

17. Let $\{x_k\}$ be bounded above and $x = \sup\{x_k : k \in \mathbb{N}\}$. Let $y_k = \max\{x_1, \ldots, x_k\}$. Show that $y_k \uparrow x$.

18. Let $x_k \to x$. Put $u_k = \sup\{x_j : j \geq k\}$, $v_k = \inf\{x_j : j \geq k\}$. Show that $u_k \downarrow x$ and $v_k \uparrow x$.

19. Let $x_1 = 1$ and $x_{k+1} = \sqrt{2x_k}$ for $k \in \mathbb{N}$. Show that $\{x_k\}$ converges to 2. *Hint:* Use Theorem 11.

20. Choose $x_1 > 0$ and set $x_{k+1} = \sqrt{x_k + 2}$. Show that $x_k \to 2$.

21. If $x_k = (k^2 + k)^{1/2} - k$, show that $x_k \uparrow \frac{1}{2}$.

22. If $x_k = k!/k^k$, does $\{x_{k+1}/x_k\}$ converge? If so, what is its limit?

23. If $x_k \to x$ and $x < y$, show that there is N such that $x_k < y \ \forall k \geq N$.

24. If $x_k \to x$, $y_k \to y$, and $x_k < y_k \ \forall k$, does it follow that $x < y$?

25. If $x \in \mathbb{Q}$ ($x \notin \mathbb{Q}$), show that there is a sequence $\{x_k\} \subseteq \mathbb{R} \setminus \mathbb{Q}(\{x_k\} \subseteq \mathbb{Q})$ such that $x_k \to x$.

26. If $0 \leq b \leq a$, find $\lim [a^k + b^k]^{1/k}$.

27. If $E \subseteq \mathbb{R}$ is bounded above (below), show that there exists a sequence $\{x_k\} \subseteq E$ such that $x_k \to \sup E$ (inf E).

28. If $x_k \to x$, show that the sequence of arithmetic means $a_k = (x_1 + \cdots + x_k)/k$ also converges to x. Does the converse hold true? *Hint:* Consider $\{(-1)^k\}$.

29. Define $x_k \to -\infty$.

30. If $x_k \to \infty(-\infty)$, show that $\{x_k\}$ is bounded below (above).

31. State and prove the analogue of Theorem 22 for decreasing sequences.

32. State and prove the analogues of Propositions 23 and 24 for limits at $-\infty$.

33. If $x_k \to \infty$ and $y_k \to -\infty$, can anything be said about $\lim (x_k + y_k)$?

34. Can the condition, $x_k > 0$, be dropped in Proposition 24?

35. For $\{x_k\}$ a real-valued sequence, set $\underline{\lim} x_k = \sup_{k \geq 1} \inf_{n \geq k} x_n$ and $\overline{\lim} x_k = \inf_{k \geq 1} \sup_{n \geq k} x_n$. $\underline{\lim} x_k$ and $\overline{\lim} x_k$ are called the *limit inferior* and *limit superior*, respectively, of $\{x_k\}$ and are also sometimes denoted by $\lim \inf x_k$ and $\lim \sup x_k$.

(a) If $x_k \to x$ (where x might be $\pm \infty$), show that

$$\underline{\lim} x_k = \overline{\lim} x_k = x.$$

(b) Show that

$$\underline{\lim} x_k \le \overline{\lim} x_k.$$

(c) If $\underline{\lim} x_k = \overline{\lim} x_k = x$, show that $x_k \to x$.

36. If $\{x_k\}$ and $\{y_k\}$ are sequences in $\mathbb{R}$, show that $\underline{\lim}(x_k + y_k) \ge \underline{\lim} x_k + \underline{\lim} y_k$ and $\overline{\lim}(x_k + y_k) \le \overline{\lim} x_k + \overline{\lim} x_k$. Give examples to show that strict inequality can occur.

37. If $|x| \ge 1$, discuss the convergence of $\{x^k\}$.

38. Let $\{x_k\}$ be a bounded sequence in $\mathbb{R}$.
   (a) If $\overline{\lim} x_k = M$, show that for every $\varepsilon > 0$, $x_k < M + \varepsilon$ for all but a finite number of values of k, and that $x_k > M - \varepsilon$ for infinitely many values of k.
   (b) If $\underline{\lim} x_k = m$, show that for every $\varepsilon > 0$, $x_k > m - \varepsilon$ for all but a finite number of values of k, and that $x_k < m + \varepsilon$ for infinitely many values of k.

39. Using the concepts of $\underline{\lim}$ and $\overline{\lim}$, show that if $\{x_k\}$ is Cauchy, then it converges.

40. If $\{x_k\} \subseteq \mathbb{R}$ is such that $|x_{k+1} - x_k| \le |x_k - x_{k-1}|/2 \ \forall\, k$, show that $\{x_k\}$ converges.

41. Show that the condition "$|x_k - x_j| < \varepsilon$ for $k, j \ge N$" in the definition of a Cauchy sequence cannot be replaced by "$|x_{k+1} - x_k| < \varepsilon$ for $k \ge N$." Hint: Consider $x_k = \ln(k + 1)$.

42. If $t_k > 0$ and $\lim_k t_k = t > 0$, show that there is $\delta > 0$ such that $t_k \ge \delta$ $\forall\, k$.

43. Show that a sequence $\{x_k\}$ in $\mathbb{R}$ converges to 0 iff given any positive sequence $\{\varepsilon_k\}$ such that $\varepsilon_k \downarrow 0$, then every subsequence of $\{x_k\}$ has a subsequence $\{x_{n_k}\}$ satisfying $|x_{n_k}| \le \varepsilon_k \ \forall\, k$.

44. Show that $\lim_{k \to \infty} \{1/\sqrt{k^2 + 1} + 1/\sqrt{k^2 + 2} + \cdots + 1/\sqrt{k^2 + k}\} = 1$.

45. If $\{x_k\}$ is such that $|x_{k+1} - x_k| \le Mr^k$ for some $M > 0$ and $0 < r < 1$, show that $\{x_k\}$ converges.

46. Let a real number $x > 0$ and an integer $k \ge 2$ be given. Let a_0 be the largest integer $\le x$ and, assuming that $a_0, a_1, \ldots, a_{n-1}$ have been defined, let a_n be the largest integer such that

$$a_0 + \frac{a_1}{k} + \frac{a_2}{k^2} + \cdots + \frac{a_n}{k^n} \le x.$$

(a) Show that $0 \le a_i \le k - 1$ for each $i = 1, 2, \ldots$.
(b) Let $r_n = \sum_{i=0}^{n} a_i / k^i$ and show that $x = \sup \{r_n : n \in \mathbb{N}\}$.
When $k = 10$, the integers $a_0, a_1, \ldots$ are the digits in the decimal representation for x; when $k = 2$, they are the digits in the binary representation for x.

47. Let $\{x_k\} \subseteq \mathbb{R}$. The set $E = \{x \in \overline{\mathbb{R}} : \exists$ a subsequence $\{x_{n_k}\}$ converging to $x\}$ is called a set of *subsequential limits* of $\{x_k\}$. Show that $\overline{\lim} x_k = \sup E$ and $\underline{\lim} x_k = \inf E$ (see Exercise 3.35).

48. Show that $x_k \to x$ iff $\exists t_k \downarrow 0$ such that $|x_k - x| \le t_k$.

49. If there is $0 < r < 1$ such that $|x_{k+1} - x| \le r|x_k - x|$, show that $x_k \to x$.

CHAPTER 4

Infinite Series

We turn now to a special kind of sequence, infinite series, or infinite sums as they were sometimes called. One of the early problems with the concept was its definition. The operation of addition of real numbers, being binary and associative, extended easily using the principle of induction to the sum of any finite number of real numbers. But, as early as Archimedes, mathematicians had to work with expressions like

$$1 + \tfrac{1}{4} + \tfrac{1}{9} + \tfrac{1}{16} + \cdots. \tag{1}$$

The question is "what does such an expression mean?," that is "what does it mean to add an infinite number of real numbers?" It can be elegantly answered by using infinite sequences.

Given an infinite sequence of real numbers $\{a_i\}_{i=1}^{\infty}$, form a new sequence, $\{s_k\}$, by defining $s_k = \sum_{i=1}^{k} a_i$. The sequence $\{s_k\}$ is called an *infinite series*, s_k is called the kth *partial sum*, and a_i is called the ith *term* of the series. If the sequence of partial sums, $\{s_k\}$, converges to a limit $s \in \mathbb{R}$, then s is called the *sum* of the sequence $\{a_k\}$; we also call s the *sum* of the infinite series generated by $\{a_k\}$ and write $s = \Sigma_{k=1}^{\infty} a_k$. We often abuse our notation and use $\Sigma_{k=1}^{\infty} a_k$ or Σa_k to denote both the series $\{s_k\}$ and its sum s. The intended meaning is usually clear from the context and so causes no confusion. The infinite series $\Sigma_{k=1}^{\infty} a_k$ is said to *converge* if the sequence $\{s_k\}$ converges in $\mathbb{R}$; otherwise the series $\Sigma_{k=1}^{\infty} a_k$ is said to *diverge*.

The expression in (1) then is an infinite series generated by the sequence $\{1/i^2\}$; the ith term is $1/i^2$ and the kth partial sum is $\Sigma_{i=1}^{k} 1/i^2$. We shall see later that this series converges to a real number s, which will be the sum of the series $\Sigma_{k=1}^{\infty} 1/k^2$.

It is usually assumed that a sequence $\{a_k\}$ is indexed by the positive integers, but it will occasionally be convenient to index a sequence beginning with an integer other than 1. Thus, if we use $\{a_k\}_{k=0}^{\infty}$, we then write $\Sigma_{k=0}^{\infty} a_k$ for the series generated by this sequence. For sequences, convergence behavior is related to what happens sufficiently far out in the sequence. Changing the values of any finite number of terms in a sequence does not alter the convergence behavior. Similarly the convergence or divergence of an infinite series is related to the "tail piece" of the series and is not affected by alteration in any finite number of its terms. Thus, changing a finite number of terms may alter the sum of a series, but it will not influence its convergence or divergence. See Exercise 2.

Note that if $\{s_k\}$ converges to s, then $\lim a_k = \lim(s_k - s_{k-1}) = s - s = 0$; so we have Proposition 1.

Proposition 1. If $\Sigma_{k=1}^{\infty} a_k$ converges, then $\lim a_k = 0$.

The following example shows that the converse of Proposition 1 is false.

EXAMPLE 2. The *harmonic series* $\Sigma_{k=1}^{\infty} 1/k$ diverges, since

$$s_{2k} = 1 + \frac{1}{2} + \cdots + \frac{1}{k} + \frac{1}{k+1} + \cdots + \frac{1}{2k}$$

$$= s_k + \frac{1}{k+1} + \cdots + \frac{1}{2k} > s_k + \frac{1}{2}$$

shows that $\{s_k\}$ is not a Cauchy sequence. Clearly, $\lim a_k = \lim 1/k = 0$. This example shows that $\lim a_k = 0$ is a necessary, but not sufficient condition for the convergence of Σa_k. $\qquad\square$

In the following example, it is assumed that the reader is familiar with the elementary properties of the natural logarithm function. We denote this function by ln or log. A working knowledge of the elementary functions exp, sin, cos, and other trigonometric functions is also assumed throughout the text.

EXAMPLE 3. Consider the series $\sum_{k=1}^{\infty} \ln(1 + 1/k)$. Since

$$s_k = \sum_{i=1}^{k} \ln\left(\frac{i+1}{i}\right) = \ln\left(\frac{2}{1} \cdot \frac{3}{2} \cdots \frac{k}{k-1} \cdot \frac{k+1}{k}\right) = \ln(k+1),$$

the series diverges but $\lim \ln(1 + 1/k) = 0$. □

EXAMPLE 4. The *geometric series*, $\sum_{k=1}^{\infty} ar^{k-1}$, $a \in \mathbb{R}$, $r \in \mathbb{R}$, is an important class of infinite series. It is both a frequently used example and an important tool for determining convergence of series, either by direct application or by comparison. The number r is called the *common ratio* of the series. Since

$$s_k - rs_k = a + ar + \cdots + ar^{k-1} - ar - \cdots - ar^{k-1} - ar^k$$

$$= a - ar^k$$

yields $s_k = a(1 - r^k)/(1 - r)$, the geometric series with $a \neq 0$ converges iff $|r| < 1$, and in this case, $\sum_{k=1}^{\infty} ar^{k-1} = a/(1 - r)$. □

Since the convergence of a series is just the convergence of its sequence of partial sums, the results of Chapter 3 are applicable and give several tests for convergence and divergence.

Proposition 5 (Cauchy Criterion). The series $\sum_{k=1}^{\infty} a_k$ converges if, and only if, $\forall \varepsilon > 0 \; \exists N$ such that $k \geq N$ implies $|\sum_{i=k}^{k+p} a_i| < \varepsilon$ for each $p \in \mathbb{N}$.

Proof. Theorem 3.17 implies that the sequence $s_n = \sum_{i=1}^{n} a_i$ converges iff it is Cauchy, that is, iff for each $\varepsilon > 0$, there is $N > 0$ such that $|s_m - s_n| < \varepsilon$ $\forall m, n \geq N$. For any $p \in \mathbb{N}$, let $m = k + p$ and $n = k - 1$. Then $m, n \geq N$ is equivalent to $p \in \mathbb{N}$ and $k > N$. So, the series converges iff

$$|s_m - s_n| = |s_{k+p} - s_{k-1}| = \left|\sum_{i=k}^{k+p} a_i\right| < \varepsilon$$

$\forall p \in \mathbb{N}$ and $k > N$. □

It's sometimes particularly easy to obtain results about series with nonnegative terms, for the sequence of partial sums in this case is monotone increasing. From 3.11, we immediately have Proposition 6.

Proposition 6. Let $a_k \geq 0$. Then $\sum_{k=1}^{\infty} a_k$ converges if, and only if, the sequence of partial sums is bounded (from above).

Corollary 7 (Comparison Test). Suppose that $0 \le a_k \le b_k$.

  (i) If $\sum_{k=1}^{\infty} b_k$ converges, then $\sum_{k=1}^{\infty} a_k$ converges.
  (ii) If $\sum_{k=1}^{\infty} a_k$ diverges, then $\sum_{k=1}^{\infty} b_k$ diverges.

  Proof. Note that $\sum_{i=1}^{k} a_i \le \sum_{i=1}^{k} b$ and apply Proposition 6.   □

Corollary 8 (Limit Form of the Comparison Test). Let $a_k \ge 0$, $b_k > 0$ and suppose that $\lim a_k/b_k = L \ne 0$. Then either $\sum_{k=1}^{\infty} a_k$ and $\sum_{k=1}^{\infty} b_k$ both converge or both diverge.

  Proof. There is N such that $k \ge N$ implies $|a_k/b_k - L| < L/2$, or $L/2 \le a_k/b_k \le 3L/2$ (Exercise 2.2), or $Lb_k/2 \le a_k \le 3Lb_k/2$. Apply Corollary 7. □

  In Exercise 6, the reader will be asked to examine the situation in Corollary 8 when $L = 0$.

  For series with decreasing positive terms, Cauchy developed an elegant convergence test that uses only a very "thin" subsequence of $\{a_k\}$.

Proposition 9 (Cauchy Condensation Test). Let $a_k \ge 0$ and $a_k \downarrow$. Then $\sum_{k=1}^{\infty} a_k$ converges iff the series $\sum_{k=0}^{\infty} 2^k a_{2^k}$ converges.

  Proof. Let $s_k = \sum_{i=1}^{k} a_i$ and $t_k = \sum_{i=0}^{k} 2^i a_{2^i}$. For $n < 2^k$,

$$s_n \le a_1 + (a_2 + a_3) + \cdots + (a_{2^k} + \cdots + a_{2^{k+1}-1})$$

$$\le a_1 + 2a_2 + \cdots + 2^k a_{2^k} = t_k.$$

Thus, the convergence of $\sum_{k=0}^{\infty} 2^k a_{2^k}$ implies the convergence of $\sum_{k=1}^{\infty} a_k$.
  For $n > 2^k$,

$$s_n \ge a_1 + a_2 + (a_3 + a_4) + \cdots + (a_{2^{k-1}+1} + \cdots + a_{2^k})$$

$$\ge a_1/2 + a_2 + 2a_4 + \cdots + 2^{k-1} a_{2^k} = t_k/2.$$

Thus, if $\sum_{k=1}^{\infty} a_k$ converges, then $\sum_{k=0}^{\infty} 2^k a_{2^k}$ converges.   □

  The next two results are applications of the Cauchy condensation test.

EXAMPLE 10 (*p*-series). The series $\sum_{k=1}^{\infty} 1/k^p$ is called a *p-series*; the series converges if $p > 1$ and diverges for $p \le 1$. If $p \le 0$, then $\lim 1/k^p \ne 0$ (Exercise 3.1); so the series $\sum_{k=1}^{\infty} 1/k^p$ diverges. For $p > 0$, the Cauchy condensation test is applicable with

$$\sum_{k=0}^{\infty} 2^k/(2^k)^p = \sum_{k=0}^{\infty} 2^{(1-p)k}.$$

This is a geometric series with common ratio $r = 2^{1-p}$. The result follows from Example 4. Exercise **9**.13 can be used to establish this result in yet another way. The series in expression (1) at the beginning of the chapter is a p-series with $p = 2$ and therefore converges.   □

As another application of Proposition 9, we establish an interesting result of the Norwegian mathematician Niels Abel (1802–1829), which asserts that for a convergent series of positive decreasing terms, the kth term goes to zero "faster" than the kth term of the harmonic series.

Corollary 11 (Abel). Let $a_k \geq 0$ and $a_k \downarrow$. If $\sum_{k=1}^{\infty} a_k$ converges, then $\lim k a_k = 0$.

Proof. By the Cauchy condensation test, $2^k a_{2^k} \to 0$. If $2^n \leq k \leq 2^{n+1}$, then $k a_k \leq 2^{n+1} a_{2^n}$ and $k a_k \to 0$.   □

Note that Corollary 11 is an improvement of Proposition 1 for positive termed series whose terms decrease.

The ratio test, when it works, is usually one of the easiest to apply.

Proposition 12 (Ratio Test). Let $a_k > 0$.

  (i) If there are N and q, $0 < q < 1$, such that $k \geq N$ implies $a_{k+1}/a_k \leq q$, then $\sum_{k=1}^{\infty} a_k$ converges.
  (ii) If there is N such that $k \geq N$ implies $a_{k+1}/a_k \geq 1$, then $\sum_{k=1}^{\infty} a_k$ diverges.

Proof. For (i), note that $a_{N+1} \leq q a_N$, $a_{N+2} \leq q a_{N+1} \leq q^2 a_N$; so $a_{N+p} \leq q^p a_N$. Now apply Corollary 7 to the convergent geometric series $\sum_{p=1}^{\infty} q^p$ and use Exercise 2.

For (ii), $0 < a_N \leq a_{N+1} \leq \cdots \leq a_{N+p}$; so $\lim a_k \neq 0$, and the series $\sum_{k=1}^{\infty} a_k$ diverges.   □

The more familiar form of the ratio test from calculus now follows easily.

Corollary 13 (Calculus Ratio Test). Let $a_k > 0$ and suppose that $\lim a_{k+1}/a_k = L$.

  (i) If $L < 1$, then $\sum_{k=1}^{\infty} a_k$ converges.
  (ii) If $L > 1$, then $\sum_{k=1}^{\infty} a_k$ diverges.
  (iii) If $L = 1$, no conclusion about convergence of the series can be made.

Proof. If $L < 1$, then $L < (L + 1)/2 < 1$. So $\exists N$ such that $k \geq N$ implies $a_{k+1}/a_k \leq (L + 1)/2$. Thus, (i) follows from Proposition 12.

If $L > 1$, then $\exists\, N$ such that $k \geq N$ implies $a_{k+1}/a_k > 1$ so (ii) follows from Proposition 12.

For (iii), consider the two p-series, $\sum_{k=1}^{\infty} 1/k$ and $\sum_{k=1}^{\infty} 1/k^2$. $\qquad\square$

Another version of the ratio test, using the limit superior, is given in Exercise 33. An example is given in Exercise 7 where the ratio test is applicable but the calculus ratio test fails.

Another useful and more powerful test is the root test. A version of it using the limit superior is given in Exercise 32.

Proposition 14 (Root Test). Let $a_k \geq 0$.

   (i) If there are N and q, $0 < q < 1$ such that $k \geq N$ implies $\sqrt[k]{a_k} \leq q$, then $\sum_{k=1}^{\infty} a_k$ converges.

   (ii) If for infinitely many k, $\sqrt[k]{a_k} \geq 1$, then $\sum_{k=1}^{\infty} a_k$ diverges.

Proof. For (i), if $k \geq N$, then $a_k \leq q^k$; so Corollary 7 is applicable using the convergent geometric series $\sum_{k=1}^{\infty} q^k$.

For (ii), we have $a_k \geq 1$ for infinitely many k; so $\sum_{k=1}^{\infty} a_k$ diverges by Proposition 1. $\qquad\square$

Corollary 15 (Limit Form of the Root Test). Suppose $a_k \geq 0$ and $\lim \sqrt[k]{a_k} = L$.

   (i) If $L < 1$, then $\sum_{k=1}^{\infty} a_k$ converges.

   (ii) If $L > 1$, then $\sum_{k=1}^{\infty} a_k$ diverges.

   (iii) If $L = 1$, no conclusion about convergence or divergence of the series can be made.

Proof. The proof is basically the same as that of Corollary 13 (Exercise 9). $\qquad\square$

We give an example in which the Root Test shows convergence, but the ratio test fails.

EXAMPLE 16. Let $a_k = \begin{cases} 1/3^k & \text{if } k \text{ is odd,} \\ 1/2^k & \text{if } k \text{ is even.} \end{cases}$ Then $\sqrt[k]{a_k} \leq 1/2$ and $\sum_{k=1}^{\infty} a_k$ converges by the root test. Note that the ratio test fails since

$$a_{k+1}/a_k = \begin{cases} 3^k/2^{k+1} & \text{if } k \text{ is odd,} \\ 2^k/3^{k+1} & \text{if } k \text{ is even.} \end{cases}$$

Note also that the limit form of the root test fails.

It is true, in general, that if the ratio test shows convergence then the root test shows convergence. For suppose (i) of the ratio test holds. Then $a_{N+k} \leq q^k a_N$ for $k \geq 1$ or $a_k \leq a_N q^{k-N}$ for $k \geq N$. So,

$$\sqrt[k]{a_k} \leq q\sqrt[k]{a_N/q^N},$$

and since

$$q\sqrt[k]{a_N/q^N} \rightarrow q,$$

(i) of the root test holds. Thus, the root test is a stronger test for convergence than the ratio test; however, the ratio test is usually easier to apply. □

Absolute Convergence

If Σa_k is a series with real terms, then $\Sigma |a_k|$, formed using the absolute values of the terms a_k, is a series of nonnegative real numbers and all the tests that we've developed for such series apply to this new series. Moreover, it will be shown that Σa_k converges whenever $\Sigma |a_k|$ converges. Examination of the convergence of $\Sigma |a_k|$ is often a useful approach, for it is usually difficult to determine the behavior of a series of real numbers with "mixed signs." The situation is even worse than it superficially appears. It's sometimes possible to "rearrange" a series of real numbers, using each term exactly once, so that the "rearranged" series converges to any chosen real number or actually diverges. These ideas are formalized and explored in the following propositions and examples.

Definition 17. The series $\sum_{k=1}^{\infty} a_k$ is *absolutely convergent* (or *converges absolutely*) if the series $\sum_{k=1}^{\infty} |a_k|$ converges.

Proposition 18. If $\sum_{k=1}^{\infty} a_k$ converges absolutely, then $\sum_{k=1}^{\infty} a_k$ converges.

Proof. For $k > j$, $|\sum_{i=j}^{k} a_i| \leq \sum_{i=j}^{k} |a_i|$; so Proposition 5 gives the result. □

Since $\sum_{k=1}^{\infty} |a_k|$ is a series of positive terms, we reemphasize that all the tests for series of positive terms apply to show absolute convergence and, therefore, convergence.

The converse of Proposition 18 is false. To give an example of a series that converges but is not absolutely convergent, we consider an *alternating series*, that is, a series in which successive terms alternate in sign.

Proposition 19 (Alternating Series Test). Let $a_k \geq 0$, $a_k \downarrow$ and $a_k \to 0$. Then the alternating series $\sum_{k=1}^{\infty}(-1)^{k+1}a_k$ converges.

Proof. Note that

$$0 \leq s_{2k}$$

$$= (a_1 - a_2) + (a_3 - a_4) + \cdots + (a_{2k-1} - a_{2k})$$

$$= s_{2k-2} + (a_{2k-1} - a_{2k})$$

$$= a_1 - (a_2 - a_3) - \cdots - (a_{2k-2} - a_{2k-1}) - a_{2k}$$

$$\leq a_1.$$

Thus, $\{s_{2k}\}$ is increasing and bounded above by a_1. Hence, $\lim s_{2k} = s$ exists with $0 \leq s \leq a_1$. Since $a_k \to 0$,

$$\lim s_{2k+1} = \lim(s_{2k} + a_{2k+1}) = s.$$

Hence, $\lim s_k = s$ by **3.20**.   □

The proof also yields the following error estimate for approximating the sum of an alternating series by its partial sum (Exercise 10).

Corollary 20. $|\sum_{k=N}^{\infty}(-1)^{k+1}a_k| \leq a_N$.

From Proposition 19 it follows that the alternating harmonic series $\sum_{k=1}^{\infty}(-1)^k/k$ converges but does not converge absolutely. For the sum of this series see the paragraph following **15.4**. A series that converges but does not converge absolutely is said to be *conditionally convergent*. The idea of conditional convergence arises in the examination of "rearrangements" of the terms of a series and is related to "nonabsolute" convergence. This relationship will unfold as we proceed.

A rearrangement of the terms of a series can be best described using a *permutation* of $\mathbb{N}$, that is, a one-one, onto function $\pi: \mathbb{N} \to \mathbb{N}$. If π is a permutation of $\mathbb{N}$, the series $\sum_{k=1}^{\infty}a_{\pi(k)}$ is called a *rearrangement* of the series $\sum_{k=1}^{\infty}a_k$.

A series may be convergent to a sum s but have a rearrangement that does not converge to s. For example, consider the following rearrangement of the alternating harmonic series $\sum_{k=1}^{\infty}(-1)^{k+1}/k$:

$$1 - \tfrac{1}{2} - \tfrac{1}{4} + \tfrac{1}{3} - \tfrac{1}{6} - \tfrac{1}{8} + \tfrac{1}{5} - \cdots.$$

If s_n, t_n, s, and t represent the partial sums and sums, respectively, of the series and its arrangement, then

$$t_{3n} = 1 - \frac{1}{2} - \frac{1}{4} + \cdots + \frac{1}{2n-1} - \frac{1}{4n-2} - \frac{1}{4n}$$

$$= \left(1 + \frac{1}{3} + \cdots + \frac{1}{2n-1}\right) - \left(\frac{1}{2} + \frac{1}{6} + \cdots + \frac{1}{4n-2}\right)$$

$$- \left(\frac{1}{4} + \frac{1}{8} + \cdots + \frac{1}{4n}\right)$$

$$= \left(1 + \frac{1}{3} + \cdots + \frac{1}{2n-1}\right) - \frac{1}{2}\left(1 + \frac{1}{3} + \cdots + \frac{1}{2n-1}\right)$$

$$- \frac{1}{2}\left(\frac{1}{2} + \frac{1}{4} + \cdots + \frac{1}{2n}\right) = \frac{1}{2}s_{2n} \to \frac{s}{2}.$$

Since $t_{3n+1} = t_{3n} + \alpha_n$ and $t_{3n+2} = t_{3n} + \beta_n$, where $\alpha_n \to 0$ and $\beta_n \to 0$, it follows that the rearrangement converges with sum $t = s/2$.

For an absolutely convergent series the situation is much better.

Theorem 21. Let $\Sigma_{k=1}^{\infty} a_k$ be absolutely convergent. Then any rearrangement of $\Sigma_{k=1}^{\infty} a_k$ is absolutely convergent and the rearrangement converges to $\Sigma_{k=1}^{\infty} a_k$.

Proof. Let $\Sigma_{k=1}^{\infty} a_{\pi(k)}$ be a rearrangement and $\beta = \Sigma_{k=1}^{\infty}|a_k|$. Then $\Sigma_{k=1}^{n}|a_{\pi(k)}| \leq \beta \; \forall \, n$; so $\Sigma_{k=1}^{\infty} a_{\pi(k)}$ is absolutely convergent to some $b \in \mathbb{R}$. Let $a = \Sigma_{k=1}^{\infty} a_k$, $s_n = \Sigma_{k=1}^{n} a_k$, $t_n = \Sigma_{k=1}^{n} a_{\pi(k)}$ and $\varepsilon > 0$. Choose N such that $|a - s_n| < \varepsilon/3$ and $\Sigma_{k=n+1}^{\infty}|a_k| < \varepsilon/3$ for $n \geq N$. Choose a partial sum t_r such that $|b - t_r| < \varepsilon/3$ and such that $a_1, \ldots, a_N$ all occur in t_r. Then

$$|a - b| \leq |a - s_N| + |s_N - t_r| + |t_r - b|.$$

Since $s_N = \Sigma_{k=1}^{N} a_k$ and each of $a_1, a_2, \ldots, a_N$ is in t_r, the minimum k in $s_N - t_r$ is $k = N + 1$; so $s_N - t_r \leq \Sigma_{k=N+1}^{\infty}|a_k| < \varepsilon/3$. It follows that

$$|a - b| < \varepsilon/3 + \varepsilon/3 + \varepsilon/3 = \varepsilon.$$

Hence, $a = b$.   □

A result of the German mathematician Bernhard Riemann (1826–1866) shows that if the series $\Sigma_{k=1}^{\infty} a_k$ is conditionally convergent and $c \in \mathbb{R}$, then there is a rearrangement of $\Sigma_{k=1}^{\infty} a_k$ which converges to c. (See Rudin, 1976, p. 76 and also Schaefer, 1981, pp. 33–40, for some interesting remarks concerning rearrangements.)

Let $a: \mathbb{N} \times \mathbb{N} \to \mathbb{R}$. Then a is called a *double sequence*. The values $a(i, j)$ are usually denoted a_{ij}. The series $\sum_{i=1}^{\infty}\sum_{i=1}^{\infty}a_{ij} = \sum_{j=1}^{\infty}(\sum_{i=1}^{\infty}a_{ij})$ and $\sum_{i=1}^{\infty}\sum_{j=1}^{\infty}a_{ij}$ are called *iterated series*. For example, if $a_{ij} = 1/2^{i+j}$, then $\sum_{i=1}^{\infty}a_{ij} = 1/2^i$ and $\sum_{i=1}^{\infty}\sum_{j=1}^{\infty}a_{ij} = \sum_{i=1}^{\infty}1/2^i = 1$. However, if $a_{ij} = 1/(i2^j)$, then $\sum_{j=1}^{\infty}a_{ij} = 1/i$ and $\sum_{i=1}^{\infty}(\sum_{j=1}^{\infty}a_{ij})$ fails to exist.

For positive termed series, we have the following result, which has important applications to the theory of Lebesgue measure on $\mathbb{R}$.

Proposition 22. Let $a_{ij} \geq 0$ and let $\{a_i\}$ be any enumeration of $\{a_{ij}: i, j \in \mathbb{N}\}$. The following are equivalent:

  (i) $\sum_{i=1}^{\infty}\sum_{j=1}^{\infty}a_{ij}$ converges.
  (ii) $\sum_{i=1}^{\infty}a_i$ converges. In this case, $\sum_{i=1}^{\infty}a_i = \sum_{i=1}^{\infty}\sum_{j=1}^{\infty}a_{ij}$.

Proof. Suppose (i) holds and put $s_n = \sum_{i=1}^{n}a_i$. Fix n. Pick i_0, j_0 such that $\{a_1, \dots, a_n\} \subseteq \{a_{ij}: i \leq i_0, j \leq j_0\}$. Then

$$s_n \leq \sum_{i=1}^{i_0}\sum_{j=1}^{j_0}a_{ij} \leq \sum_{i=1}^{\infty}\sum_{j=1}^{\infty}a_{ij}.$$

Hence,

$$\sum_{i=1}^{\infty}a_i \leq \sum_{i=1}^{\infty}\sum_{j=1}^{\infty}a_{ij}.$$

Suppose (ii) holds. First we show that $\sum_{j=1}^{\infty}a_{ij}$ converges for each i. Given n pick m such that $\{a_{i1}, \dots, a_{in}\} \subseteq \{a_1, \dots, a_m\}$. Then $\sum_{j=1}^{n}a_{ij} \leq \sum_{j=1}^{m}a_j \leq \sum_{j=1}^{\infty}a_j$ so that $\sum_{j=1}^{\infty}a_{ij} \leq \sum_{j=1}^{\infty}a_j$ for each i.

Given m, n, pick N such that $\{a_{ij}: i \leq m, j \leq n\} \subseteq \{a_i: i \leq N\}$. Then $\sum_{i=1}^{m}\sum_{j=1}^{n}a_{ij} \leq \sum_{i=1}^{N}a_i \leq \sum_{i=1}^{\infty}a_i$; so for each m, $\sum_{i=1}^{m}\sum_{j=1}^{\infty}a_{ij} \leq \sum_{i=1}^{\infty}a_i$ and $\sum_{i=1}^{\infty}\sum_{j=1}^{\infty}a_{ij} \leq \sum_{i=1}^{\infty}a_i$. $\square$

For series with terms that are not positive see Exercise 17.

Finally we give a series representation for the number e. Recall that $e = \lim(1 + 1/k)^k$ (Example 3.13).

Proposition 23. $e = \sum_{k=0}^{\infty}1/k!$.

Proof. The series $\sum_{k=0}^{\infty}1/k!$ clearly converges since $1/k! \leq 1/2^{k-1}$ (Exercise 3.2). Now

$$\left(1 + \frac{1}{k}\right)^k = 1 + 1 + \frac{1}{2!}\left(1 - \frac{1}{k}\right) + \frac{1}{3!}\left(1 - \frac{1}{k}\right)\left(1 - \frac{2}{k}\right) + \cdots$$

$$+ \frac{1}{k!}\left(1 - \frac{1}{k}\right)\cdots\left(1 - \frac{k-1}{k}\right) \leq \sum_{i=0}^{k}\frac{1}{i!};$$

so $e \leq \sum_{k=0}^{\infty}1/k!$.

If $k > n$,

$$\left(1 + \frac{1}{k}\right)^k \geq 1 + 1 + \frac{1}{2!}\left(1 - \frac{1}{k}\right) + \cdots + \frac{1}{n!}\left(1 - \frac{1}{k}\right)\cdots\left(1 - \frac{n-1}{k}\right);$$

now fix n and let $k \to \infty$ to obtain $e \geq \sum_{k=0}^{n} 1/k!$. Hence, $e \geq \sum_{k=0}^{\infty} 1/k!$.   □

Corollary 24. If $s_n = \sum_{k=0}^{n} 1/k!$, then $0 < e - s_n < 1/n(n!)$.

Proof.

$$e - s_n = \sum_{k=n+1}^{\infty} \frac{1}{k!} < \frac{1}{(n+1)!}\left\{1 + \frac{1}{n+1} + \frac{1}{(n+1)^2} + \cdots\right\} = \frac{1}{n(n!)}.$$

□

Using the series representation, we can easily show that e is irrational.

Proposition 25. The number e is irrational.

Proof. Suppose that $e = p/q$ with $p, q \in \mathbb{N}$. By Corollary 24, $0 < q!(e - s_q) < 1/q$. By assumption, $q!e \in \mathbb{N}$. Since $q!s_q = q!(1 + 1 + 1/2! + \cdots + 1/q!)$ is an integer, $q!(e - s_q)$ is also an integer. Thus, we have an integer which lies between 0 and $1/q$; this is clearly impossible.   □

For a generalization of the method of proof used in Proposition 25, see N. J. Lord, 1985, 213–214, and also A. E. Parks, 1986, 722–723.

EXERCISES 4

1. Let Σa_k and Σb_k be convergent series. Show that
   (a) $\Sigma(a_k \pm b_k) = \Sigma a_k \pm \Sigma b_k$
   (b) $\Sigma M a_k = M \Sigma a_k$ for every $M \in \mathbb{R}$.

2. Show that $\sum_{k=1}^{\infty} a_k$ converges if, and only if, $\sum_{k=N}^{\infty} a_k$ converges for every $N \in \mathbb{N}$.

3. Can "nonnegativity" be dropped in Proposition 6? In Corollary 7?

4. Use Example 3 and Corollary 7 to show that $\sum_{k=1}^{\infty} 1/k$ diverges.

5. Check the following series for convergence or divergence:

   (a) $\sum_{k=1}^{\infty} 1/(2k-1)(2k+1)$ (b) $\sum_{k=1}^{\infty} k/(k+1)$

   (c) $\sum_{k=1}^{\infty} (\sqrt{k+1} - \sqrt{k})/k$ (d) $\sum_{k=1}^{\infty} 3^k/5^{k+1}$

   (e) $\sum_{k=1}^{\infty} \left(\frac{100k+1}{k}\right)/k^2$ (f) $\sum_{k=1}^{\infty} 1/(2k+1)$

6. If $a_k \geq 0$, $b_k > 0$ and $\lim a_k/b_k = 0$, can anything be said about the relation of convergence and/or divergence between $\sum_{k=1}^{\infty} a_k$ and $\sum_{k=1}^{\infty} b_k$?

7. Let

$$0 < r < 1/2, \qquad a_k = \begin{cases} r^k & \text{if } k \text{ is odd,} \\ r^k/2 & \text{if } k \text{ is even.} \end{cases}$$

Show that $\sum_{k=1}^{\infty} a_k$ converges by using the ratio test. Show that the calculus ratio test is not applicable.

8. Check the following series for convergence or divergence:

(a) $\displaystyle\sum_{k=1}^{\infty} k^p p^k,\ p > 0$ (b) $\displaystyle\sum_{k=1}^{\infty} (\sqrt[k]{k} - 1)^k$ (c) $\displaystyle\sum_{k=1}^{\infty} k^{-1-1/k}$

(d) $\displaystyle\sum_{k=1}^{\infty} 1/(p^k - q^k),\ 0 < q < p$ (e) $\displaystyle\sum_{k=2}^{\infty} 1/k(\log k)^k$

(f) $\displaystyle\sum_{k=2}^{\infty} 1/(\log k)^k$ (g) $\displaystyle\sum_{k=1}^{\infty} k!/k^k$

(h) $\displaystyle\sum_{k=2}^{\infty} 1/(\log k)^p,\ p > 0$ (i) $\displaystyle\sum_{k=2}^{\infty} (\log k)/k^p,\ p > 0$

9. Prove Corollary 15.

10. Prove Corollary 20.

11. Show that if $\sum a_k$ converges, then any series formed by inserting parentheses in $\sum a_k$ also converges. Show that the converse of this is not true.

12. Give an example of a divergent series $\sum_{k=1}^{\infty} a_k$ such that $(a_1 + a_2) + (a_3 + a_4) + (a_5 + a_6) + \cdots$ converges.

13. If $\{a_{n_k}\}$ is a subsequence of the sequence $\{a_k\}$, the series $\sum_{k=1}^{\infty} a_{n_k}$ is called a *subseries* of $\sum_{k=1}^{\infty} a_k$. Give an example of a convergent series that has a divergent subseries.

14. Show that a series $\sum_{k=1}^{\infty} a_k$ is absolutely convergent iff every subseries is convergent.

15. Let $\{a_k\}_{k=1}^{\infty}$ be a sequence such that $a_k \geq 0$, $k \geq N$, for some $N \geq 1$. Show that $\sum_{k=1}^{\infty} a_k$ converges if, and only if, it converges absolutely. Some or all of the a_k, $k = 1, \ldots, N - 1$ may be negative.

16. An iterated series $\sum_{i=1}^{\infty}\sum_{j=1}^{\infty} a_{ij}$ is said to *converge absolutely* if the series $\sum_{i=1}^{\infty}\sum_{j=1}^{\infty} |a_{ij}|$ converges, or, equivalently by Proposition 22, if the series $\sum_{i=1}^{\infty} a_i$ converges absolutely for every enumeration $\{a_i\}$ of $\{a_{ij}\}$. Analogous to the situation in Theorem 21, if $\sum_{i=1}^{\infty}\sum_{j=1}^{\infty} a_{ij}$ converges absolutely, show that it converges. Moreover, if $\{a_i\}$ is any enumeration of $\{a_{ij}\}$, show that $\sum a_i$ converges with $\sum_{i=1}^{\infty} a_i = \sum_{i=1}^{\infty}\sum_{j=1}^{\infty} a_{ij} = \sum_{j=1}^{\infty}\sum_{i=1}^{\infty} a_{ij}$. For an example of a nonabsolutely convergent iterated series, see Exercise 17.

17. Let

$$a_{ij} = \begin{cases} +1 & \text{if } i - j = 1 \\ -1 & \text{if } i - j = -1 \\ 0 & \text{otherwise.} \end{cases}$$

Show that $\sum_{i=1}^{\infty}\sum_{j=1}^{\infty}a_{ij}$ and $\sum_{j=1}^{\infty}\sum_{i=1}^{\infty}a_{ij}$ both exist, but are not equal. Find an enumeration $\{a_i\}$ of $\{a_{ij}\}$ such that $\sum_{i=1}^{\infty}a_i$ diverges.

18. Show that $\sum_{k=1}^{\infty}1/k(k+1) = 1$, and $\sum_{k=1}^{\infty}1/(\alpha+k-1)(\alpha+k) = 1/\alpha$ for $\alpha > 0$.
    Hint: Use partial fractions.

19. Show that $\sum_{k=1}^{\infty}x_k$ converges absolutely iff $\sum_{k=1}^{\infty}x_k y_k$ converges for every bounded sequence $\{y_k\}$.

20. Show that if $a_k \geq 0$ and $\sum_{k=1}^{\infty}a_k$ converges, then $\sum_{k=1}^{\infty}a_k^2$ converges. Does the converse hold? Can the nonnegativity conditions be dropped?

21. Does $\sum_{k=1}^{\infty} \sin^2 \pi(k + 1/k)$ converge?

22. (Dirichlet's Test) If $\sum_{k=1}^{\infty}b_k$ is a series whose partial sums are bounded and if $a_k \downarrow 0$, show that $\sum_{k=1}^{\infty}a_k b_k$ converges.
    Hint: Establish the "partial summation" formula

$$\sum_{k=1}^{n} a_k b_k = a_n B_n + \sum_{k=1}^{n-1} B_k(a_k - a_{k+1}), \text{ where } B_k = \sum_{j=1}^{k} b_j.$$

23. Use Dirichlet's test to show that the series $\sum_{k=1}^{\infty}a_k \sin kx$ converges for all x if $a_k \downarrow 0$.
    Hint: $\displaystyle\sum_{k=1}^{n} \sin kx = \sin \frac{nx}{2} \sin \frac{(n+1)x}{2} \left/ \sin \frac{x}{2} \right.$

24. Use the Dirichlet test to establish the alternating series test 4.19.

25. (Abel's Test) If $\sum_{k=1}^{\infty}b_k$ converges and $\{a_k\}$ is bounded and monotonic, show that $\sum_{k=1}^{\infty}a_k b_k$ converges.
    Hint: Use the hint in problem 22.

26. Show that $\sum_{k=1}^{\infty}x_k$ converges absolutely iff the series $\sum_{k=1}^{\infty}\varepsilon_k x_k$ converges for every sequence $\{\varepsilon_k\}$ with $\varepsilon_k = \pm 1$.

27. If Σa_k converges absolutely, show that Σa_k^2 converges. Can "converges absolutely" be replaced by "converges"?

28. If Σa_k^2 and Σb_k^2 converge, show that $\Sigma a_k b_k$ converges absolutely.

29. Show that the monotonicity condition in the alternating series test (Proposition 19) cannot be dropped.

Hint: Let

$$a_k = \frac{1}{\sqrt{k}} + \frac{(-1)^{k-1}}{k}$$

and consider $\Sigma(-1)^k a_k$.

30. Let $x_k > 0$. Show that Σx_k and $\Sigma x_k/(1 + x_k)$ both converge or both diverge.

31. Show that if $a_n \to 0$, then $\{a_n\}$ has a subsequence $\{a_{n_k}\}$ such that the series Σa_{n_k} is absolutely convergent.

32. Let $\{a_k\}$ be a sequence in $\mathbb{R}$ and let $r = \overline{\lim} |a_k|^{1/k}$. Show that Σa_k converges absolutely if $r < 1$ and diverges if $r > 1$. See Exercise 3.35 for the definition of $\overline{\lim}$.

33. Let $\{a_k\}$ be a sequence of nonzero real numbers and let $r = \overline{\lim} |a_{k+1}|/|a_k|$. Show that Σa_k converges absolutely if $r < 1$ and diverges if $r > 1$.

CHAPTER 5

Euclidean Space

We next direct our attention to the extension of the ideas of convergence of sequences of real numbers to sequences whose values lie in higher dimensional Euclidean spaces. This will greatly facilitate our discussion of functions and sequences of functions with domains and/or ranges in these spaces. The reader has probably already encountered many of these ideas during the study of functions of several variables and vector-valued functions in the calculus.

Most of the properties of the cartesian plane extend neatly to *n-dimensional Euclidean space* $\mathbb{R}^n$, which is the set of all ordered *n*-tuples $\mathbf{x} = (x_1, \ldots, x_n)$ of real numbers, that is, $\mathbb{R}^n = \mathbb{R} \times \cdots \times \mathbb{R}$, where there are n factors $\mathbb{R}$ in the cartesian product. The *n*-tuples $\mathbf{x} = (x_1, \ldots, x_n)$ are called *vectors* and, in this context, the elements of $\mathbb{R}$ are called *scalars*. The definitions of addition and multiplication by scalars extend directly from $\mathbb{R}^2$ and $\mathbb{R}^3$ to $\mathbb{R}^n$. If $\mathbf{x} = (x_1, \ldots, x_n)$ and $\mathbf{y} = (y_1, \ldots, y_n)$ are vectors in $\mathbb{R}^n$, then *addition* and *scalar multiplication* (or, multiplication by a scalar) are defined by

$$\mathbf{x} + \mathbf{y} = (x_1 + y_1, \ldots, x_n + y_n) \qquad \text{and} \qquad a\mathbf{x} = (ax_1, \ldots, ax_n),$$

where $a \in \mathbb{R}$. Under these operations, $\mathbb{R}^n$ is a *vector space* over the scalar field $\mathbb{R}$, that is, the following properties hold.

Proposition 1. Let

$$\mathbf{x} = (x_1, \ldots, x_n), \mathbf{y} = (y_1, \ldots, y_n), \mathbf{z} = (z_1, \ldots, z_n) \in \mathbb{R}^n.$$

Then

   (i) $\mathbf{x} + \mathbf{y} = \mathbf{y} + \mathbf{x}$.
   (ii) $(\mathbf{x} + \mathbf{y}) + \mathbf{z} = \mathbf{x} + (\mathbf{y} + \mathbf{z})$.
   (iii) if $\mathbf{0} = (0, \ldots, 0)$, then $\mathbf{x} + \mathbf{0} = \mathbf{x}$.
   (iv) if $-\mathbf{x} = (-1)\mathbf{x}$, then $\mathbf{x} + (-\mathbf{x}) = \mathbf{0}$.
   (v) $1\mathbf{x} = \mathbf{x}$.
   (vi) $a(b\mathbf{x}) = (ab)\mathbf{x}$ for $a, b \in \mathbb{R}$.
   (vii) $a(\mathbf{x} + \mathbf{y}) = a\mathbf{x} + a\mathbf{y}$ and $(a + b)\mathbf{x} = a\mathbf{x} + b\mathbf{x}$ for $a, b \in \mathbb{R}$.

Similarly, it is easy to extend the definitions of dot product, norm, and distance between vectors from the plane and 3-space to $\mathbb{R}^n$. If $\mathbf{x} \in \mathbb{R}^n$, the (Euclidean) *norm* of $\mathbf{x}$ is

$$\|\mathbf{x}\| = \left(\sum_{i=1}^{n} x_i^2 \right)^{1/2}$$

For $\mathbf{x}, \mathbf{y} \in \mathbb{R}^n$, the (Euclidean) *distance* between $\mathbf{x}$ and $\mathbf{y}$ is

$$d(\mathbf{x}, \mathbf{y}) = \|\mathbf{x} - \mathbf{y}\|;$$

note that this is the usual distance in $\mathbb{R}$, $\mathbb{R}^2$, and $\mathbb{R}^3$. The *dot product* (*inner product, scalar product*) of $\mathbf{x}$ and $\mathbf{y}$ is

$$\mathbf{x} \cdot \mathbf{y} = \sum_{i=1}^{n} x_i y_i;$$

the dot product is a map from $\mathbb{R}^n \times \mathbb{R}^n$ into $\mathbb{R}$.

The basic properties of norm, distance, and dot product are summarized in Propositions 2 through 4.

Proposition 2. For $\mathbf{x}, \mathbf{y}, \mathbf{z} \in \mathbb{R}^n$ and $a \in \mathbb{R}$

   (i) $\mathbf{x} \cdot \mathbf{y} = \mathbf{y} \cdot \mathbf{x}$.
   (ii) $\mathbf{x} \cdot (\mathbf{y} + \mathbf{z}) = \mathbf{x} \cdot \mathbf{y} + \mathbf{x} \cdot \mathbf{z}$.

(iii) $\mathbf{x} \cdot (a\mathbf{y}) = a(\mathbf{x} \cdot \mathbf{y})$.

(iv) $\mathbf{x} \cdot \mathbf{x} = \|\mathbf{x}\|^2$ and $\mathbf{x} \cdot \mathbf{x} = 0$ iff $\mathbf{x} = \mathbf{0}$.

(v) $|\mathbf{x} \cdot \mathbf{y}| \le \|\mathbf{x}\|\,\|\mathbf{y}\|$ (Cauchy–Schwartz inequality).

Proof. We prove (v); the proofs of the other properties are easy. For (v), note that

$$0 \le \sum_{i=1}^{n} \sum_{j=1}^{n} (x_i y_j - x_j y_i)^2$$

$$= \sum_{i=1}^{n} x_i^2 \sum_{j=1}^{n} y_j^2 - 2 \sum_{i=1}^{n} x_i y_i \sum_{j=1}^{n} y_j x_j + \sum_{j=1}^{n} x_j^2 \sum_{i=1}^{n} y_i^2$$

$$= 2\|\mathbf{x}\|^2\|\mathbf{y}\|^2 - 2(\mathbf{x} \cdot \mathbf{y})^2.$$

It follows that $(\mathbf{x} \cdot \mathbf{y})^2 \le \|\mathbf{x}\|^2\|\mathbf{y}\|^2$. Taking principal square roots of both sides of this last inequality yields (v).   □

Proposition 3. For $\mathbf{x}, \mathbf{y} \in \mathbb{R}^n$ and $a \in \mathbb{R}$,

(i) $\|\mathbf{x}\| \ge 0$ and $\|\mathbf{x}\| = 0$ iff $\mathbf{x} = \mathbf{0}$.

(ii) $\|a\mathbf{x}\| = |a|\,\|\mathbf{x}\|$.

(iii) $\|\mathbf{x} + \mathbf{y}\| \le \|\mathbf{x}\| + \|\mathbf{y}\|$ (triangle inequality).

(iv) $\|\mathbf{x} - \mathbf{y}\| \ge |\,\|\mathbf{x}\| - \|\mathbf{y}\|\,|$.

Proof. We prove (iii) and (iv). For (iii), note that $\|\mathbf{x} + \mathbf{y}\|^2 = (\mathbf{x} + \mathbf{y}) \cdot (\mathbf{x} + \mathbf{y}) = \|\mathbf{x}\|^2 + 2\mathbf{x} \cdot \mathbf{y} + \|\mathbf{y}\|^2 \le \|\mathbf{x}\|^2 + 2\|\mathbf{x}\|\,\|\mathbf{y}\| + \|\mathbf{y}\|^2 = (\|\mathbf{x}\| + \|\mathbf{y}\|)^2$ by the Cauchy–Schwarz inequality.

For (iv), the triangle inequality gives $\|\mathbf{x}\| = \|\mathbf{x} - \mathbf{y} + \mathbf{y}\| \le \|\mathbf{x} - \mathbf{y}\| + \|\mathbf{y}\|$, which implies that $\|\mathbf{x} - \mathbf{y}\| \ge \|\mathbf{x}\| - \|\mathbf{y}\|$. By symmetry, $\|\mathbf{y} - \mathbf{x}\| = \|\mathbf{x} - \mathbf{y}\| \ge \|\mathbf{y}\| - \|\mathbf{x}\|$. It follows that $\|\mathbf{x} - \mathbf{y}\| \ge |\,\|\mathbf{x}\| - \|\mathbf{y}\|\,|$.   □

Proposition 4. For $\mathbf{x}, \mathbf{y}, \mathbf{z} \in \mathbb{R}^n$,

(i) $d(\mathbf{x}, \mathbf{y}) = d(\mathbf{y}, \mathbf{x}) \ge 0$.

(ii) $d(\mathbf{x}, \mathbf{y}) = 0$ iff $\mathbf{x} = \mathbf{y}$.

(iii) $d(\mathbf{x}, \mathbf{y}) \le d(\mathbf{x}, \mathbf{z}) + d(\mathbf{z}, \mathbf{y})$ (triangle inequality).

The proof is left to the exercises.

To define the convergence of sequences in $\mathbb{R}$, it is only necessary to have the notion of distance between real numbers. Since we now have a natural distance function in $\mathbb{R}^n$, it is a straightforward matter to define convergence of sequences in $\mathbb{R}^n$. Let $\{\mathbf{x}_k\}$ be a sequence in $\mathbb{R}^n$. Then $\{\mathbf{x}_k\}$ *converges* to $\mathbf{x} \in \mathbb{R}^n$ iff $\forall \varepsilon > 0 \; \exists N$ such that $k \geq N$ implies $\|\mathbf{x}_k - \mathbf{x}\| < \varepsilon$. We write $\lim \mathbf{x}_k = \lim_k \mathbf{x}_k = \lim_{k \to \infty} \mathbf{x}_k = \mathbf{x}$ or $\mathbf{x}_k \to \mathbf{x}$. Note that $\mathbf{x}_k \to \mathbf{x}$ iff $\lim \|\mathbf{x}_k - \mathbf{x}\| = 0$.

A statement equivalent to the definition of convergence of $\{x_k\}$ to x in $\mathbb{R}$ is that, given any open interval about x, eventually all remaining terms of the sequence (i.e, for k sufficiently large) lie in that interval. See Definition 3.1 and particularly Figure 3.1. In $\mathbb{R}^2$ "any open interval about x" is replaced by "the interior of any circle about x"; in $\mathbb{R}^3$, "circle" is replaced by "sphere"; and in $\mathbb{R}^n$, $n > 3$, "sphere" is replaced by "hypersphere." A *hypersphere* of radius r in $\mathbb{R}^n$, $n > 3$, is the set of all n-tuples $(x_1, \ldots, x_n)$ such that $x_1^2 + \cdots + x_n^2 = r^2$. The essential idea is to have a statement that precisely describes what we mean when we say that x_k comes arbitrarily near to x for k sufficiently large.

The following criterion reduces convergence in $\mathbb{R}^n$ to convergence in $\mathbb{R}$ and makes it possible to compute the limits of sequences in $\mathbb{R}^n$ by using the techniques of Chapter 3.

Proposition 5. Let $\mathbf{x}_k = (x_{k1}, \ldots, x_{kn})$ and $\mathbf{x} = (x_1, \ldots, x_n)$. Then $\lim \mathbf{x}_k = \mathbf{x}$ if, and only if, $\lim_k x_{ki} = x_i$ for each $i, \ldots, n$.

Proof. The sufficiency follows from the standard limit theorems of Chapter 3 and the definition of the distance function. The necessity follows from the inequalities

$$\|\mathbf{x}_k - \mathbf{x}\| \geq |x_{ki} - x_i| \qquad \text{for } i = 1, \ldots, n. \qquad \square$$

The usual limit theorems about uniqueness, sums, and multiplication by a scalar follow from Proposition 5. We consider the dot product in Proposition 8 below. For this we need the following definition.

Definition 6. A subset $E \subseteq \mathbb{R}^n$ is *bounded* if there is $r > 0$ such that $\|\mathbf{x}\| \leq r$ $\forall \mathbf{x} \in E$.

Note that this agrees with the previous definition of a bounded set in $\mathbb{R}$ (3.6).

Proposition 7. A convergent sequence in $\mathbb{R}^n$ is bounded.

The proof is left to Exercise 5.

Proposition 8. If $\mathbf{x}_k \to \mathbf{x}$ and $\mathbf{y}_k \to \mathbf{y}$, then $\mathbf{x}_k \cdot \mathbf{y}_k \to \mathbf{x} \cdot \mathbf{y}$.

Proof. The result follows from Proposition 7 and the inequalities

$$|\mathbf{x}_k \cdot \mathbf{y}_k - \mathbf{x} \cdot \mathbf{y}| \leq |\mathbf{x}_k \cdot (\mathbf{y}_k - \mathbf{y})| + |(\mathbf{x}_k - \mathbf{x}) \cdot \mathbf{y}|$$

$$\leq \|\mathbf{x}_k\| \|\mathbf{y}_k - \mathbf{y}\| + \|\mathbf{x}_k - \mathbf{x}\| \|\mathbf{y}\|. \qquad \square$$

The distance function enables us to extend the definition of a *Cauchy sequence* to $\mathbb{R}^n$. A sequence $\{x_k\}$ is *Cauchy* if for each $\varepsilon > 0$, there is N such that $k, j \geq N$ imply $\|x_k - x_j\| < \varepsilon$.

The following is an analog in $\mathbb{R}^n$ of the basic Cauchy convergence criterion given in **3.17**:

Proposition 9. A sequence $\{x_k\}$ in $\mathbb{R}^n$ converges if, and only if, it is Cauchy.

Proof. Let $x_k = (x_{k1}, \ldots, x_{kn})$. Since $\|x_k - x_j\| \geq |x_{ki} - x_{ji}|$ for each $i = 1, \ldots, n$, the sequence $\{x_{ki}\}_{k=1}^{\infty}$ is Cauchy in $\mathbb{R}$ if $\{x_k\}$ is Cauchy. Since it is clear that a convergent sequence is Cauchy, the result now follows from **3.17** and Proposition 5. $\square$

Finally, note that if a function $\mathbf{f}$ defined on a set S has values in $\mathbb{R}^n$, then $\mathbf{f}$ induces *component functions* $f_i \colon S \to \mathbb{R}$ $(i = 1, \ldots, n)$ by defining $f_i(x) = e_i \cdot \mathbf{f}(x)$, where e_i is the vector in $\mathbb{R}^n$ with a 1 in the ith coordinate and 0 in all the other coordinates. $f_i(x)$ is the ith coordinate of $\mathbf{f}(x)$. Thus $\mathbf{f}(x) = (f_1(x), \ldots, f_n(x))$ and we write $\mathbf{f} = (f_1, \ldots, f_n)$.

EXERCISES 5

1. Prove Propositions 1 and 4 and complete the proofs of Propositions 2 and 3.

2. Show that if $x_k \to x$, then $\|x_k\| \to \|x\|$. Does the converse hold?

3. Use the inner product to define the angle between two vectors. Give a condition for perpendicularity.
   Hint: Recall the dot product from calculus.

4. Prove or disprove: If $x_k \to x$, then the angle between x_k and x converges to zero.

5. Prove Proposition 7.

6. Show that a Cauchy sequence is bounded.

7. Let $\{x_k\}$ be a sequence in $\mathbb{R}^n$. The (formal) series, $\sum_{k=1}^{\infty} x_k$, is said to *converge* if the sequence of partial sums $s_k = \sum_{i=1}^{k} x_i$ converges. If $s_k \to s$, we write $s = \sum_{k=1}^{\infty} x_k$. Does the series $\sum_{k=1}^{\infty}((-1)^k/k, 1/k^2, (-1)^k/\ln(k+1))$ converge in $\mathbb{R}^3$?

8. The series $\sum_{k=1}^{\infty} x_k$ *converges absolutely* if the series $\sum_{k=1}^{\infty} \|x_k\|$ converges. Does the series in Exercise 7 converge absolutely?

9. Show that if $\sum_{k=1}^{\infty} x_k$ converges, then $x_k \to 0$.

10. State and prove the analogue of **4.18** for series in $\mathbb{R}^n$.

11. State and prove the analogue of **4.21** for series in $\mathbb{R}^n$.

12. Prove the parallelogram law:

$$2\|\mathbf{x}\|^2 + 2\|\mathbf{y}\|^2 = \|\mathbf{x} + \mathbf{y}\|^2 + \|\mathbf{x} - \mathbf{y}\|^2 \quad \text{for all } \mathbf{x}, \mathbf{y} \in \mathbb{R}^n.$$

13. If $\{\mathbf{x}_k\}$ is such that the series $\sum_{k=1}^{\infty} \|\mathbf{x}_k - \mathbf{x}_{k+1}\|$ converges, show that $\{\mathbf{x}_k\}$ is Cauchy.

14. If $\sum_{k=1}^{\infty} \mathbf{x}_k$ converges absolutely and $\{\mathbf{y}_k\}$ is a bounded sequence, show that $\sum_{k=1}^{\infty} \mathbf{x}_k \cdot \mathbf{y}_k$ converges absolutely.

CHAPTER 6

Limits of Functions

In Chapters 3 and 4, convergence or limit of a sequence was studied. In this chapter we introduce the limit of a function and relate it to the convergence of sequences. We will deal directly with functions defined on $\mathbb{R}^n$; this will allow us in subsequent chapters to treat such topics as finding the extreme values of functions of several variables.

Most of our work will involve a study or analysis of functions defined on subsets of $\mathbb{R}$ or $\mathbb{R}^n$ with ranges in $\mathbb{R}$ or $\mathbb{R}^n$ with occasional diversions to more general settings such as metric spaces (see Chapters 19 and 20). Such analysis often begins with an examination of the behavior of functions "close to" particular points and so we need to describe what we mean by "x being near to x_0." To this end, we introduce the notion of "neighborhood." We'll find that sometimes when "x is near to x_0" in the domain of a function, all the values of the function exhibit some particular kind of behavior. The discussion of such behavior will be greatly facilitated by the introduction of "limit of a function." We'll discover soon that the introduction of these and related ideas leads to a structure, called a topology, on the supporting spaces that can be described in a variety of ways, for example, through the use of convergence of sequences, neighborhoods, or yet-to-be-introduced concepts like "open sets" and "closed sets." Some of the properties of "topological spaces" will be considered in Chapter 20, but the subject is so rich that it generally requires a separate course for adequate examination. We begin with a definition of neighborhood.

An *ε-neighborhood* of a point $\mathbf{x} \in \mathbb{R}^n$ is the sphere $S(\mathbf{x}, \varepsilon) = \{\mathbf{y} \in \mathbb{R}^n : \|\mathbf{x} - \mathbf{y}\| < \varepsilon\}$. Only in $\mathbb{R}^3$ is an ε-neighborhood actually a sphere of radius ε about x; in $\mathbb{R}$ it's an interval, in $\mathbb{R}^2$ an open disk, and in $\mathbb{R}^n$, $n \geq 4$, it's a "hypersphere." We accommodate all of these under the appelation sphere. If $D \subseteq \mathbb{R}^n$, then $\mathbf{x}$ is an *accumulation point* of D if every ε-neighborhood of $\mathbf{x}$ contains infinitely many points of D. Intuitively, a point $\mathbf{x}$ is an accumulation point of a set D if there are "lots of points of D" arbitrarily close to $\mathbf{x}$; however, it should be noted that for $\mathbf{x}$ to be an accumulation point of D it suffices for each neighborhood of $\mathbf{x}$ to contain a single point of D distinct from $\mathbf{x}$ (Exercise 1). See Exercise 2 for some examples.

Finally, we address one of the fundamental topics of real analysis; defining the limit of a real-valued function. For a function f to have a limit $L \in \mathbb{R}$ at the point $\mathbf{x}_0$, the values $f(\mathbf{x})$ should be arbitrarily close to L for $\mathbf{x}$ sufficiently close to $\mathbf{x}_0$. For this to be meaningful, without being trivial, it is necessary that $\mathbf{x}_0$ be an accumulation point of the domain of f. With this requirement, we give the formal definition.

Definition 1. Let $\mathbf{x}_0$ be an accumulation point of D and $f \colon D \to \mathbb{R}$. Then f has a *limit L* at $\mathbf{x}_0$ if for each $\varepsilon > 0$ there is $\delta > 0$ such that $0 < \|\mathbf{x} - \mathbf{x}_0\| < \delta$, $\mathbf{x} \in D$ implies $|f(\mathbf{x}) - L| < \varepsilon$. We write

$$\lim_{\mathbf{x} \to \mathbf{x}_0} f(\mathbf{x}) = L.$$

Figure 6.1 illustrates the definition. It is not required that $\mathbf{x}_0$ belong to the domain of f, and since we require that $0 < \|\mathbf{x} - \mathbf{x}_0\| < \delta$, the value of f at $\mathbf{x}_0$ (if $\mathbf{x}_0 \in D$) is irrelevant to the existence of the limit.

Below are several examples applying the definition to establish the existence and values of some limits.

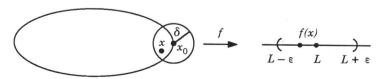

Figure 6.1

EXAMPLE 2. Let $f: \mathbb{R} \to \mathbb{R}$ be defined by $f(t) = t^2$. We claim that

$$\lim_{t \to t_0} f(t) = t_0^2 \quad \text{for each} \quad t_0 \in \mathbb{R}.$$

Let $\varepsilon > 0$. We must show that $|t^2 - t_0^2| < \varepsilon$ for $|t - t_0|$ sufficiently small. Since $|t^2 - t_0^2| = |t + t_0| \cdot |t - t_0|$, it's clear that we can make the left side of this equation small by making $|t - t_0|$ small, provided that we can control the size of the t-dependent quantity $|t + t_0|$. However, $|t + t_0| \le |t - t_0| + 2|t_0|$. So, if we require that $|t - t_0| < 1$ (t has to approach t_0 anyway), then $|t + t_0| \le 1 + 2|t_0|$. Thus, if we set

$$\delta = \min \left\{1, \varepsilon/(1 + 2|t_0|)\right\},$$

then $|t - t_0| < \delta$ implies that

$$|t^2 - t_0^2| = |t + t_0| \, |t - t_0| < (1 + 2|t_0|)\varepsilon/(1 + 2|t_0|) = \varepsilon. \qquad \square$$

In this example, δ depends on both ε and t_0. This is an important observation and we'll have more to say about it later.

EXAMPLE 3. Let $f(t) = \begin{cases} 1 & t = 0, \\ t \sin(1/t), & t \ne 0. \end{cases}$ Since $|t \sin(1/t)| \le |t|$, it follows that $\lim_{t \to 0} f(t) = 0$ (take $\delta = \varepsilon$ in Definition 1). Note that $\lim_{t \to 0} f(t) \ne f(0)$. $\qquad \square$

EXAMPLE 4. Let $f(x_1, x_2) = x_1 x_2 (x_1^2 - x_2^2)/(x_1^2 + x_2^2)$. Then

$$|f(x_1, x_2)| \le |x_1 x_2| \, |x_1^2 - x_2^2|/(x_1^2 + x_2^2) \le |x_1 x_2| < \varepsilon$$

whenever $\|x\| < \delta = \sqrt{\varepsilon}$ implies $\lim_{x \to 0} f(x) = 0$. $\qquad \square$

EXAMPLE 5. Let $f(x_1, x_2) = (x_1 + x_2)/(x_1 - x_2)$ for $(x_1, x_2) \in D = \{(x_1, x_2): x_1 \ne x_2\}$. Since the function f has the constant value $(1 + m)/(1 - m)$ along the line $x_2 = mx_1$, $\lim_{x \to 0} f(x)$ fails to exist. $\qquad \square$

Although the limit of a function and the limit of a sequence have different definitions, there is a useful relationship between $\lim f(x)$ as x approaches x_0 "continuously" and as it approaches x_0 "sequentially." This is made explicit in Proposition 6.

Proposition 6. Let $f: D \to \mathbb{R}$ and x_0 be an accumulation point of D. Then $\lim_{x \to x_0} f(x) = L$ if, and only if, for every sequence $\{x_k\} \subseteq D$ such that $x_k \to x_0, x_k \ne x_0$ we have $f(x_k) \to L$.

Proof. Suppose $\lim_{x \to x_0} f(x) = L$. Let $x_k \to x_0$, $x_k \neq x_0$, $x_k \in D$ and let $\varepsilon > 0$. There is $\delta > 0$ such that $0 < \|x - x_0\| < \delta$, $x \in D$ implies $|f(x) - L| < \varepsilon$. There also is N such that $k \geq N$ implies $\|x_k - x_0\| < \delta$. Hence, $k \geq N$ implies $|f(x_k) - L| < \varepsilon$, that is, $f(x_k) \to L$.

Next suppose that $\lim_{x \to x_0} f(x) \neq L$. Then there is an $\varepsilon > 0$ such that for each $k \in \mathbb{N}$ $\exists x_k \in D$ with $0 < \|x_k - x_0\| < 1/k$ and $|f(x) - L| \geq \varepsilon$. But then $x_k \to x_0$ while $f(x_k)$ does not converge to L. $\square$

Corollary 7. $\lim_{x \to x_0} f(x)$ exists if, and only if, for every sequence $\{x_k\} \subseteq D$ such that $x_k \to x_0, x_k \neq x_0$, $\lim f(x_k)$ exists.

Proof. The "only if" part is clear from Proposition 6. To show the "if" part, let $x_k \to x_0$ and $y_k \to x_0$ with $x_k \neq x_0$, $y_k \neq x_0$ and $f(x_k) \to L$, $f(y_k) \to L'$. Define the sequence $\{z_i\}$ by $z_{2k} = x_k$, $z_{2k-1} = y_k$. Then $z_k \to x_0$ (3.20 and 5.5); so $\lim f(z_k)$ exists. Hence, $L = L'$ and Proposition 6 now gives the result. $\square$

The idea of a function that gives information on whether or not an object belongs to a particular set is of great utility in many branches of mathematics. If E is a subset of a set S, the *characteristic function* of E, denoted by C_E, is the function $C_E \colon S \to \mathbb{R}$ defined by $C_E(t) = 1$ if $t \in E$ and $C_E(t) = 0$ if $t \notin E$.

EXAMPLE 8. Proposition 6, **2.2**, and **2.3** show that the characteristic function of the rationals $C_\mathbb{Q}$ does not have a limit at any point of $\mathbb{R}$. $\square$

If f and g are real-valued functions defined on a set S, the sum, product, quotient, and scalar multiples of these functions are defined respectively by

$$(f + g)(t) = f(t) + g(t), \qquad t \in S,$$

$$(fg)(t) = f(t)g(t), \qquad t \in S,$$

$$(f/g)(t) = f(t)/g(t), \qquad t \in S \text{ provided } g(t) \neq 0,$$

$$(af)(t) = af(t), \qquad t \in S, a \in \mathbb{R}.$$

The usual properties of limits such as uniqueness, sums, products, quotients, and scalar multiples now follow immediately from Proposition 6 and the corresponding limit properties for sequences.

Recall that a convergent sequence is bounded. For limits of functions, the following analogous property might be termed "local boundedness." A property of a function is *local* at x_0 if it holds only in some ε-neighborhood of x_0. If a property of a function f holds on the whole domain of f, the property is said to be *global*. We say that a function f is *bounded on a set S* if $f(S)$ is a bounded set.

Proposition 9. Suppose that $\lim_{x \to x_0} f(x) = L$ exists. Then $\exists\, M > 0$ and a δ-neighborhood of x_0 such that $|f(x)| \leq M \; \forall\, x \in S(x_0, \delta) \cap D$.

Proof. $\exists\, \delta > 0$ such that $0 < \|x - x_0\| < \delta$, $x \in D$, implies $|f(x) - L| < 1$. Let $M = |L| + 1$, if $x_0 \notin D$, and $M = \max\{|f(x_0)|, |L| + 1\}$, if $x_0 \in D$. $\square$

Similarly, if a function has a nonzero limit, then the function is "locally bounded away from 0."

Proposition 10. Suppose $\lim_{x \to x_0} f(x) = L \neq 0$. Then $\exists\, m > 0$ and a δ-neighborhood of x_0 such that $|f(x)| \geq m \; \forall\, x \in S(x_0, \delta) \cap D$, $x \neq x_0$.

Proof. There is $\delta > 0$ such that $0 < \|x - x_0\| < \delta$, $x \in D$, implies $|f(x) - L| < |L|/2$. Thus, $|f(x)| > |L|/2$ for such x. $\square$

$\mathbb{R}^m$-Valued Functions

Next, we consider limits for vector-valued functions defined on subsets of $\mathbb{R}^n$. Again, since we have a distance function defined, the definition of limit given in Definition 1 can be generalized to the case of $\mathbb{R}^m$-valued functions.

Let x_0 be an accumulation point of $D \subseteq \mathbb{R}^n$ and

$$f = (f_1, \ldots, f_m) : D \to \mathbb{R}^m$$

Definition 11. $\lim_{x \to x_0} f(x) = L \in \mathbb{R}^m$ if $\lim_{x \to x_0} \|f(x) - L\| = 0$.

Proposition 6 and 5.5, yield the following useful criterion for convergence in $\mathbb{R}^m$:

Proposition 12. $\lim_{x \to x_0} f(x) = L = (L_1, \ldots, L_m)$ if, and only if, $\lim_{x \to x_0} f_i(x) = L_i$ for $i = 1, \ldots, m$.

The usual properties of limits regarding uniqueness, sums and scalar multiples follow immediately from Proposition 12. Note that the notions of product and quotient are not applicable. However, Proposition 6 and 5.8 easily yield a "product formula" for the inner product.

Proposition 13. If $\lim_{x \to x_0} f(x) = L$ and $\lim_{x \to x_0} g(x) = M$, then

$$\lim_{x \to x_0} f(x) \cdot g(x) = L \cdot M.$$

Proof. Proposition 6 and 5.8. $\square$

One-Sided Limits

Continuing the study of various types of limits, we direct our attention to "one-sided limits." The existence of an order on the real numbers and the impossibility of one on $\mathbb{R}^n$ for $n \geq 2$, restricts consideration of a limit from one side or another to functions defined on subsets of $\mathbb{R}$.

Definition 14. Let $D \subseteq \mathbb{R}$ and $\mathbf{f} = (f_1, \ldots, f_m): D \to \mathbb{R}^m$. Let x_0 be an accumulation point of $D \cap (x_0, \infty)$. Then $\mathbf{f}$ has a *right-hand limit* at x_0 if the restriction $\mathbf{f}_0$, of $\mathbf{f}$ to $D \cap (x_0, \infty)$, has a limit at x_0. We write $\lim_{x \to x_0^+} \mathbf{f}(x) = \lim_{x \to x_0} \mathbf{f}_0(x) = \mathbf{R}$ or $\mathbf{f}(x) \to \mathbf{R}$ as $x \to x_0^+$. This means that the function $\mathbf{f}$ has a right-hand limit $\mathbf{R}$ at x_0 if, and only if, for each $\varepsilon > 0$, there is a $\delta > 0$ such that $0 < x - x_0 < \delta$ and $x \in D$ implies that $\|\mathbf{f}(x) - \mathbf{R}\| < \varepsilon$.

Left-hand limits are defined in a similar fashion and are denoted by $\lim_{x \to x_0^-} \mathbf{f}(x) = \mathbf{L}$ or $\mathbf{f}(x) \to \mathbf{L}$ as $x \to x_0^-$ (Exercise 10). Since right- and left-hand limits are special cases of the limit of a function, the results on sums, scalar products, etc. follow immediately and will not be explicitly stated.

The results on convergence of monotone sequences (3.11) have analogues for limits of functions. A real-valued function f defined on an interval I of $\mathbb{R}$ is said to be *increasing* (*strictly increasing*) if $x, y \in I$ and $x < y$ imply $f(x) \leq f(y)$ ($f(x) < f(y)$). The function is *decreasing* (*strictly decreasing*) if $-f$ is increasing (strictly increasing).

Proposition 15. Let $a < x_0 < b$ and $I = (a, b)$. If $f: I \to \mathbb{R}$ is increasing on the interval I, then it has both right- and left-hand limits at x_0.

Proof. Let $R = \inf \{ f(x): x_0 < x, x \in I \}$. Let $\varepsilon > 0$. Then there is $x \in I$, $x_0 < x$ such that $R + \varepsilon > f(x)$. Set $\delta = x - x_0$. If $0 < y - x_0 < \delta$, then $R + \varepsilon > f(x) \geq f(y) \geq R$; so $\lim_{x \to x_0^+} f(x) = R$.

The case of left-hand limit is similar. The left-hand limit at x_0 is $L = \sup \{ f(x): x < x_0, x \in I \}$. $\qquad\qquad\square$

Limits at Infinity

For functions defined on $\mathbb{R}$, the definition of the extended real numbers in Chapter 2 leads to a useful extension of the limit concept to limits at the points $\pm\infty$.

Definition 16. Let $\mathbf{f} = (f_1, \ldots, f_m): [a, \infty) \to \mathbb{R}^m$. The function $\mathbf{f}$ has a *limit* $\mathbf{L} \in \mathbb{R}^m$ at ∞ if $\forall \varepsilon > 0$ there is M such that $x \geq M$ implies $\|\mathbf{f}(x) - \mathbf{L}\| < \varepsilon$. We write

$$\lim_{x \to \infty} \mathbf{f}(x) = \mathbf{L} \qquad \text{or } \mathbf{f}(x) \to \mathbf{L} \qquad \text{as } x \to \infty.$$

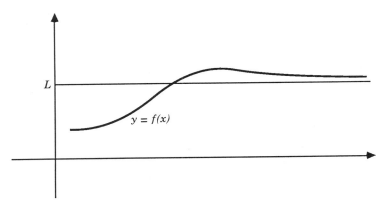

Figure 6.2

Limits at $-\infty$ are defined similarly (Exercise 12).

Intuitively, a function has a limit **L** at ∞ if the function values $\mathbf{f}(x)$ become and remain arbitrarily close to **L** for all sufficiently large values of x. If the range of f is the set of real numbers and f has limit L at ∞, the graph of f approaches the horizontal line $y = L$ as $x \to \infty$. See Figure 6.2.

The following is an analogue of **3.11**.

Proposition 17. Let $f: [a, \infty) \to \mathbb{R}$ be increasing. Then $\lim_{x \to \infty} f(x)$ exists iff f is bounded on $[b, \infty)$ for some $b \geq a$. In this case $\lim_{x \to \infty} f(x) = \sup\{f(x): x \geq a\}$.

Proof. Suppose that $\lim_{x \to \infty} f(x) = L$. Then there is b such that $x \geq b$ implies $|f(x) - L| < 1$ or $|f(x)| \leq |L| + 1$ for $x \geq b$.

Suppose f is bounded on $[b, \infty)$. Since f is increasing, $\sup\{f(x): x \geq b\} = \sup\{f(x): x \geq a\} = A$. If $\varepsilon > 0$, there is M such that $f(M) > A - \varepsilon$. Hence, if $x \geq M$, $A \geq f(x) \geq f(M) \geq A - \varepsilon$. It follows that $\lim_{x \to \infty} f(x) = A$. $\square$

Infinite Limits

Another type of "infinite limit" for real-valued functions is one in which the function values increase or decrease without bound as $x \to x_0$. If the domain of the function is also in $\mathbb{R}$, the point x_0 may be $\pm\infty$. We'll consider the case in which the function has domain in $\mathbb{R}^n$ and range in $\mathbb{R}$.

Let $D \subseteq \mathbb{R}^n$, $\mathbf{x}_0$ be an accumulation point of D, and $f: D \to \mathbb{R}$.

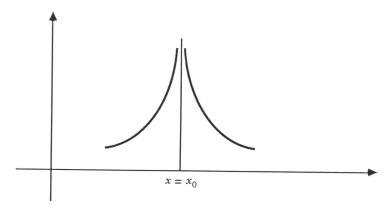

Figure 6.3

Definition 18. $\lim_{\mathbf{x} \to \mathbf{x}_0} f(\mathbf{x}) = \infty$ if $\forall M \in \mathbb{R}$, there is $\delta > 0$ such that

$$f(\mathbf{x}) \geq M \qquad \text{for } 0 < \|\mathbf{x} - \mathbf{x}_0\| < \delta, \mathbf{x} \in D.$$

Limits with values $-\infty$ are defined similarly. See Exercise 13.

If the function f has as its domain an interval (a, b) in $\mathbb{R}$, $a < x_0 < b$, and $\lim_{x \to x_0} f(x) = \infty$, then the graph of f has the vertical line $x = x_0$ as an asymptote. See Figure 6.3.

Development of the properties of infinite limits will be left as an exercise (Exercise 17); however, we will establish the following proposition.

Proposition 19. Let $\lim_{\mathbf{x} \to \mathbf{x}_0} f(\mathbf{x}) = \infty$. Suppose that $g: D \to \mathbb{R}$ is such that g is bounded from below on D. Then $\lim_{\mathbf{x} \to \mathbf{x}_0} (f(\mathbf{x}) + g(\mathbf{x})) = \infty$.

Proof. Let b be such that $g(\mathbf{x}) \geq b$ for $\mathbf{x} \in D$. Let $M \in \mathbb{R}$. Then $\exists \delta > 0$ such that $0 < \|\mathbf{x} - \mathbf{x}_0\| < \delta$ implies $f(\mathbf{x}) \geq M - b$. Thus, $f(\mathbf{x}) + g(\mathbf{x}) \geq M$ for such $\mathbf{x}$.   □

After working through this seemingly endless maze of various types of limits, it's natural to inquire about a general theory that would encompass all these limits and for which the basic properties of limits could be established once and for all. The good news is that there is such a theory. A readable exposition of a general theory is given in E. J. McShane's article, "A Theory of Limits," in *Studies in Modern Analysis*, Vol. I, R. C. Buck, editor.

EXERCISES 6

1. Show that x_0 is an accumulation point of D iff every ε-neighborhood of x_0 contains a point of D distinct from x_0.

2. Find the accumulation points of:
   (a) $\mathbb{N} \subseteq \mathbb{R}$ (b) $\mathbb{Q} \subseteq \mathbb{R}$ (c) $\mathbb{R} \setminus \mathbb{Q} \subseteq \mathbb{R}$
   (d) $\{1/2^k \colon k \in \mathbb{N}\}$ (e) $\{i + 1/j \colon i, j \in \mathbb{N}\}$
   (f) $\{\mathbf{x} \in \mathbb{R}^m \colon 0 < \|\mathbf{x}\| < 1\}$ (g) $\{\mathbf{x} \in \mathbb{R}^n \colon 1 \le \|\mathbf{x}\| < 2\}$
   (h) $\{\mathbf{x} \in \mathbb{R}^n \colon \|\mathbf{x}\| = 1 \text{ or } \|\mathbf{x}\| = 2\}$

3. Where does the function $f(t) = tC_{\mathbb{Q}}(t)$ have a limit?

4. Using only the definition, show that $\lim_{x \to x_0} \sin x = \sin x_0 \; \forall \, x_0 \in \mathbb{R}$.
   Hint: Use trigonometric formulas.

5. Establish the familiar calculus limit $\lim_{x \to 0} x/\sin x = 1$.
   Hint: For $0 < x < \pi/2$, $\sin x < x < \tan x$.

6. Using only the definition, show that $\lim_{x \to x_0} 1/x^2 = 1/x_0^2$.

7. Let $f(x) = \begin{cases} 0, & x = 0, \\ \sin(1/x), & x \ne 0, \end{cases}$ and $h(x) = xf(x)$. Do $\lim_{x \to 0} f(x)$ and/or $\lim_{x \to 0} h(x)$ exist?

8. If $\lim_{x \to x_0} \mathbf{f}(\mathbf{x}) = \mathbf{L}$, show that $\lim_{x \to x_0} \|\mathbf{f}(\mathbf{x})\| = \|\mathbf{L}\|$. Is the converse true?

9. If $D \subseteq \mathbb{R}^n$, $\mathbf{f} \colon D \to \mathbb{R}^m$, $g \colon D \to \mathbb{R}$ and $\lim_{x \to x_0} \mathbf{f}(\mathbf{x}) = \mathbf{L}$, $\lim_{x \to x_0} g(\mathbf{x}) = a$, show that $\lim_{x \to x_0} g(\mathbf{x})\mathbf{f}(\mathbf{x}) = a\mathbf{L}$.

10. Write out a formal definition of left-hand limit.

11. Show that a function $\mathbf{f} \colon (a, b) \to \mathbb{R}^m$ has a limit at $x_0 \in (a, b)$ iff $\mathbf{f}$ has both a left- and right-hand limit at x_0 and they are equal. Is it important that the left- and right-hand limits be the same?

12. Give a formal definition of $\lim_{x \to -\infty} f(x)$.

13. Define $\lim_{x \to x_0} f(x) = -\infty$, $\lim_{x \to x_0^+} f(x) = \pm\infty$, $\lim_{x \to x_0^-} f(x) = \pm\infty$, $\lim_{x \to \pm\infty} f(x) = \pm\infty$. Make sketches illustrating the behavior of the function in each of these situations.

14. If $f \colon [a, \infty) \to \mathbb{R}$ is increasing, show that $\lim_{x \to \infty} f(x)$ exists (it might be infinite).

15. Let $f \colon (a, b) \to \mathbb{R}$ be increasing. Show that $\lim_{x \to a^+} f(x)$ and $\lim_{x \to b^-} f(x)$ exist.

16. Let $f = (f_1, \ldots, f_m) \colon \mathbb{R} \to \mathbb{R}^m$. Show that $\lim_{x \to \infty} f(x) = \mathbf{L}$ iff $\lim_{x \to \infty} f_i(x) = L_i$ for $i = 1, \ldots, m$.

17. Show that $\lim_{x \to x_0} f(x) = \infty$ iff for every sequence $\{\mathbf{x}_k\} \subseteq D$, $\mathbf{x}_k \ne \mathbf{x}_0$, $\mathbf{x}_k \to \mathbf{x}_0$, we have $\lim f(\mathbf{x}_k) = \infty$.

18. Let $\xi = (\xi_1, \xi_2)$, $\mathbf{e}_1 = (1,0)$, $\mathbf{e}_2 = (0,1)$ and $\mathbf{x}, \mathbf{x}_1, \mathbf{x}_2 \in \mathbb{R}^n$. Find the following limits, if they exist:

  (a) $\lim\limits_{\xi \to 0} \dfrac{\xi_1^4 + \xi_2^4}{\xi_1^2 + \xi_2^2}$ (b) $\lim\limits_{\xi \to 0} \dfrac{\xi_1 \xi_2^2}{\xi_1^2 + \xi_2^4}$

  (c) $\lim\limits_{\xi \to 0} \xi_1^2 \xi_2^2 / (\xi_1^2 \xi_2^2 + (\xi_1 - \xi_2)^2)$

  (d) $\lim\limits_{\xi \to \xi_0} \xi_1 \xi_2 / (\xi_1^2 + \xi_2^2)$, $\xi_0 = \mathbf{e}_1 + \mathbf{e}_2$

  (e) $\lim\limits_{\|\mathbf{x}\| \to \infty} \dfrac{\|\mathbf{x} - \mathbf{x}_1\|}{\|\mathbf{x} - \mathbf{x}_2\|}$ (f) $\lim\limits_{\|\mathbf{x}\| \to \infty} \dfrac{(\mathbf{x} \cdot \mathbf{x}_1)(\mathbf{x} \cdot \mathbf{x}_2)}{\mathbf{x} \cdot \mathbf{x}}$

19. Find $\lim_{x \to 0} \sin(x^2)/x$ and $\lim_{x \to 1} (x^n - 1)/(x - 1)$, $n \in \mathbb{N}$.

20. Does the function $f(x, y) = |x|^y$ have a limit at $(0,0)$?

21. Show that $\lim_{\mathbf{x} \to \mathbf{x}_0} f(\mathbf{x}) = \infty$ iff $\lim_{\mathbf{x} \to \mathbf{x}_0} 1/f(\mathbf{x}) = 0$.

22. Give a sequential characterization of $\lim_{x \to \infty} \mathbf{f}(x)$.

23. Show that $\lim_{x \to \infty} f(x) = L$ (possibly $\pm \infty$) iff $\lim_{t \to 0^+} f(1/t) = L$.

24. Let $\mathbf{x}_0$ be an accumulation point of $D \subseteq \mathbb{R}^n$ and $f \colon D \to \mathbb{R}^m$. Prove that $\lim_{\mathbf{x} \to \mathbf{x}_0} \mathbf{f}(\mathbf{x}) = \mathbf{L}$ iff $\forall \{\mathbf{x}_k\} \subseteq D$ such that $\mathbf{x}_k \to \mathbf{x}_0$, $\mathbf{x}_k \neq \mathbf{x}_0$, we have $\lim_{k \to \infty} \mathbf{f}(\mathbf{x}_k) = \mathbf{L}$.

CHAPTER 7

Continuity and Uniform Continuity

An understanding of the class of continuous functions is essential to any study in mathematical analysis. For example, the continuous functions enjoy many "nice" properties that make them indispensable to the study of such subjects as optimization theory. There, the existence of extreme values of real-valued functions can be guaranteed by suitably invoking conditions of continuity.

We begin with a definition of continuity at an elementary level and build the tools that will be needed for the continued development of the properties of real-valued functions.

Definition 1. Let $D \subseteq \mathbb{R}^n$, $\mathbf{f}\colon D \to \mathbb{R}^m$ and $\mathbf{x}_0 \in D$. The function f is *continuous at* $\mathbf{x}_0$ if $\forall\, \varepsilon > 0\, \exists\, \delta > 0$ such that $\mathbf{x} \in D$ and $\|\mathbf{x} - \mathbf{x}_0\| < \delta$ implies $\|\mathbf{f}(\mathbf{x}) - \mathbf{f}(\mathbf{x}_0)\| < \varepsilon$. The function $\mathbf{f}$ is *continuous on D* if $\mathbf{f}$ is continuous at every point of D.

Note that δ in general depends on both ε and $\mathbf{x}_0$. In the event that $\mathbf{x}_0$ is an accumulation point of D, the essential ingredients for continuity at $\mathbf{x}_0$ are (i) f is defined at $\mathbf{x}_0$; (ii) $\lim_{\mathbf{x} \to \mathbf{x}_0} f(\mathbf{x})$ exists; and (iii) $\lim_{\mathbf{x} \to \mathbf{x}_0} f(\mathbf{x}) = f(\mathbf{x}_0)$.

For accumulation points of the domain of a function, the following proposition gives some equivalents to continuity.

Proposition 2. If x_0 is an accumulation point of D, the following are equivalent:

    (i) $\mathbf{f}$ is continuous at x_0.

    (ii) $\lim_{x \to x_0} \mathbf{f}(x) = \mathbf{f}(x_0)$.

    (iii) for every sequence $\{x_k\} \subseteq \mathbf{D}$ such that $x_k \to x_0$, we have $\mathbf{f}(x_k) \to \mathbf{f}(x_0)$.

Proof. The equivalence of (i) and (ii) is clear and (iii) implies (ii) by **6.6**. That (ii) implies (iii) follows exactly as in the proof of **6.6**. (Note that the condition $x_k \neq x_0$ can be dropped from **6.6**.)   □

A point $x_0 \in D$ that is not an accumulation point of D is called an *isolated point* of D. Equivalently, a point $x_0 \in D$ is an isolated point of D iff $\exists \delta > 0$ such that $D \cap S(x_0, \delta) = \{x_0\}$. Thus, every function is continuous at the isolated points of its domain of definition.

The reader is undoubtedly familiar with many continuous functions, such as the polynomials, the trigonometric functions, the absolute value function, the exponential and logarithmic functions, etc. encountered in any reasonable investigation of the calculus. A more exotic function is given in the next example.

EXAMPLE 3. Let $f \colon (0,1] \to \mathbb{R}$ be defined by

$$f(x) = \begin{cases} 0, & x \notin \mathbb{Q} \\ 1/q, & x = p/q, \end{cases}$$

where p and q are relatively prime integers. Then f is certainly not continuous at any point $x \in \mathbb{Q}$ (2.3). We claim, however, that f is continuous at every point $x_0 \notin \mathbb{Q}$. Let $\varepsilon > 0$. Then there is $q_0 \in \mathbb{N}$ such that $1/q_0 < \varepsilon$. In the interval $(0,1]$ there is at most a finite number of rationals of the form p/q with $0 < q < q_0$; which we denote by $r_1, \ldots, r_N$. Put $\delta = \min\{|x_0 - r_i| \colon 1 \le i \le N\}$. If $|x - x_0| < \delta$, then x irrational implies $f(x) = 0$, while $x = p/q$ implies $q \ge q_0$ and $1/q \le 1/q_0 < \varepsilon$. In either case, $|x - x_0| < \delta$ implies $|f(x) - f(x_0)| < \varepsilon$.   □

For later use, we consider a more familiar example.

EXAMPLE 4. Let $f \colon (0,1) \to \mathbb{R}$ be defined by $f(x) = 1/x$. If $x_0 \in (0,1)$ and $\varepsilon > 0$, set $\delta = \min\{x_0/2, \varepsilon x_0^2/2\}$. Then $|x - x_0| < \delta$ implies $|f(x) - f(x_0)|$

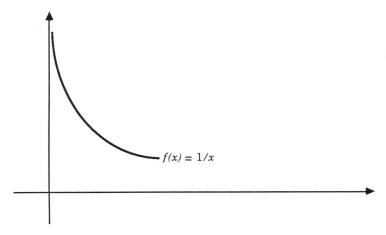

Figure 7.1

$= |x - x_0|/x x_0 < \varepsilon$, since $x > x_0/2$. Note that δ depends on both x_0 and ε and as x_0 approaches 0, δ must be taken smaller for the same fixed ε. See the sketch of the graph of $y = 1/x$ in Figure 7.1.   □

A real-valued function f defined on a subset $D \subseteq \mathbb{R}$, has a *jump discontinuity* at $x_0 \in (a, b)$ if both the right- and left-hand limits of f exist at x_0 but are not equal. For example, an increasing (decreasing) function has only jump discontinuities (**6.15**). Because they are usually easy to define and work with, jump discontinuities are the most common type of discontinuity encountered in elementary analysis; however, there are discontinuities that are not in this category.

EXAMPLE 5. Let $f(x) = \begin{cases} 0, & x = 0, \\ \sin(1/x), & x \neq 0. \end{cases}$ Then f is not continuous at 0 and 0 is not a jump discontinuity. (See Figure 7.2).   □

Application of the algebraic operations of addition, subtraction, multiplication, and division (a zero denominator is, of course, not allowed) to continuous functions f and g again yield a continuous function. This is made explicit in the next proposition.

Proposition 6. Suppose $D \subset \mathbb{R}^n$ and $\mathbf{f}, \mathbf{g}: D \to \mathbb{R}^m$ and $\mathbf{f}, \mathbf{g}$ are continuous at $\mathbf{x}_0 \in D$. Then

    (i) $\mathbf{f} + \mathbf{g}$ is continuous at $\mathbf{x}_0$.
    (ii) $a\mathbf{f}$ is continuous at $\mathbf{x}_0$ for $a \in \mathbb{R}$.

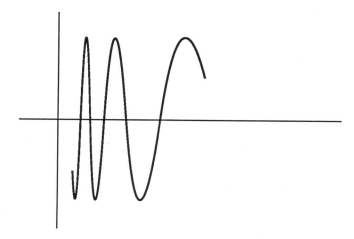

Figure 7.2

(iii) if $m = 1$, fg is continuous at x_0.

(iv) if $m = 1$ and $g(x) \neq 0 \; \forall \, x$, then f/g is continuous at x_0.

(v) $\mathbf{f} \cdot \mathbf{g} : D \to \mathbb{R}$ is continuous at x_0.

Each of these is a consequence of a similar result in Chapter 6 on limits of functions. From this proposition and our previous examples it follows that the familiar functions of calculus, polynomials, rational functions, etc., wherever they are defined, are continuous. We assume that the reader is familiar with the elementary functions of the calculus and their continuity properties.

Let $D \subseteq \mathbb{R}^n$, $E \subseteq \mathbb{R}^m$, $\mathbf{f} \colon D \to E$ and $\mathbf{g} \colon E \to \mathbb{R}^k$. The *composition* of $\mathbf{g}$ with $\mathbf{f}$ is the function $\mathbf{g} \circ \mathbf{f} \colon D \to \mathbb{R}^k$ defined by $\mathbf{g} \circ \mathbf{f}(\mathbf{x}) = \mathbf{g}(\mathbf{f}(\mathbf{x}))$.

Proposition 7. If $\mathbf{f}$ is continuous at $x_0 \in D$ and $\mathbf{g}$ is continuous at $\mathbf{f}(x_0)$, then $\mathbf{g} \circ \mathbf{f}$ is continuous at x_0.

Proof. Let $\varepsilon > 0$. There is $\delta_1 > 0$ such that $\|\mathbf{y} - \mathbf{f}(x_0)\| < \delta_1$, $\mathbf{y} \in E$, implies $\|\mathbf{g}(\mathbf{y}) - \mathbf{g}(\mathbf{f}(x_0))\| < \varepsilon$. There is also $\delta > 0$ such that $\|\mathbf{x} - \mathbf{x}_0\| < \delta$, $\mathbf{x} \in D$, implies $\|\mathbf{f}(\mathbf{x}) - \mathbf{f}(\mathbf{x}_0)\| < \delta_1$. Hence, $\|\mathbf{x} - \mathbf{x}_0\| < \delta$, $\mathbf{x} \in D$, implies $\|\mathbf{g}(\mathbf{f}(\mathbf{x})) - \mathbf{g}(\mathbf{f}(\mathbf{x}_0))\| < \varepsilon$. ☐

Proposition 8. The function $\mathbf{x} \to \|\mathbf{x}\|$ from $\mathbb{R}^n$ into $\mathbb{R}$ is continuous.

Proof. By 5.3, $|\,\|\mathbf{x}\| - \|\mathbf{x}_0\|\,| \leq \|\mathbf{x} - \mathbf{x}_0\|$. ☐

Corollary 9. If $\mathbf{f}\colon D \to \mathbb{R}^m$ is continuous at $\mathbf{x}_0 \in D$, then $\exists\, M > 0$ and a δ-neighborhood of $\mathbf{x}_0$ such that $\|\mathbf{f}(\mathbf{x})\| \leq M\ \forall\, \mathbf{x} \in D \cap S(\mathbf{x}_0, \delta)$.

Proof. The function $\mathbf{x} \to \|\mathbf{f}(\mathbf{x})\|$ is continuous by Propositions 7 and 8. Apply **6.9**. $\qquad\square$

Recall that when the definition of continuity was applied to the function $f(x) = 1/x$ for $0 < x < 1$ in Example 4, the δ in Definition 1 depended on both ε and the proximity of the point x_0 to 0. On the other hand, consider the function $g(x) = x^2$, $0 < x < 1$. For $\varepsilon > 0$ we can take $\delta = \varepsilon/2$ in Definition 1, because $|x - x_0| < \delta$ implies $|g(x) - g(x_0)| = |x + x_0| \cdot |x - x_0| < \varepsilon$. So, for this particular function there is a $\delta = \delta(\varepsilon)$, which works "uniformly" for all points x_0 in the interval $(0,1)$. The lack of dependence of δ on x_0 will later yield some very nice results, and so, we distinguish this case with a definition.

Definition 10. Let $D \subseteq \mathbb{R}^n$ and $\mathbf{f}\colon D \to \mathbb{R}^m$. $\mathbf{f}$ is *uniformly continuous* on D if for each $\varepsilon > 0$, there is $\delta > 0$ such that $\|\mathbf{x} - \mathbf{y}\| < \delta$, $\mathbf{x}, \mathbf{y} \in D$ implies $\|\mathbf{f}(\mathbf{x}) - \mathbf{f}(\mathbf{y})\| < \varepsilon$.

Some examples should help to illustrate the concept.

EXAMPLE 11. The function $f(x) = 1/x$, $0 < x < 1$, of Example 4 was such that $\delta(x, \varepsilon) = \min\{x/2, \varepsilon x^2/2\} \to 0$ as $x \to 0^+$. This apparent dependence of δ on x suggests that f might fail to be uniformly continuous on $(0,1)$. This is the case, since $|f(1/m) - f(1/n)| = |m - n|\ \forall\, m, n \in \mathbb{N}$. If we take $m = 2n$, then $1/2n$ and $1/n$ can be made arbitrarily close by taking n sufficiently large, and yet the difference in the function values is n and this does not decrease in size in this process. Thus, for any pair of values, $\varepsilon > 0$ and $\delta > 0$, there are points x and y in $(0,1)$ such that $|x - y| < \delta$ and $|f(x) - f(y)| \geq \varepsilon$, and so, f is not uniformly continuous on $(0,1)$. The function is continuous on $(0,1)$ as we demonstrated in Example 4. $\qquad\square$

EXAMPLE 12. The function $f\colon \mathbb{R} \to \mathbb{R}$ defined by $f(x) = x^2$ is continuous with $\delta(x, \varepsilon) = \min\{1, \varepsilon/(1 + 2|x|)\}$ (Example 6.2). Since $\delta(x, \varepsilon) \to 0$ as $x \to \infty$, this suggests a dependence of δ on x and thus f might not be uniformly continuous on $\mathbb{R}$. Again, this is the case because

$$|f(m + 1/m) - f(m)| \geq 2\ \forall\, m \in \mathbb{N}.$$

Note that this function is uniformly continuous on $(0,1)$; in fact, it's uniformly continuous on every finite interval $[a, b]$. $\qquad\square$

EXAMPLE 13. The function $\mathbf{x} \to \|\mathbf{x}\|$ is uniformly continuous on $\mathbb{R}^m$. (See the proof of Proposition 8.) $\qquad\square$

Joint and Separate Continuity

A function of two variables, $f(x, y)$, can be considered as a function of x alone, of y alone, or of x and y together. It is important to distinguish between these concepts, and so, we introduce "separate" and "joint" continuity.

Let $D \subseteq \mathbb{R}^n$, $E \subseteq \mathbb{R}^m$ and $\mathbf{f}: D \times E \to \mathbb{R}^k$. For $\mathbf{x} \in D$ ($\mathbf{y} \in E$) we denote by $\mathbf{f}(\mathbf{x}, \cdot): E \to \mathbb{R}^k$ ($\mathbf{f}(\cdot, \mathbf{y}): D \to \mathbb{R}^k$) the function $\mathbf{f}(\mathbf{x}, \cdot)(\mathbf{y}) = \mathbf{f}(\mathbf{x}, \mathbf{y})$ ($\mathbf{f}(\cdot, \mathbf{y})(\mathbf{x}) = \mathbf{f}(\mathbf{x}, \mathbf{y})$). The function $\mathbf{f}$ is *separately continuous* at $(\mathbf{x}_0, \mathbf{y}_0) \in D \times E$ if the function $\mathbf{f}(\mathbf{x}_0, \cdot)$ is continuous at $\mathbf{y}_0$ and the function $\mathbf{f}(\cdot, \mathbf{y}_0)$ is continuous at $\mathbf{x}_0$; $\mathbf{f}$ is said to be *jointly continuous* (or simply *continuous*) at $(\mathbf{x}_0, \mathbf{y}_0)$ if $\mathbf{f}$ is continuous at $(\mathbf{x}_0, \mathbf{y}_0)$. If a function $\mathbf{f}$ is jointly continuous at $(\mathbf{x}_0, \mathbf{y}_0)$, then it is certainly separately continuous at $(\mathbf{x}_0, \mathbf{y}_0)$, but the converse is false.

EXAMPLE 14. Define

$$f(x, y) = \begin{cases} 2xy/(x^2 + y^2), & (x, y) \neq (0,0), \\ 0, & (x, y) = (0,0). \end{cases}$$

f is clearly separately continuous at $(0,0)$ but is not continuous at $(0,0)$ since, along the line $y = mx$, f has the constant value $2m/(1 + m^2)$. □

EXERCISES 7

1. Examine the function

$$f(x) = \begin{cases} 8x, & x \in \mathbb{Q}, \\ 2x^2 + 8, & x \notin \mathbb{Q}, \end{cases}$$

for continuity at $x = 1, 2$.

2. If $f: \mathbb{R} \to \mathbb{R}$ is continuous on $\mathbb{R}$ and $f(x) = 0 \; \forall x \in \mathbb{Q}$, show that $f = 0$.

3. Give an example of a function on $\mathbb{R}$ that is continuous at exactly one point.

4. If $\mathbf{f}: D \to \mathbb{R}^m$ is continuous at $\mathbf{x}_0 \in D$ and $\mathbf{f}(\mathbf{x}_0) \neq \mathbf{0}$, show that there is $m > 0$ and a δ-neighborhood of $\mathbf{x}_0$ such that $\|\mathbf{f}(\mathbf{x})\| \geq m \; \forall \mathbf{x} \in D \cap S(\mathbf{x}_0, \delta)$.

5. Show that the sine function is uniformly continuous on $\mathbb{R}$.

6. Let $\mathbf{f}, \mathbf{g}: D \to \mathbb{R}^m$ be uniformly continuous on D. Show that $a\mathbf{f} + b\mathbf{g}$ is uniformly continuous for $a, b \in \mathbb{R}$.

7. If $f, g: D \to \mathbb{R}$ are uniformly continuous on D, is fg uniformly continuous on D?

8. If $f: D \to \mathbb{R}$ is uniformly continuous on D and $f(x) \neq 0$ for $x \in D$, is $1/f$ uniformly continuous on D?

9. Is the composition of uniformly continuous functions necessarily uniformly continuous?

10. Is the function in Example 5 uniformly continuous?

11. Is a bounded, continuous function necessarily uniformly continuous?

12. If $\mathbf{f}: D \to \mathbb{R}^m$ is uniformly continuous and $\{\mathbf{x}_k\}$ is a Cauchy sequence in D, show that $\{\mathbf{f}(\mathbf{x}_k)\}$ is Cauchy. Can uniform continuity be replaced by continuity? Is the converse true?

13. If $f: (a, b) \to \mathbb{R}$ is uniformly continuous, show that $\lim_{x \to a} f(x)$ exists.

14. If $F \subseteq \mathbb{R}^n$, $G \subseteq \mathbb{R}^m$ and $f: F \to \mathbb{R}$, $g: G \to \mathbb{R}$ are continuous, show that the function $(\mathbf{x}, \mathbf{y}) \to f(\mathbf{x})g(\mathbf{y})$ is continuous on $F \times G \subseteq \mathbb{R}^{n+m}$.

15. Suppose that $f: \mathbb{R} \to \mathbb{R}$ is continuous at some $x_0 \in \mathbb{R}$ and satisfies the functional equation $f(x + y) = f(x) + f(y) \; \forall \, x, y \in \mathbb{R}$. Show that there is $a \in \mathbb{R}$ such that $f(x) = ax \; \forall \, x \in \mathbb{R}$.
Hint: $f(n) = nf(1)$ for $n \in \mathbb{N}$; extend this to $\mathbb{Q}$ and then to $\mathbb{R}$.

16. Let $f, g: D \to \mathbb{R}$ be continuous at $\mathbf{x}_0 \in D \subseteq \mathbb{R}^n$. Show that the functions $f \wedge g$ and $f \vee g$ defined by $f \vee g(\mathbf{x}) = \max\{f(\mathbf{x}), g(\mathbf{x})\}$ and $f \wedge g(\mathbf{x}) = \min\{f(\mathbf{x}), g(\mathbf{x})\}$ are both continuous at $\mathbf{x}_0$.
Hint: $f \vee g = (f + g + |f - g|)/2$ and $f \wedge g = (f + g - |f - g|)/2$.

17. Show that an increasing function $f: [a, b] \to \mathbb{R}$ can have at most countably many points of discontinuity.
Hint: If x is a point of discontinuity, pick a rational r_x between $\lim_{t \to x^-} f(t)$ and $\lim_{t \to x^+} f(t)$ and consider the map $x \to r_x$.

CHAPTER 8

Topology in $\mathbb{R}^n$

There are some elegant and useful results, particularly for continuous functions, that can be obtained most easily by introducing a structure, called a topology, on $\mathbb{R}^n$. As we indicated in Chapter 6, there are many ways in which a topological structure can be introduced. In the present situation, we'll define our topology via a distinguished class of subsets, called the open sets, of $\mathbb{R}^n$. A set E is *open* if for each $\mathbf{x} \in E$ there is $\delta > 0$ such that $S(\mathbf{x}, \delta) \subseteq E$. The family of all open subsets of $\mathbb{R}^n$ defines a *topology* for $\mathbb{R}^n$ and can be used to describe many things of interest, for example, neighborhood, convergence of sequences and functions, continuity, differentiability, etc. A subset E of $\mathbb{R}^n$ is *closed* if E contains all of its accumulation points.

In $\mathbb{R}$, closed intervals are closed sets and open intervals are open sets. The converses are not true; for example, every set with just a finite number of elements is closed, and the sets $\mathbb{R}$ and $\varnothing$ are both open and closed. See Exercise 1 for more examples.

It's easy to see that the closed sets in $\mathbb{R}^n$ are precisely those whose complements are open, and conversely.

Proposition 1. A subset E of $\mathbb{R}^n$ is closed if and only if its complement $\mathbb{R}^n \setminus E$ is open.

Proof. Suppose E is closed. If $\mathbf{x}_0 \in \mathbb{R}^n \setminus E$, then $\mathbf{x}_0$ is not an accumulation point of E. Consequently, there is a δ-neighborhood $S(\mathbf{x}_0, \delta)$ of $\mathbf{x}_0$ that contains no point of E except possibly $\mathbf{x}_0$. But $\mathbf{x}_0 \notin E$. Therefore $S(\mathbf{x}_0, \delta) \subset \mathbb{R}^n \setminus E$, and it follows that $\mathbb{R}^n \setminus E$ is open.

Next, suppose $\mathbb{R}^n \setminus E$ is open and let $\mathbf{x}_0$ be an accumulation point of E. Then every neighborhood of $\mathbf{x}_0$ contains points of E, and so $\mathbf{x}_0 \notin \mathbb{R}^n \setminus E$. It follows that $\mathbf{x}_0 \in E$, and that E is closed. $\quad\square$

Since $E = \mathbb{R}^n \setminus (\mathbb{R}^n \setminus E)$, it follows from Proposition 1 that E is open if and only if $\mathbb{R}^n \setminus E$ is closed.

Remark 2. For each α in a nonempty set I, let there correspond a set G_α in $\mathbb{R}^n$. Then the set I is called an *index set* for the family of sets $\{G_\alpha : \alpha \in I\}$. If each G_α is open, the set $\bigcup_{\alpha \in I} G_\alpha$ is open. If F_α is a closed set for each $\alpha \in I$, then the set $\bigcap_{\alpha \in I} F_\alpha$ is closed. It is possible to have a family of open sets whose intersection is not open and a family of closed sets whose union is not closed. The verification of all of this is left to the exercises (see Exercise 8). The following proposition characterizes the notion of a closed set in terms of sequences.

Proposition 3. A subset E of $\mathbb{R}^n$ is closed if and only if every convergent sequence of points in E has its limit in E.

Proof. Suppose E is closed. If $\mathbf{x}_0 \notin E$, then the fact that $\mathbb{R}^n \setminus E$ is open implies that there is a δ-neighborhood $S(\mathbf{x}_0, \delta)$ contained in $\mathbb{R}^n \setminus E$. Thus, no sequence in E can converge to $\mathbf{x}_0$.

Next, suppose every convergent sequence of points in E has its limit in E and let $\mathbf{x}_0$ be an accumulation point of E. Then for each $k \in \mathbb{N}$, there is $\mathbf{x}_k \in E \cap S(\mathbf{x}_0, 1/k)$. It follows that $\mathbf{x}_k \to \mathbf{x}_0$ and $\mathbf{x}_0 \in E$, that is, E is closed. $\quad\square$

Recall that a set E in $\mathbb{R}^n$ is *bounded* if there is $M > 0$ such that $\|\mathbf{x}\| \leq M$ for all $\mathbf{x} \in E$.

It is clear that a sequence in a bounded set is bounded; it is also easy to construct examples of nonconvergent sequences in bounded sets. The following theorem shows that every sequence in a bounded set has a convergent subsequence or, equivalently, every bounded sequence has a convergent subsequence.

Theorem 4 (Bolzano–Weierstrass). If a subset E of $\mathbb{R}^n$ is bounded, then every sequence in E has a convergent subsequence.

Proof. The proof is by induction on n. For $n = 1$, let $\{x_k\}$ be a sequence in E. If the set $\{x_k : k \in \mathbb{N}\}$ is finite, the existence of a convergent subsequence is immediate (Exercise 5). So assume that $\{x_k: k \in \mathbb{N}\}$ has an infinite number of elements. Since E is bounded, there is a closed interval $I_0 = [a, b]$, $a < b$, containing E. Bisect I_0 to obtain two closed intervals I_0' and I_0''. At least one of I_0', I_0'' contains infinitely many terms of the sequence $\{x_k\}$. Call it I_1. Repeat the process on I_1. Continue inductively to obtain a sequence $\{I_j\}$ of nested closed intervals with the properties (i) $\ell(I_j) = (b - a)/2^j$, and (ii) I_j contains infinitely many terms of $\{x_k\}$ for $j = 1, 2, \ldots$. By the nested interval theorem (3.16), there is exactly one x common to all I_j. Now, choose any x_{k_1} in $\{x_k\}$ such that $x_{k_1} \in I_1$. Assume $x_{k_1}, \ldots, x_{k_m}$ have been chosen such that $x_{k_i} \in I_i$ and $k_i < k_{i+1}$ for $i = 1, 2, \ldots, m - 1$. Then choose $k_{m+1} > k_m$ so that $x_{k_{m+1}} \in I_{m+1}$. Then $\{x_{k_m}\}$ converges to x, since

$$|x_{k_m} - x| \le (b - a)/2^{k_m}$$

and $k_m \to \infty$.

Suppose that the result holds for $n = k$ and let $\{\mathbf{x}_j\}$ be a sequence in $E \subseteq \mathbb{R}^{k+1}$. Then we can write $\mathbf{x}_j = (\mathbf{a}_j, b_j)$, where $\{\mathbf{a}_j\} \subseteq \mathbb{R}^k$ and $\{b_j\} \subseteq \mathbb{R}$. Since $\{\mathbf{x}_j\}$ is bounded, both $\{\mathbf{a}_j\}$ and $\{b_j\}$ are bounded. By the induction hypothesis, $\{\mathbf{a}_j\}$ has a convergent subsequence, say $\{\mathbf{a}_{j_i}\}$, and by the first part of this proof, $\{b_{j_i}\}$ has a convergent subsequence $\{b_{j_{i_m}}\}$. Thus, $\{\mathbf{x}_{j_{i_m}}\}$ is a convergent subsequence of $\{\mathbf{x}_j\}$. □

We next describe a class of subsets of $\mathbb{R}^n$ that are of importance in some of the applications and that enable us to describe some important and attractive results in the theory of functions. The definition that we give is in terms of convergent subsequences; the usual definition is given in terms of the characterization given in Theorem 7. This is the definition that we use in Chapter 25.

Definition 5. A set K in $\mathbb{R}^n$ is *compact* if every sequence in K contains a subsequence that converges to a point in K. For subsets of $\mathbb{R}^n$, we have the following useful and descriptive characterization of compactness.

Theorem 6 (Heine–Borel). A subset of $\mathbb{R}^n$ is compact if and only if it is closed and bounded.

Proof. Suppose K is a closed and bounded subset of $\mathbb{R}^n$. Then Theorem 4 implies that every sequence in K has a convergent subsequence and, since K is closed, this subsequence converges to a point in K. It follows that K is compact.

Suppose next that K is compact. Then every convergent sequence in K converges to a point in K. It follows from Proposition 3 that K is closed. If K is not bounded, then for each $k \in \mathbb{N}$, there is $\mathbf{x}_k \in K$ such that $\|\mathbf{x}_k\| > k$. Consequently, no subsequence of $\{\mathbf{x}_k\}$ can converge and so K is not compact. □

We now give another characterization of compactness. Since this characterization is not used again until Chapter 25, the reader interested only in the properties of continuous functions on compact sets can skip ahead to Theorem 8. In order to give this characterization, we introduce the notion of an open cover of a set E in $\mathbb{R}^n$. Let E be a subset of $\mathbb{R}^n$. A family $\mathscr{F} = \{ G_\alpha : \alpha \in I \}$ of subsets of $\mathbb{R}^n$ is called a *cover* of E if $E \subset \bigcup_{\alpha \in I} G_\alpha$. A cover $\mathscr{F}$ is called an *open cover* of E if each G_α is an open set. If I' is a subset of the index set I, then $\mathscr{F}' = \{ G_\alpha : \alpha \in I' \}$ is a subfamily of $\mathscr{F}$; if $\mathscr{F}'$ also covers E, it is called a *subcover* for E.

The family $\mathscr{F} = \{ (1/k, 2) : k \in \mathbb{N} \}$ is an open cover of $(0, 1]$; the family $\mathscr{F}' = \{ (2^{-k}, 2) : k \in \mathbb{N} \}$ is a subfamily of $\mathscr{F}$. Each point of $(0, 1]$ belongs to $(2^{-k}, 2)$ for k sufficiently large, and so $\mathscr{F}'$ is a subcover for $(0, 1]$. If $\mathscr{F}''$ is a finite subfamily of $\mathscr{F}$, there is a largest value of k, say k_1 such that $(2^{-k_1}, 2) \in \mathscr{F}''$. Since $2^{-k_1 - 1} \in (0, 1]$ and does not belong to any element of $\mathscr{F}''$, $\mathscr{F}''$ is not a subcover for $(0, 1]$. Close examination of this example reveals that the problem lies in the fact that $(0, 1]$ is not closed. Other examples, which the reader should construct, show that similar problems can arise with unbounded sets. However, when the set being considered is both closed and bounded, that is, compact, we obtain the characterization given in the following theorem.

Theorem 7. A subset K of $\mathbb{R}^n$ is compact if and only if every open cover of K has a finite subcover.

Proof. Suppose first that every open cover of K has a finite subcover. We'll show that K is then closed and bounded and, therefore, compact. To show that K is bounded, set $G_{\mathbf{x}} = S(\mathbf{x}, 1)$. Then $\{ G_{\mathbf{x}} : \mathbf{x} \in K \}$ is an open cover of K (Exercise 1(c)) and so has a finite subcover, say, $G_{\mathbf{x}_1}, \ldots, G_{\mathbf{x}_k}$. Let $M = \max \{ \|\mathbf{x}_j\| : j = 1, \ldots, k \}$. If $\mathbf{x} \in K$, then $\mathbf{x} \in G_{\mathbf{x}_i}$ for some i and $\|\mathbf{x}\| \leq \|\mathbf{x}_i\| + 1 \leq M + 1$. Thus, K is bounded.

To show that K is closed, consider an $\mathbf{x}_0 \in \mathbb{R}^n \setminus K$. For $\mathbf{x} \in K$, there are disjoint neighborhoods $U_{\mathbf{x}}$ and $V_{\mathbf{x}}$ of $\mathbf{x}$ and $\mathbf{x}_0$, respectively, which generally will depend on $\mathbf{x}$. The family $\{ U_{\mathbf{x}} : \mathbf{x} \in K \}$ is an open cover of K and so has a finite subcover, say, $U_{\mathbf{x}_1}, \ldots, U_{\mathbf{x}_k}$. Then $K \subseteq \bigcup_{j=1}^k U_{\mathbf{x}_j}$, which is disjoint from the open sphere $\bigcap_{i=1}^k V_{\mathbf{x}_i}$. The latter is a neighborhood of $\mathbf{x}_0$ (in fact, it is a sphere with center x_0) contained entirely in $\mathbb{R}^n \setminus K$. It follows that $\mathbb{R}^n \setminus K$ is open and so K is closed.

In order to keep things simple and easily understood, the remainder of the proof will be given only for $n = 1$; a more general result is established in Chapter 25 (see also Exercise 34). Suppose that K is a compact subset of $\mathbb{R}$ and $\mathscr{F}$ is an open cover of K such that no finite subfamily of $\mathscr{F}$ is a cover for K. Since K is bounded, there is a closed interval $I_0 = [a, b]$ such that $K \subset I_0$. Bisect I_0 to obtain two closed subintervals I_0', I_0''. Then either $K \cap I_0'$ or $K \cap I_0''$ is not covered by a finite subfamily of $\mathscr{F}$. Label such a subinterval I_1.

Continue this process to obtain a sequence $\{I_k\}$ of nested closed intervals whose lengths approach zero and each with the property that no finite subfamily of $\mathscr{F}$ will cover $K \cap I_k$, $k = 0, 1, \ldots$. Let $\{\mathbf{x}\} = \cap_{k=0}^{\infty} I_k$ and for each k, choose $\mathbf{x}_k \in K \cap I_k$. Then $\mathbf{x}_k \to \mathbf{x}$. Since K is closed, it follows that $\mathbf{x} \in K$. Therefore $\mathbf{x} \in G$ for some $G \in \mathscr{F}$. Since G is open, there is k_0 such that $\mathbf{x} \in I_{k_0} \subseteq G$. Thus $\{G\}$ is a finite subcover of $I_{k_0} \cap K$. This is a contradiction. $\qquad\qquad\qquad\qquad\qquad\qquad\qquad\qquad\qquad\qquad\qquad\square$

An advantage of this last characterization of compactness is that it remains meaningful when the discussion is extended to topological spaces much more general than $\mathbb{R}^n$, and thus serves as the basis for a definition of compactness in these more general spaces. These ideas will be explored further in Chapter 19.

Properties of Continuous Functions on Compact Sets

The importance of the concept of "compact set" is illustrated by some of the properties enjoyed by continuous functions defined on compact sets.

Theorem 8. Let K be a compact subset of $\mathbb{R}^n$ and $\mathbf{f}: K \to \mathbb{R}^m$ a continuous function on K. Then $\mathbf{f}$ is uniformly continuous.

Proof. Suppose the conclusion is not true. Then there is $\varepsilon > 0$ and sequences $\{\mathbf{x}_k\}$ and $\{\mathbf{y}_k\}$ of points in K such that $\|\mathbf{x}_k - \mathbf{y}_k\| < 1/k$ and $\|\mathbf{f}(\mathbf{x}_k) - \mathbf{f}(\mathbf{y}_k)\| \geq \varepsilon$. Since K is compact, $\{\mathbf{x}_k\}$ has a subsequence $\{\mathbf{x}_{k_j}\}$, which converges to a point $\mathbf{x} \in K$. Then $\mathbf{y}_{k_j} \to \mathbf{x}$, since $\mathbf{y}_{k_j} - \mathbf{x} = (\mathbf{y}_{k_j} - \mathbf{x}_{k_j}) + (\mathbf{x}_{k_j} - \mathbf{x})$. Since $\mathbf{f}$ is continuous, we have

$$\lim_{j \to \infty} \left\| \mathbf{f}(\mathbf{x}_{k_j}) - \mathbf{f}(\mathbf{y}_{k_j}) \right\| = 0.$$

This is a contradiction. $\qquad\qquad\qquad\qquad\qquad\qquad\qquad\qquad\qquad\qquad\qquad\square$

The compactness of the domain is important. For example, the function f, defined by $f(x) = 1/x$ on $(0, 1]$, is continuous but is not uniformly continuous on the noncompact set $(0, 1]$. The problem, of course, lies at the left endpoint of the interval. The function values are arbitrarily large in every neighborhood of 0.

Let E be a subset of $\mathbb{R}^n$. Recall that a function $\mathbf{f}: E \to \mathbb{R}^m$ is *bounded* on E if $\{\mathbf{f}(\mathbf{x}): \mathbf{x} \in E\}$ is a bounded subset of $\mathbb{R}^m$.

Proposition 9. If E is a bounded subset of $\mathbb{R}^n$ and $\mathbf{f}: E \to \mathbb{R}^m$ is uniformly continuous, then $\mathbf{f}$ is bounded.

Proof. Suppose $\mathbf{f}$ is not bounded. Then there is a sequence $\{\mathbf{x}_k\}$ in E such that $\|f(\mathbf{x}_k)\| \geq k$ for each k. By the Bolzano–Weierstrass theorem, $\{\mathbf{x}_k\}$ has a convergent subsequence $\{\mathbf{x}_{k_j}\}$. It follows from Exercise 7.12 that $\{f(\mathbf{x}_{k_j})\}$ is a Cauchy sequence and therefore is bounded. This is a contradiction. □

Since a continuous function on a compact set is uniformly continuous, we immediately have the following corollary to Proposition 9.

Corollary 10. If K is a compact subset of $\mathbb{R}^n$ and $\mathbf{f}: K \to \mathbb{R}^m$ is continuous, then $\mathbf{f}$ is bounded on K.

The example $f(x) = x$ on $(0, 1)$ shows that a bounded continuous real-valued function need not assume either its least upper bound or its greatest lower bound on its domain of definition. If, however, the domain is compact, then these values are actually taken on by the continuous function.

Theorem 11. Suppose K is a compact subset of $\mathbb{R}^n$ and $f: K \to \mathbb{R}$ is continuous. Then

  (i) $\sup\{f(\mathbf{x})\colon \mathbf{x} \in K\} = M$ and $\inf\{f(\mathbf{x})\colon \mathbf{x} \in K\} = m$ exist.
  (ii) there are $\mathbf{a}, \mathbf{b} \in K$ such that $f(\mathbf{a}) = M$ and $f(\mathbf{b}) = m$.

Proof. (i) follows directly from Corollary 10.

To prove (ii), for each k, there is $\mathbf{x}_k \in K$ such that $M - 1/k < f(\mathbf{x}_k) \le M$. Since K is compact, $\{\mathbf{x}_k\}$ has a convergent subsequence $\{\mathbf{x}_{k_j}\}$, which converges to a point $\mathbf{a} \in K$. Since f is continuous, $f(\mathbf{a}) = \lim f(\mathbf{x}_{k_j}) = M$. The argument for the existence of $\mathbf{b}$ such that $f(\mathbf{b}) = m$ is similar. □

If a real-valued function $f: E \to \mathbb{R}$ is such that there exists a point $\mathbf{x} \in E$ with $f(\mathbf{x}) = \sup\{f(\mathbf{y})\colon \mathbf{y} \in E\}$ $[f(\mathbf{x}) = \inf\{f(\mathbf{y})\colon \mathbf{y} \in E\}]$, we say that f attains its *maximum* (*minimum*) at $\mathbf{x}$ and usually write $f(\mathbf{x}) = \max\{f(\mathbf{y})\colon \mathbf{y} \in E\}$ $[f(\mathbf{x}) = \min\{f(\mathbf{y})\colon \mathbf{y} \in E\}]$. Thus, from Theorem 11, a continuous real-valued function on a compact set attains both its maximum and minimum.

The image of a bounded, closed, or open set under a continuous mapping is not necessarily bounded, closed or open, respectively, (Exercise 21). However, in the case of compact sets, we have the following:

Proposition 12. If K is a compact subset of $\mathbb{R}^n$ and $\mathbf{f}: K \to \mathbb{R}^m$ is continuous, then $\mathbf{f}(K) = \{\mathbf{f}(\mathbf{x})\colon \mathbf{x} \in K\}$ is compact.

Proof. Boundedness of $\mathbf{f}(K)$ follows from Corollary 10. To show that $\mathbf{f}(K)$ is also closed, and therefore compact, let $\{\mathbf{y}_k\}$ be a convergent sequence in $\mathbf{f}(K)$, and $\mathbf{x}_k \in K$ such that $\mathbf{f}(\mathbf{x}_k) = \mathbf{y}_k$ for each k. Then $\{\mathbf{x}_k\}$ has a subsequence $\{\mathbf{x}_{k_j}\}$, which converges to a point $\mathbf{x} \in K$. Since $\mathbf{f}$ is continuous, the sequence $\{\mathbf{f}(\mathbf{x}_{k_j})\}$ converges to $\mathbf{f}(\mathbf{x}) \in \mathbf{f}(K)$. Hence, $\mathbf{f}(K)$ is closed. □

We show next that a continuous one-one (1-1) function on a compact set has a continuous inverse.

Recall that if f is a one-to-one function from the set A onto the set B, then the function $f^{-1}: B \to A$ defined by $f^{-1}(b) = a$ iff $f(a) = b$ is called the *inverse* of f.

Theorem 13. If K is a compact subset of $\mathbb{R}^n$ and $\mathbf{f}: K \to \mathbb{R}^m$ is continuous and 1-1, then $\mathbf{f}^{-1}: \mathbf{f}(K) \to \mathbb{R}^n$ is continuous.

Proof. Let $\mathbf{y}_0 \in \mathbf{f}(K)$ and $\{\mathbf{y}_k\}$ be a sequence in $\mathbf{f}(K)$ converging to $\mathbf{y}_0$. If $\mathbf{f}(\mathbf{x}_k) = \mathbf{y}_k$ for $k = 0, 1, \ldots$, then there is a subsequence $\{\mathbf{x}_{k_j}\}$ of $\{\mathbf{x}_k\}$ converging to some $\mathbf{x} \in K$. Then $\mathbf{f}(\mathbf{x}_{k_j}) = \mathbf{y}_{k_j}$ converges to $\mathbf{f}(\mathbf{x}) = \mathbf{y}_0$ since $\mathbf{f}$ is continuous. Hence $\mathbf{x}_0 = \mathbf{x}$ and $\{\mathbf{x}_{k_j}\}$ converges to $\mathbf{x}_0$. This argument can be applied to every subsequence of $\{\mathbf{x}_k\}$ and so 3.19 implies that $\{\mathbf{x}_k\}$ converges to $\mathbf{x}_0$. Thus, $\mathbf{x}_k = f^{-1}(\mathbf{y}_k)$ converges to $\mathbf{x}_0 = f^{-1}(y_0)$ and the continuity of $\mathbf{f}^{-1}$ follows. $\qquad\square$

To see that the compactness condition cannot be eliminated in Theorem 13, consider the function $f: [0, 2\pi) \to \mathbb{R}^2$ defined by $f(\theta) = (\cos\theta, \sin\theta)$. The inverse of this function fails to be continuous at the point $(1, 0)$ because in every neighborhood of $(1, 0)$, there are points whose images under f^{-1} are near zero and points whose images are near 2π. Consequently, the limit at $(1, 0)$ cannot exist. A sketch of the domain and range of f may help illustrate this more clearly. In this counterexample, it is necessary that $m > 1$; see Exercise 29.

Finally, we consider the intermediate value property, which is enjoyed by continuous real-valued functions defined on intervals in $\mathbb{R}$. We first establish a special case and then use it to prove the more general statement in Theorem 15.

Lemma 14. Let $f: [a, b] \to \mathbb{R}$ be continuous and such that $f(a)$ and $f(b)$ are nonzero and of opposite sign. Then there is $\zeta \in (a, b)$ such that $f(\zeta) = 0$.

Proof. Consider the case where $f(a) < 0$ and $f(b) > 0$. Set $E = \{x: f(x) < 0\}$ and $\zeta = \sup E$. Since $f(a) < 0$ and $f(b) > 0$, $a < \zeta < b$ (Exercise 23). Choose a sequence $\{x_k\} \subseteq E$ such that $x_k \to \zeta$. Since f is continuous, we have $f(\zeta) \leq 0$. If $f(\zeta) < 0$, then Exercise 23 implies that there is an open interval J containing ζ and such that f is strictly negative on J. This contradicts $\zeta = \sup E$. Consequently, $f(\zeta) = 0$. $\qquad\square$

Theorem 15 (Intermediate Value Property). Let $f: [a, b] \to \mathbb{R}$ be continuous with $f(a) < f(b)$. If $f(a) < y < f(b)$, then there is $\zeta \in [a, b]$ such that $f(\zeta) = y$.

Proof. Apply Lemma 14 to the function $F(x) = y - f(x)$. $\qquad\square$

It may appear that the intermediate value property characterizes continuity; however, this is not the case (see Exercise **8**.30). It is also shown in Chapter 9 that any derivative, even if discontinuous, has the intermediate value property (see Proposition **9**.18).

A more detailed study of continuous functions in a more general setting is given in Chapter 21. We have derived the more important properties of continuous functions here based on the sequential methods developed in the earlier chapters.

EXERCISES 8

1. Determine which of the following sets are open, closed, or neither:
   (a) (in $\mathbb{R}$) $\mathbb{Q}, \mathbb{N}$, (a, b), $[a, b]$, $(a, b]$, $\{1/n: \; n \in \mathbb{N}\}$, $\{n + 1/m: \; m, n \in \mathbb{N}\}$.
   (b) (in $\mathbb{R}^2$) $\mathbb{Q} \times \mathbb{Q}, \mathbb{N} \times \mathbb{N}, \mathbb{N} \times \mathbb{Q}, \mathbb{R} \times \{0\}$.
   (c) (in $\mathbb{R}^n$) $S(\mathbf{x}_0, \delta)$, $\{\mathbf{x}: 0 < \|\mathbf{x}\| \le 1\}$.

2. If $E \subset \mathbb{R}$ is bounded above and closed, show that $\sup E \in E$. Is it necessary that E be closed?

3. Show that $[0, 1)$ and $[0, \infty)$ are not compact subsets of $\mathbb{R}$.

4. Show that any finite subset in $\mathbb{R}^n$ is compact.

5. Show that if the sequence $\{\mathbf{x}_k\}$ has finite range, that is, if $\{\mathbf{x}_k: k \in \mathbb{N}\}$ is finite, then $\{\mathbf{x}_k\}$ has a convergent subsequence.

6. Using only the definition, show that the closed interval $[a, b]$, $-\infty < a < b < \infty$ is a compact subset of $\mathbb{R}$.

7. If A and B are open and closed sets, respectively, of $\mathbb{R}^n$, show that $A \setminus B$ is open and $B \setminus A$ is closed.

8. If E_α is closed $\forall \alpha \in I$, show that $\bigcap_{\alpha \in I} E_\alpha$ is closed. If E_α is compact $\forall \alpha \in I$, show that $\bigcap_{\alpha \in I} E_\alpha$ is compact. If E_α is open $\forall \alpha \in I$, show that $\bigcup_{\alpha \in I} E_\alpha$ is open.

9. If E_j is closed for $j = 1, 2, \ldots, k$, show that $\bigcup_{j=1}^k E_j$ is closed. If E_j is compact for $j = 1, 2, \ldots, k$, show that $\bigcup_{j=1}^k E_j$ is compact. If E_j is open for $j = 1, 2, \ldots, k$, show that $\bigcap_{j=1}^k E_j$ is open. What is the situation if $j \in \mathbb{N}$?

10. Show that if E is a closed subset of a compact set K, then E is compact.

11. Let $Y \subset \mathbb{R}^n$ be compact and $f: Y \to \mathbb{R}$ continuous. If $Z = \{\mathbf{z} \in Y: f(\mathbf{z}) = \sup \{f(\mathbf{y}): \mathbf{y} \in Y\}\}$, show that $Z \ne \varnothing$ and Z is compact.

12. Suppose $f: [a, b] \to \mathbb{R}$ is continuous. Show that the function $g(x) = \max \{f(t): a \le t \le x\}$ is continuous.

13. Show that if $K_i \neq \varnothing$ is compact and $K_{i+1} \subseteq K_i$ for $i = 1, 2, \ldots$, then $\bigcap_{i=1}^{\infty} K_i \neq \varnothing$. Can "compact" be replaced by "closed"?

14. Let $f: [\delta, \infty) \to \mathbb{R}$, $\delta > 0$, be defined by $f(x) = 1/x$. Show that f is uniformly continuous.

15. Can "uniform continuity" in Proposition 9 be replaced by "continuity"?

16. Give an example of a function $f: \mathbb{R} \to \mathbb{R}$ that is continuous and bounded, but not uniformly continuous.

17. Can the compactness condition be eliminated in Corollary 10?

18. Let $X \subset \mathbb{R}^n$ and $Y \subset \mathbb{R}^n$ be compact and $f: X \times Y \to \mathbb{R}$ be continuous. Show that the function $F(x) = \sup \{ f(x, y): y \in Y \}$ is continuous on X.

19. Is the statement of Proposition 12 true if "compact" is replaced by "closed"?
    [*Hint:* Consider $f: \mathbb{R}^2 \to \mathbb{R}$ defined by $f(x, y) = x$ and the set $E = \{(x, y): xy = 1, x > 0\}$.] What if "compact" is replaced by "open"?

20. If K is a compact subset of $\mathbb{R}^n$, E is a closed subset of K, and $\mathbf{f}: K \to \mathbb{R}^m$ is continuous, show that $\mathbf{f}(E)$ is closed. Can the compactness of K be dropped?

21. Give an example of a continuous function f and a closed (bounded, open) set E such that $f(E)$ is not closed (bounded, open).

22. Show that there is $\mathbf{b}$ such that $f(\mathbf{b}) = m$ in Theorem 11.

23. If K is a compact subset of $\mathbb{R}^n$, $f: K \to \mathbb{R}$ is continuous, and $f(\mathbf{x}) > 0$ $\forall \mathbf{x} \in K$, show that $\exists \delta > 0$ such that $f(\mathbf{x}) \geq \delta$ $\forall \mathbf{x} \in K$.

24. Let $f: [a, b] \to \mathbb{R}$ be continuous. Show that the graph of f, $G(f) = \{(x, f(x)): x \in [a, b]\}$, is a compact subset of $\mathbb{R}^2$. Can the continuity be dropped? Can you generalize?

25. If $f: \mathbb{R} \to \mathbb{R}$ is continuous and $\lim_{x \to \infty} f(x) = \lim_{x \to -\infty} f(x) = 0$, show that f is uniformly continuous and bounded on $\mathbb{R}$.

26. Let $f: \mathbb{R}^n \to \mathbb{R}$ be continuous. For $c \in \mathbb{R}$, show that $\{x: f(x) \leq c\}$ and $\{x: f(x) \geq c\}$ are closed and that $\{x: f(x) < c\}$ and $\{x: f(x) > c\}$ are open.

27. If $f: [0, 1] \to [0, 1]$ is continuous, show that there is $x \in [0, 1]$ such that $f(x) = x$.
    Hint: Make a sketch.

28. Suppose that $f: [0, 1] \to \mathbb{R}$ satisfies (a) f is continuous, (b) $f(t)$ is irrational for all t, and (c) $f(0) = e$. Describe f.

29. Suppose $f: [a, b] \to \mathbb{R}$ is continuous and one-one. Show that f is either strictly increasing or strictly decreasing. Show that f^{-1} is continuous.

30. Show that the function $f: [0,1] \to [0,1]$ defined by $f(x) = x$ for x rational and $f(x) = 1 - x$ for x irrational is onto and thus satisfies the intermediate value property. Show, however, that f is continuous only at $x = \frac{1}{2}$.

31. If $E \subseteq \mathbb{R}^n$, $f: E \to \mathbb{R}$ is continuous, and E is closed, show that the graph of f, $G(f) = \{(x, f(x)): x \in E\}$ is closed in $\mathbb{R}^{n+1}$. Give an example of a continuous function whose graph is not closed.

32. Can the interval $[a, b]$ in Theorem 15 be replaced by an arbitrary subset of $\mathbb{R}$?

33. If $f: [0,1] \to \mathbb{R}^2$ is continuous and $f(0) = (0,0)$, $f(1) = (0,1)$, does there necessarily exist a point $t \in [0,1]$ such that $f(t) = (0, \frac{1}{2})$? In other words, is there an analogue to Theorem 15 for vector valued functions?

34. Give a proof of the "if" part of Theorem 7 for $n = 2$.
    Hint: K is contained in a rectangle $I_0 = [a, b] \times [c, d]$. Divide I_0 into four parts by bisecting the sides of I_0 and proceed as in the proof of Theorem 7. Do you see how to proceed for a general proof in $\mathbb{R}^n$?

35. A function $\mathbf{f}: D \to \mathbb{R}^m$ satisfies a *Lipschitz condition* of order $\alpha > 0$ if there is L such that $\|\mathbf{f}(\mathbf{x}) - \mathbf{f}(\mathbf{y})\| \le L\|\mathbf{x} - \mathbf{y}\|^{\alpha}$ for all $\mathbf{x}, \mathbf{y} \in D$. Show that such a function is uniformly continuous.

36. Show that a function $\mathbf{f}: D \to \mathbb{R}^m$ is uniformly continuous iff for each pair of sequences $\{\mathbf{x}_k\}$ and $\{\mathbf{y}_k\}$ in D such that $\mathbf{x}_k - \mathbf{y}_k \to \mathbf{0}$, we have $\mathbf{f}(\mathbf{x}_k) - \mathbf{f}(\mathbf{y}_k) \to \mathbf{0}$.

37. Show that every bounded infinite subset of $\mathbb{R}$ has at least one accumulation point.
    Hint: Consider the proof of Theorem 4.

38. A subset $E \subseteq \mathbb{R}^n$ is called *discrete* if all of its points are isolated points. Give a characterization of compact discrete sets. Give an example of a noncompact discrete set.

39. Show that there is a sequence $\{n_k\}$ of distinct integers such that $\lim_k \sin(n_k)$ exists.

40. If E is an uncountable subset of $\mathbb{R}$, show that E has an accumulation point.

41. Show that the function f defined by $f(x) = \sqrt{x}$, $x \ge 0$, is uniformly continuous on $[0, \infty)$.

42. Let $D \subset \mathbb{R}^n$ and $f: D \to \mathbb{R}^m$. Show that f is continuous iff f carries convergent sequences into convergent sequences.
    Hint: If $x_n \to x$, consider the sequence $x_1, x, x_2, x, \ldots$.

43. Give a sequential characterization for a set to be open.
    Hint: If G is open and $\mathbf{x}_k \to x$, $\mathbf{x} \in G$, then there is N such that $\mathbf{x}_k \in G$ for $k \ge N$.

44. If $f: [a, b] \to \mathbb{R}$ is continuous and $\varepsilon > 0$, show that there is a piecewise linear function g such that $|f(t) - g(t)| < \varepsilon$ for all $t \in [a, b]$.

45. Let $f: (0, 1] \to \mathbb{R}$ be uniformly continuous. If $\lim_{k \to \infty} f(1/k) = L$, show that $\lim_{x \to 0} f(x) = L$. Can "uniformly continuous" be replaced by "continuous"?

46. Let $f: I \to \mathbb{R}$ have the property that for every $\varepsilon > 0$, there is a uniformly continuous function $g: I \to \mathbb{R}$ such that $|f(t) - g(t)| < \varepsilon$. Show that f is uniformly continuous.

47. A function $f: \mathbb{R} \to \mathbb{R}$ is said to be *periodic* with *period p* if $f(t + p) = f(t)$ $\forall t \in \mathbb{R}$. Show that a continuous periodic function is uniformly continuous and bounded.

48. Show that if $f: [a, b] \to \mathbb{R}$ is monotonic and has the intermediate value property, then f is continuous.

49. Let $f: \mathbb{R} \to \mathbb{R}$ be continuous and $\lim_{|x| \to \infty} f(x) = 0$. Show that f has either a maximum or a minimum value on $\mathbb{R}$.

50. Let $f: [0, \infty) \to \mathbb{R}$ be continuous, $f(0) = 0$, and $\lim_{x \to \infty} f(x) = 1$. If $0 < \zeta < 1$, show that there is x such that $f(x) = \zeta$.

51. Prove or disprove: A function $f: \mathbb{R} \to \mathbb{R}$ is uniformly continuous on $\mathbb{R}$ if, and only if, f is uniformly continuous on every bounded interval.

52. Let $f: [0, 1] \times [0, 1] \to \mathbb{R}$ be continuous. Set $g(x) = \sup \{ f(x, y): 0 \leq y \leq x \}$. Show that g is continuous.

53. Let $f, g: \mathbb{R} \to \mathbb{R}$ be continuous and $f(t) \neq 0$ for all t. If $f^2(t) = g^2(t)$ for all t, show that either $f - g = 0$ or $f + g = 0$. Can continuity be dropped?

54. Let $\mathscr{F}$ be a family of nonvoid pairwise disjoint open intervals in $\mathbb{R}$. Show that $\mathscr{F}$ is countable.

55. Let I be an interval in $\mathbb{R}$ and $f: I \to \mathbb{R}$. Show that f is uniformly continuous on I iff for every $\varepsilon > 0$, there exists $M > 0$ such that $|(f(x) - f(y))/(x - y)| > M$ implies $|f(x) - f(y)| < \varepsilon$.
Hint See D. Paine, "Visualizing Uniform Continuity, " *American Mathematical Monthly 75* (1968), 44–45.

56. Let $\mathbf{f}: \mathbb{R}^n \to \mathbb{R}^m$. Show that $\mathbf{f}$ is continuous iff $\mathbf{f}^{-1}(G)$ is open $\forall G$ open in $\mathbb{R}^m$ iff $\mathbf{f}^{-1}(F)$ is closed $\forall F$ closed in $\mathbb{R}^m$.

CHAPTER 9

Differentiation in $\mathbb{R}$

One of the fundamental problems leading to the development of the calculus was that of defining the tangent line to a curve in $\mathbb{R}^2$. Given a curve C in $\mathbb{R}^2$ and a point P on C, how do we find a tangent line to the curve at the point P? There are two basic questions here: first, "what is a curve?"; and second, "what is a tangent line?." We'll be content with the definition of a curve that is usually given in the calculus: A *curve* in $\mathbb{R}^2$ is a point set that can be represented as the range of a continuous function defined on an interval in $\mathbb{R}$. To define what is meant by the tangent line, we need first to decide what the fundamental property of a tangent line ought to be. A good starting point for this is the case where the curve C is a circle, a situation encountered early in the study of plane geometry. The tangent line to a circle is that line which is perpendicular to the circle's radius at the point of tangency. In this particular case, the tangent line gives an excellent approximation to the circle near the point of tangency, and this is the fundamental property that leads to the definition of tangent line for more general curves.

Suppose the curve C is the graph of a function $y = f(x)$, $a \leq x \leq b$, and $a \leq x_0 \leq b$. If h is small, the secant line from the point $(x_0, f(x_0))$ to the

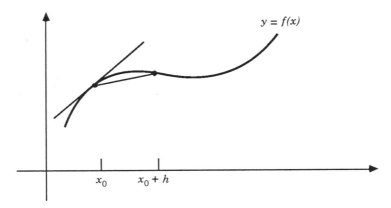

Figure 9.1

point $(x_0 + h, f(x_0 + h))$ should give a good approximation to the curve near $(x_0, f(x_0))$ (see Figure 9.1). The slope of this secant line is $(f(x_0 + h) - f(x_0))/h$. Intuitively, as h becomes smaller, this secant line should give increasingly better approximations to the curve C. This suggests that we should define the tangent line to C at $(x_0, f(x_0))$ to be the line that passes through the point $(x_0, f(x_0))$ having slope $\lim_{h \to 0} (f(x_0 + h) - f(x_0))/h$, provided this limit exists. This same limit appears in attempts to solve other problems, such as those related to velocity or radioactive decay, in the general area of rate of change. This, of course, leads to the examination of this limit as an abstract idea and to the introduction of the concept of derivative.

Definition 1. The function $f: [a, b] \to \mathbb{R}$ is *differentiable* at $x_0 \in [a, b]$ if

$$\lim_{x \to x_0} \frac{f(x) - f(x_0)}{x - x_0} \tag{1}$$

exists; in this case, the limit is called the *derivative* of f at x_0 and is denoted by $f'(x_0)$. f is *differentiable* on $[a, b]$ if it is differentiable at each point of $[a, b]$. When $x_0 = a$ or b, the limit above is understood to be a one-sided limit and we refer to the *right-* or *left-hand derivative* at a or b, respectively.

The notation $f'(x)$ is a descendant of notation introduced by Isaac Newton (1642–1727), one of two inventors of the calculus. The other inventor of the calculus, Gottfried Wilhelm Leibniz (1646–1716), working independently of Newton, used the notation $(df/dx)(x_0)$ for $f'(x_0)$. This notation is still commonly used and, in many applications, is more descriptive of the processes involved.

Familiarity with the elementary calculus is assumed here, and so the task of establishing the elementary computational properties of the derivative is left to Exercise 1. We do, however, give a proof of the chain rule since the treatment of this important result in the elementary texts is often incomplete.

Since the denominator in (1) approaches zero, the limit will exist only if the numerator does the same. This is, of course, equivalent to saying that f is continuous at x_0, and so we have the following.

Proposition 2. If f is differentiable at x_0, then it is continuous there.

The simple example $\phi(x) = |x|$, $x_0 = 0$, shows that the converse of Proposition 2 is not true. More complicated examples show that there are continuous functions that are differentiable nowhere (see Example **11**.20).

Theorem 3 (The Chain Rule). If

   i. $f: [a, b] \to [c, d]$ and $g: [c, d] \to \mathbb{R}$,
   ii. $x_0 \in [a, b]$,
   iii. f and g are differentiable at x_0 and $f(x_0)$ respectively, then the composite function $g \circ f$ is differentiable at x_0 and has derivative $(g \circ f)'(x_0) = g'[f(x_0)]f'(x_0)$.

Proof. Set

$$h(y) = \frac{g(y) - g[f(x_0)]}{y - f(x_0)} - g'[f(x_0)] \qquad \text{if } y \neq f(x_0), \, y \in [c, d]$$

$$= 0 \qquad \text{if } y = f(x_0), \, y \in [c, d].$$

Then $h: [c, d] \to \mathbb{R}$ is continuous at $f(x_0)$. Now f is continuous at x_0; so $h \circ f$ is continuous at x_0. Hence

$$\lim_{x \to x_0} \{h \circ f(x) + g'[f(x_0)]\} \cdot \frac{f(x) - f(x_0)}{x - x_0}$$

$$= \lim_{x \to x_0} \frac{g[f(x)] - g[f(x_0)]}{x - x_0}$$

$$= (g \circ f)'(x_0) = g'[f(x_0)] f'(x_0). \qquad \square$$

The following two theorems are of considerable importance in the study of real-valued functions. Their applications yield results in optimization theory, Taylor's series, location of zeros of a function, and many other areas.

90 Introduction to Real Analysis

Theorem 4 (Rolle's Theorem). If $f: [a, b] \to \mathbb{R}$ is continuous on $[a, b]$ and differentiable on (a, b) and $f(a) = f(b)$, then there is $\zeta \in (a, b)$ such that $f'(\zeta) = 0$.

Proof. If $f(x) = f(a)$ for all $x \in (a, b)$, then $f'(x) \equiv 0$. If f is not constant, then the continuity of f implies that there are $x_M, x_m \in [a, b]$ such that $f(x_M) = \sup \{ f(x): a \le x \le b \}$, $f(x_m) = \inf \{ f(x): a \le x \le b \}$. Since f is not constant, either x_M or x_m is in (a, b). Suppose it is x_M. Then $[f(x) - f(x_M)]/(x - x_M) \le 0$ for $x > x_M$ implies $f'(x_M) \le 0$. Similarly, $[f(x) - f(x_M)]/(x - x_M) \ge 0$ for $x < x_M$ implies $f'(x_M) \ge 0$. Consequently $f'(x_M) = 0$. □

Geometrically, Rolle's theorem states that there is a point in the interval (a,b) at which the tangent line to the graph of f is parallel to the x-axis.

Theorem 5 (Mean Value Theorem). If $f: [a, b] \to \mathbb{R}$ is continuous on $[a, b]$ and differentiable on (a, b), then there is $\zeta \in (a, b)$ such that $f'(\zeta) = [f(b) - f(a)]/(b - a)$.

Proof. Set

$$F(x) = f(x) - f(a) - (x - a)\frac{f(b) - f(a)}{b - a}.$$

Apply Rolle's theorem to F. □

Again, the geometrical description of the mean value theorem is that there is a point in (a, b) at which the tangent line to the graph of f is parallel to the chord connecting the end points $(a, f(a))$ and $(b, f(b))$ of the graph.

As an application of the mean value theorem, we'll establish the inequality $(1 + h)^\alpha > 1 + \alpha h$, for $h > 0$ and $\alpha > 1$. Consider the function $f(x) = x^\alpha$, $\alpha > 1$. Applying the mean value theorem with $a = 1$ and $b = 1 + h$, $h > 0$, yields, for some ζ between 1 and $1 + h$, $(1 + h)^\alpha - 1 = \alpha \zeta^{\alpha - 1} h > \alpha h$.

Another application of the mean value theorem yields the following familiar relationships between a function and its derivatives.

Corollary 6. If $f: [a, b] \to \mathbb{R}$ is continuous on $[a, b]$ and differentiable on (a, b), then

    (i) $f'(x) \ne 0 \; \forall x \in (a, b)$ implies f is one-one on $[a, b]$.
    (ii) $f'(x) = 0 \; \forall x \in (a, b)$ implies $f \equiv$ constant.
    (iii) $f'(x) > 0 \; (< 0) \; \forall x \in (a, b)$ implies f is strictly increasing (decreasing) on $[a, b]$.
    (iv) $f'(x) \ge 0 \; (\le 0) \; \forall x \in (a, b)$ implies f is increasing (decreasing) on $[a,b]$.

Proof. For (i), (iii), and (iv): if $a \le x < y \le b$, then there is $\zeta_{x,y} = \zeta \in$ (x, y) such that $f(x) - f(y) = f'(\zeta)(x - y)$. The results follow.

To prove (ii), note that for $x \in (a, b)$ there is $\zeta_x \in (a, x)$ such that $f(x) - f(a) = f'(\zeta_x)(x - a) = 0$. ☐

Corollary 7. Let f and g be real-valued functions that are continuous on $[a, b]$ and differentiable on (a, b). If $f' = g'$ on (a, b) then there is $k \in \mathbb{R}$ such that $f(x) = g(x) + k \ \forall x \in [a, b]$.

Proof. Apply Corollary 6(ii) to the function $f - g$. ☐

Cauchy extended the mean value theorem to the following:

Theorem 8 (The Generalized Mean Value Theorem). Let f and g be real-valued functions that are continuous on $[a, b]$ and differentiable on (a, b). Then there is $\zeta \in (a, b)$ such that

$$[f(b) - f(a)]g'(\zeta) = [g(b) - g(a)]f'(\zeta).$$

Proof. Apply Rolle's theorem to the function defined by

$$h(x) = [f(b) - f(a)]g(x) - [g(b) - g(a)]f(x).$$ ☐

Note that the case $g(x) = x$ gives the mean value theorem.

A consequence of the generalized mean value theorem is L'Hospital's rule, which is useful for the evaluation of certain kinds of limits. We establish two versions of the rule; others are contained in the exercises.

Theorem 9 (L'Hospital's Rule). Let $f, g: (a, b) \to \mathbb{R}$ be differentiable with $g'(x) \neq 0$ for $x \in (a, b)$. Suppose that

$$\lim_{x \to a^+} \frac{f'(x)}{g'(x)} = L.$$

If either

(i) $\lim_{x \to a^+} f(x) = \lim_{x \to a^+} g(x) = 0$,

or

(ii) $\lim_{x \to a^+} f(x) = \lim_{x \to a^+} g(x) = \infty$,

then

$$\lim_{x \to a^+} \frac{f(x)}{g(x)} = L.$$

Proof. Let $0 < \varepsilon < 1$. There is $c > a$ such that $|f'(x)/g'(x) - L| < \varepsilon$ for $a < x < c$. If $a < x < c$, $a < y < c$ and $x < y$, then Theorem 8 implies that there is a point ζ between x and y such that

$$L - \varepsilon < \frac{f(x) - f(y)}{g(x) - g(y)} = \frac{f'(\zeta)}{g'(\zeta)} < L + \varepsilon. \tag{2}$$

If (i) holds, then letting $x \to a$ in (2) gives

$$\left| \frac{f(y)}{g(y)} - L \right| \leq \varepsilon$$

for $0 < y < c$, and the conclusion holds.

If (ii) holds, fix y in (2) and set

$$\psi(x) = \frac{1 - g(y)/g(x)}{1 - f(y)/f(x)}.$$

By (ii), there is a d, $a < d < c$, such that $|\psi(x) - 1| < \varepsilon$ for $a < x < d$. Note that $|\psi(x)| \leq 2$ for $a < x < d$. From (2),

$$\left| \frac{f(x)}{g(x)} - L \right| = \left| \psi(x) \frac{f'(\zeta)}{g'(\zeta)} - L \right|$$

$$\leq |\psi(x)| \left| \frac{f'(\zeta)}{g'(\zeta)} - L \right| + |L| \, |\psi(x) - 1|$$

$$< (2 + |L|)\varepsilon,$$

and the conclusion follows.   □

Clearly, a similar result holds for left-hand limits, and the results for the one-sided limits easily yield a similar statement for limits. The cases of infinite limits and limits at infinity are left to Exercise 7.

For counter-examples related to the converse of L'Hospital's rule see N. W. Rickert, 1968, 166 and R. P. Boas, 1986, 644–645. Some examples of computation of limits using L'Hospital's rule are given in Exercises 16 to 18.

Differentiation of Inverse Functions

Suppose f is a 1-1 real-valued function on the interval $[a, b]$. Then, of course, f has an inverse function and if f is also differentiable, one naturally inquires about the existence of the derivative of the inverse function.

Theorem 10. Let $f: [a, b] \to \mathbb{R}$ be differentiable on $[a, b]$ with $f'(x) \neq 0$ for all $x \in [a, b]$. Then f^{-1} exists and is differentiable on the range of f with $(f^{-1})'[f(x)] = 1/f'(x) \; \forall \, x \in [a, b]$.

Proof. The existence of f^{-1} follows from Corollary 6. If $m = \inf \{ f(x): a \leq x \leq b\}$ and $M = \sup \{ f(x): a \leq x \leq b\}$, then $f([a, b]) = [m, M]$ by **8.11** and **15**. Let $y_0 \in [m, M]$ and $\{ y_k \}$ be a sequence in $[m, M]$ such that $y_k \neq y_0$ and $y_k \to y_0$. Set $x_k = f^{-1}(y_k)$. Then **8.13** implies that $x_k \to f^{-1}(y_0) = x_0$. Since f is 1-1, we also have $x_k \neq x_0$ for all k. Consequently,

$$\frac{f^{-1}(y_k) - f^{-1}(y_0)}{y_k - y_0} = \frac{x_k - x_0}{f(x_k) - f(x_0)} \to \frac{1}{f'(x_0)}. \qquad \square$$

Taylor's Theorem

If f is a real-valued differentiable function on $[a, b]$, then it's possible that f' is also a differentiable function at some or all points of $[a, b]$. If x_0 is such a point, the derivative of f' at x_0, denoted by $f''(x_0)$, is called the *second* (*order*) *derivative* of f at x_0. Note that for $f''(x_0)$ to exist, $f'(x)$ must exist at all points of some neighborhood of x_0. Derivatives of higher order are defined inductively by $f^{(0)} = f$ and $f^{(k+1)} = (f^{(k)})'$ for $k = 0, 1, 2, \ldots$.

If $P(x) = a_0 + a_1 x + \cdots + a_k x^k + \cdots + a_n x^n$ is a polynomial of degree $n \geq 0$, then $P^{(k)}(0) = k! a_k$. Solving for a_k we obtain $a_k = P^{(k)}(0)/k!$. So, $P(x)$ has the representation

$$P(x) = \sum_{k=0}^{n} \frac{P^{(k)}(0)}{k!} x^k.$$

In a similar fashion, an expansion of $P(x)$ in powers of $(x - x_0)$ can be obtained:

$$\sum_{k=0}^{n} \frac{P^{(k)}(x_0)}{k!} (x - x_0)^k.$$

If $f: [a, b] \to \mathbb{R}$ has derivatives through order n at $x_0 \in (a, b)$, the same expression can be written formally to obtain

$$\sum_{k=0}^{n} \frac{f^{(k)}(x_0)}{k!} (x - x_0)^k.$$

This last expression, discovered by its eponym, the English mathematician Brook Taylor (1685–1731), is called the *Taylor polynomial* (with respect to f)

of degree n about x_0. The derivatives of this polynomial up to order n coincide with those of f at the point $x = x_0$. If P_n is the Taylor polynomial of degree n for f about x_0, the function $R_n = f - P_n$ is called the *remainder* at x_0. A representation for the remainder is given by the following.

Theorem 11 (Taylor's Theorem). Let $f \colon [a, b] \to \mathbb{R}$ have derivatives through order $n + 1$ on (a, b) and let $x_0 \in (a, b)$. Then for each $x \in (a, b)$, $R_n(x) = f^{(n+1)}(\zeta_x)(x - x_0)^{n+1}/(n + 1)!$, for some ζ_x between x and x_0.

Proof. Set $Q(t) = (t - x_0)^{n+1}$. Then R_n and Q are n-times differentiable and $R_n^{(k)}(x_0) = Q^{(k)}(x_0) = 0$ for $k = 0, 1, \ldots, n$. Application of the generalized mean value theorem n-times yields

$$\frac{R_n(x)}{Q(x)} = \frac{R_n(x) - R_n(x_0)}{Q(x) - Q(x_0)} = \frac{R_n'(\zeta_1)}{Q'(\zeta_1)} = \frac{R_n''(\zeta_2)}{Q''(\zeta_2)} = \cdots = \frac{R_n^{(n)}(\zeta_n)}{Q^{(n)}(\zeta_n)},$$

where ζ_i lies between x and x_0. But $R_n^{(n)}(t) = f^{(n)}(t) - f^{(n)}(x_0)$, and $Q^{(n)}(t) = (n + 1)!(t - x_0)$. So

$$\frac{R_n(x)}{(x - x_0)^{n+1}} = \frac{1}{(n + 1)!} \frac{f^{(n)}(\zeta_n) - f^{(n)}(x_0)}{\zeta_n - x_0}.$$

Applying the mean value theorem to this last expression gives

$$R_n(x)/(x - x_0)^{n+1} = f^{(n+1)}(\zeta_x)/(n + 1)!,$$

where ζ_x lies between x and x_0. $\square$

Note that the case $n = 0$ is just the mean value theorem.

Taylor's theorem suggests the following question: Suppose $f \colon [a, b] \to \mathbb{R}$ has derivatives of all orders in (a, b) [such functions are said to be *infinitely differentiable* in (a, b)]. For $x, x_0 \in (a, b)$, what is the relationship between $f(x)$ and the infinite series $\sum_{k=0}^{\infty} f^{(k)}(x_0)(x - x_0)^k/k!$? This latter series is called the *Taylor series* for f about x_0. Taylor's theorem then states that the series and the function are equal if $\lim_{n \to \infty} R_n(x) = 0$ for all $x \in (a, b)$. Clearly, equality always holds at $x = x_0$.

Definition 12. The function f is said to be *analytic* at x_0 if the Taylor series for f about x_0 converges to $f(x)$ for all x in some neighborhood of x_0.

In general, an infinitely differentiable function need not be analytic (see the example following Proposition 13), but we do have the following criterion for analyticity.

Proposition 13. Suppose that f has derivatives of all orders in (a, b) and that there is $M > 0$ such that $|f^{(k)}(x)| \le M$ for $k = 0, 1, \ldots$, and $x \in (a, b)$. If $x_0 \in (a, b)$, then

$$f(x) = \sum_{k=0}^{\infty} \frac{f^{(k)}(x_0)}{k!}(x - x_0)^k, \qquad \forall\, x \in (a, b).$$

Proof.

$$|R_n(x)| = \left| \frac{f^{(n+1)}(\xi_x)}{(n+1)!}(x - x_0)^{n+1} \right| \le \frac{M|x - x_0|^{n+1}}{(n+1)!}.$$

The last expression is the $(n + 1)$st term of the convergent series

$$\sum_{n=1}^{\infty} \frac{M|x - x_0|^n}{n!}.$$

Therefore $\lim_{n \to \infty} R_n(x) = 0 \,\forall\, x \in (a, b)$.   □

This result shows that many of the elementary functions of calculus are analytic (see Exercise 20). However, the next example shows that it's possible for a function to be infinitely differentiable, but not analytic.

Let the function f be defined by

$$f(t) = \begin{cases} e^{-t^{-2}} & \text{if } t \ne 0, \\ 0 & \text{if } t = 0. \end{cases}$$

Then $f^{(k)}(0) = 0$ for $k = 0, 1, \ldots$. (Exercise 30). Consequently the Taylor series for f about $x = 0$ converges for all x; but it converges to $f(x)$ only at $x = 0$. There are examples of functions whose Taylor series about a point x_0 diverge at all points of some deleted neighborhood of x_0. In fact, there is a remarkable result of E. Borel that states that if $\{a_k\}_{k=0}^{\infty}$ is any real sequence, then there is an infinitely differentiable function f such that $f^{(k)}(0) = a_k$ for all k. (See M. D. Meyerson, 1981, 51–52.)

Optimization

Let D be a subset of $\mathbb{R}^n$ and $f: D \to \mathbb{R}$. The function f has a *local maximum* (*strict*) at $\mathbf{x}_0 \in D$ if there is $S(\mathbf{x}_0, \delta)$ such that $f(\mathbf{x}_0) \ge f(\mathbf{x})$, $(f(\mathbf{x}_0) > f(\mathbf{x}))$ for all $\mathbf{x} \in D \cap S(\mathbf{x}_0, \delta) \setminus \{\mathbf{x}_0\}$. *Local and strict local minima* are defined similarly. The function f has an *extremum* at x_0 if it has either a maximum or a minimum there.

We have the following first order (i.e., involving the first derivative) necessary condition for an extremum.

Proposition 14. Suppose $f: [a, b] \to \mathbb{R}$ has a local extremum at $x_0 \in (a, b)$. If f is differentiable at x_0, then $f'(x_0) = 0$.

Proof. Suppose that f has a local minimum at x_0. For x sufficiently near x_0 and $x > x_0$, we have $[f(x) - f(x_0)]/(x - x_0) \geq 0$. Consequently, $f'(x_0) \geq 0$. For $x < x_0$, we have $[f(x) - f(x_0)]/(x - x_0) \leq 0$ and so, $f'(x_0) \leq 0$. Since f is differentiable at x_0, if follows that $f'(x_0) = 0$.

The proof for a local maximum is similar.   □

The condition of Proposition 14 is only necessary; it is not, in general, sufficient, as the example $f(x) = x^3$ at $x_0 = 0$ shows. A first order sufficient condition for a minimum is given in Exercise 38.

A second order necessary condition (i.e., involving the second derivative) is given by the following.

Proposition 15. If $f: [a, b] \to \mathbb{R}$ has a local maximum at $x_0 \in (a, b)$, and f'' exists in (a, b) and is continuous at x_0, then $f''(x_0) \leq 0$.

Proof. If $x \in (a, b)$ is sufficiently near x_0, but $x \neq x_0$, then Taylor's theorem with $n = 1$ implies that

$$f'(x_0)(x - x_0) + f''(\zeta_x)\frac{(x - x_0)^2}{2} = f(x) - f(x_0) \leq 0,$$

where ζ_x lies between x and x_0. However, $f'(x_0) = 0$ and $(x - x_0)^2 > 0$, so that $f''(\zeta_x) \leq 0$. Since f'' is continuous at x_0, we have $\lim_{x \to x_0} f''(x) = f''(x_0)$. It follows that $f''(x_0) \leq 0$.   □

Again, the condition in Proposition 15 is only a necessary condition. The example $f(t) = t^5$, $t_0 = 0$, shows that, in general, the condition is not sufficient. However, by strengthening the conclusion of Proposition 15 to $f''(x) < 0$, we obtain the following second order sufficient condition. A first order sufficient condition can be obtained from Corollary 6(iii) (see Exercise 38).

Proposition 16. Suppose that $f: [a, b] \to \mathbb{R}$ is twice differentiable in (a, b) and that f'' is continuous at $x_0 \in (a, b)$. If $f'(x_0) = 0$ and $f''(x_0) < 0$, then f has a strict local maximum at x_0.

Proof. For $x \in (a, b)$, Taylor's theorem again implies that

$$f'(x_0)(x - x_0) + f''(\zeta_x)\frac{(x - x_0)^2}{2} = f(x) - f(x_0),$$

where ζ_x lies between x and x_0. Since f'' is continuous, we have $f''(x) < 0$ for x sufficiently near x_0. Thus, $f(x) - f(x_0) < 0$ and so f has a strict local maximum at x_0. □

Examination of the function $f(x) = -x^4$ shows that the conditions in Proposition 16 are sufficient, but not necessary, for f to have a strict local maximum at x_0.

We remarked in Chapter 8 that a function can be discontinuous and still possess the intermediate value property. As an application of the first order optimization condition, we show that, in fact, every derivative has the intermediate value property, although a derivative may fail to be continuous. For example, the function

$$f(t) = \begin{cases} t^2 \sin(1/t), & t \neq 0 \\ 0, & t = 0 \end{cases}$$

is differentiable everywhere, but the derivative is not continuous at $t = 0$.

Lemma 17. Let $f: [a, b] \to \mathbb{R}$ be differentiable on $[a, b]$ with $f'(a) > 0$ and $f'(b) < 0$. Then there is $\zeta \in (a, b)$ such that $f'(\zeta) = 0$.

Proof. Since $f'(a) > 0$, there is x in a neighborhood of a such that $f(x) > f(a)$. Similarly, there is y in a neighborhood of b such that $f(y) > f(b)$. Therefore, if $\zeta \in [a, b]$ is such that $f(\zeta) = \max\{f(t): a \leq t \leq b\}$, then $\zeta \in (a, b)$. By Proposition 14, $f'(\zeta) = 0$. □

Proposition 18. Let $f: [a, b] \to \mathbb{R}$ be differentiable on $[a, b]$. If $f'(a) < f'(b)$ and η is such that $f'(a) < \eta < f'(b)$, then there is $\zeta \in (a, b)$ such that $f'(\zeta) = \eta$.

Proof. Apply Lemma 17 to the function $g(t) = \eta t - f(t)$. □

EXERCISES 9

1. State and prove the familiar calculus rules for differentiation of sums, products, and quotients of functions.

2. If $f: \mathbb{R} \to \mathbb{R}$ is such that $|f(x) - f(y)| \leq |x - y|^p$ for some $p > 1$, show that f is constant.

3. Let $f: (a, b) \to \mathbb{R}$ be differentiable on (a, b) and suppose there is $M > 0$ such that $|f'(x)| \leq M \; \forall x \in (a, b)$.
   (a) Show that f is uniformly continuous on (a, b).
   (b) Show that $\lim_{x \to x_0} f(x)$ exists, where $x_0 = a$ or b.

4. Use the mean value theorem to show that $\sqrt{1 + h} < 1 + h/2$ for $h > 0$. Is there an easier way to do this?

5. Show that the equation $x^3 - 3x + b = 0$ has at most one root in $[-1, 1]$.

6. Suppose $f: \mathbb{R} \to \mathbb{R}$ is differentiable, $f(0) = 0$, and $|f'| \leq |f|$. Show that $f = 0$.

7. Prove Theorem 9 when (a) $L = \pm\infty$, (b) $\lim_{x \to \infty} f(x) = \lim_{x \to \infty} g(x) = 0$, and, (c) $\lim_{x \to \infty} f(x) = \lim_{x \to \infty} g(x) = \infty$.

8. Suppose that $f: [a, b] \to \mathbb{R}$ is differentiable and has a local maximum at both x_1 and x_2. Show that there is a point between x_1 and x_2 at which f has a local minimum.

9. Let f, g have continuous derivatives on an interval I. Show that if $fg' - f'g \neq 0$ on I, then the zeros of f and g on I separate each other [x is a zero of f if $f(x) = 0$].

10. Let $f: [0, \infty) \to \mathbb{R}$ be continuous and differentiable on $[0, \infty)$. If $f(0) = 0$ and $\lim_{x \to \infty} f(x) = 0$, show that there is $c \in (0, \infty)$ such that $f'(c) = 0$.

11. Can the function

$$f(x) = \begin{cases} 1 & \text{if } x \geq 0, \\ 0 & \text{if } x < 0, \end{cases}$$

be the derivative of any function?

12. Let $f: (a, b) \to \mathbb{R}$ be twice differentiable. If f vanishes at three distinct points, show that there is $c \in (a, b)$ such that $f''(c) = 0$.

13. Let $f: [1, \infty) \to \mathbb{R}$ be decreasing and positive. Let g be any antiderivative for f, that is, $g' = f$. Show that $\sum_{k=1}^{\infty} f(k)$ converges iff g is bounded. [*Hint:* Apply the mean value theorem to $g(k + 1) - g(k)$.] Use this result to establish **4.10**.

14. If $f: [1, \infty) \to \mathbb{R}$ is differentiable and $\lim_{x \to \infty} f'(x) = a$, $a > 0$, show that $\lim_{x \to \infty} f(x) = \infty$.

15. Define $f: \mathbb{R} \to \mathbb{R}$ by

$$f(x) = \begin{cases} x^2 & \text{if } x \in \mathbb{Q}, \\ 0 & \text{if } x \notin \mathbb{Q}. \end{cases}$$

Does $f'(0)$ exist? Does $f'(x)$ exist for $x \neq 0$? Does $f''(0)$ exist?

16. Show that $\lim_{x \to \infty} x^p/e^x = 0$ and $\lim_{x \to \infty} \ln x/x^p = 0$ for every positive integer p.

17. If $f: [a, b] \to \mathbb{R}$ is differentiable on (a, b) and $x_0 \in (a, b)$, find

$$\lim_{h \to 0} \frac{f(x_0 + h) - f(x_0 - h)}{2h}.$$

18. Compute (a) $\lim_{x \to 0} (\cot x - 1/x)$ and (b) $\lim_{x \to 0} x \ln x$.

19. Find the Taylor polynomial P_3 for f at x_0:
  (a) $f(x) = ax^3 + bx^2 + cx + d$, $x_0 = 1$.
  (b) $f(x) = e^x$, $x_0 = 0$.
  (c) $f(x) = \sin x$, $x_0 = 0$.

20. Show that Proposition 13 is applicable to e^x, $\sin x$, and $\cos x$ at $x_0 = 0$.

21. Define local minimum and strict local minimum.

22. Give statements analogous to Propositions 15 and 16 for minima.

23. If $f: [a, b] \to \mathbb{R}$ is differentiable and has a local minimum (maximum) at a, is $f'(a)$ necessarily 0? If not, state and prove a necessary condition for local minimum (maximum) at a. What is the situation at b?

24. Show that $e^x \le 1/(1 - x)$ when $x < 1$.
  Hint: Consider $f(x) = (1 - x)e^x$.

25. Suppose that f''' is continuous on (a, b) and $x_0 \in (a, b)$. If $f'(x_0) = f''(x_0) = 0$ and $f'''(x_0) \ne 0$, show that f does not have a local extremum at x_0.

26. Let $f: \mathbb{R} \to \mathbb{R}$ be differentiable at t. Set $f^+(s) = \max \{ f(s), 0 \}$. Show that the function $g = (f^+)^2$ is differentiable at t, with $g'(t) = 2f^+(t)f'(t)$.
  Hint: The function

$$\phi(t) = \begin{cases} t^2, & t \ge 0, \\ 0, & t < 0, \end{cases}$$

is useful.

27. A function $f: [a, b] \to \mathbb{R}$ is said to be *convex* if

$$f(\lambda x + (1 - \lambda)y) \le \lambda f(x) + (1 - \lambda)f(y)$$

whenever $a \le x \le b$, $a \le y \le b$, and $0 < \lambda < 1$.
  (a) If f is differentiable, show that f is convex iff $f(y) - f(x) \ge f'(x)(y - x)$ for all x, y.
  (b) Show that every increasing function of a convex function is convex.
  (c) If f is convex and $a < s < t < u < b$, show that

$$\frac{f(t) - f(s)}{t - s} \le \frac{f(u) - f(s)}{u - s} \le \frac{f(u) - f(t)}{u - t}.$$

  (d) Show that a convex function f on $[a, b]$ has a left-hand and right-hand derivative at every point. Moreover, the subset where f' does not exist is countable.

28. Prove Leibniz rule: $(fg)^{(n)} = \sum_{i=0}^{n}\binom{n}{i} f^{(i)}g^{(n-i)}$.

29. Let $f: [a, b] \to \mathbb{R}$ be continuous on $[a, b]$ and differentiable on (a, b) except possibly at $x_0 \in (a, b)$. If $\lim_{x \to x_0} f'(x) = L$ exists, show that f is differentiable at x_0 and $f'(x_0) = L$.

30. If

$$f(x) = \begin{cases} e^{-1/x^2}, & x \neq 0, \\ 0, & x = 0, \end{cases}$$

show that $f^{(k)}(0) = 0 \; \forall \, k \geq 1$.
Hint: See Exercise 29.

31. Let

$$f_k(x) = \begin{cases} x^k \sin(1/x), & x \neq 0, \\ 0, & x = 0, \end{cases}$$

$k \in \mathbb{N}$. Discuss the differentiation of f_k.

32. Why doesn't $f(x) = 1 - |x - 1|$ satisfy the conclusion in Rolle's theorem on $[0, 2]$?

33. Let $f, g: [0, \infty) \to \mathbb{R}$ be differentiable with $f(0) = g(0)$ and $f'(x) \geq g'(x)$ $\forall \, x > 0$. Show that $f(x) \geq g(x) \; \forall \, x > 0$.

34. Can the function f defined by

$$f(x) = \begin{cases} 1 & \text{if } x \text{ is rational,} \\ 0 & \text{if } x \text{ is irrational,} \end{cases}$$

be the derivative of any function?

35. Let $f: [0, 1] \to \mathbb{R}$ be differentiable and such that there is no x for which $f(x) = f'(x) = 0$. Show that $Z = \{x: f(x) = 0\}$ is finite.

36. Let $f: [a, b] \to [a, b]$ be differentiable in (a, b) and suppose that there is α, $0 < \alpha < 1$, such that $|f'(t)| \leq \alpha$ for $t \in (a, b)$. Show that f has a unique fixed point in $[a, b]$.
Hint: Pick $x_0 \in [a, b]$ and set $x_{k+1} = f(x_k)$. Show that $|x_{k+1} - x_k| \leq \alpha^k(b - a)$.

37. Let $f: (0, 1) \to \mathbb{R}$ be differentiable and $|f'(x)| \leq 1$ for all x. If $x_k = f(1/k)$, show that $\{x_k\}$ converges.

38. (First order sufficient condition for minima.) Let $f: (a, b) \to \mathbb{R}$ be differentiable and $x \in (a, b)$. If $f'(z) \geq 0$ for $z \geq x$ and $f'(z) \leq 0$ and $z \leq x$, show that f has a minimum at x.

CHAPTER 10

Differentiation in $\mathbb{R}^n$

The goal of the present chapter is to generalize the notion of differentiation in Chapter 9 to real-valued functions defined on $\mathbb{R}^n$, to real-valued functions of several variables. The problem of differentiating vector-valued functions will be discussed in Chapter 29. Throughout this chapter, D will denote an open subset of $\mathbb{R}^n$ and $f: D \to \mathbb{R}$ a real-valued function on D. If $\mathbf{u}$ is a point in $\mathbb{R}^n$ such that $\|\mathbf{u}\| = 1$, then $\mathbf{u}$ is called a *unit vector*. The unit vector in $\mathbb{R}^n$ with jth coordinate 1 and all other coordinates 0 will be called a *unit coordinate vector* and will be denoted by $\mathbf{e}_j$. This last set of unit vectors is particularly useful because every vector in $\mathbb{R}^n$ can be represented as a linear combination of the $\mathbf{e}_j$, namely,

$$(x_1, x_2, \ldots, x_n) = x_1\mathbf{e}_1 + x_2\mathbf{e}_2 + \cdots + x_n\mathbf{e}_n.$$

Differentiation of functions of several variables is a considerably more complicated affair than that of real-valued functions of one real variable. The reader has undoubtedly had some exposure to this topic in his or her study of the calculus. We'll begin with a derivative that is a generalization of the

101

derivative defined on the real line to one that is defined on a line whose direction is specified by a unit vector **u** in $\mathbb{R}^n$.

Definition 1. Let $x \in D$. The *directional derivative* of f at x in the direction u is

$$D_u f(x) = \lim_{h \to 0} \frac{f(x + hu) - f(x)}{h},$$

provided the limit exists. If u is a unit coordinate vector $u = e_j$, the corresponding directional derivative is denoted by $D_j f(x)$ and called the *partial derivative* of f at x with respect to the jth coordinate. Other common notations for $D_j f(x)$ are $f_{x_j}(x)$ and $\partial f / \partial x_j(x)$.

It's possible for a function to have partial derivatives, but fail to have directional derivatives in all directions. For example, the function

$$f(x, y) = \begin{cases} x + y, & \text{if } x = 0 \text{ or } y = 0, \\ 1, & \text{otherwise,} \end{cases}$$

has partial derivatives $D_1 f(0,0) = D_2 f(0,0) = 1$ at $(0,0)$. If $u = (a, b)$, $ab \neq 0$, then $[f(0 + hu) - f(0)]/h = 1/h$; so, $D_u f(0)$ doesn't exist. Consequently, f has partial derivatives at $(0,0)$, but does not have directional derivatives in every direction there.

For functions on $\mathbb{R}$, differentiability at a point implies continuity there. The following example shows that this situation does not carry over to functions on $\mathbb{R}^n$:

$$f(x, y) = \begin{cases} 2xy/(x^2 + y^2), & (x, y) \neq (0,0), \\ 0, & (x, y) = (0,0). \end{cases}$$

This function has partial derivatives $D_1 f(0,0) = D_2 f(0,0) = 0$ at $(0,0)$. For $y = x \neq 0$, $f(x, y) = 1$. Consequently, f has partial derivatives at $(0,0)$, but is not continuous there.

Even if a function has directional derivatives in all directions at a point, it may still fail to be continuous there. Consider the function

$$f(x, y) = \begin{cases} xy^2/(x^2 + y^4), & (x, y) \neq (0,0), \\ 0, & (x, y) = (0,0). \end{cases}$$

For $u = (a, b)$ with $a \neq 0$,

$$[f(0 + hu) - f(0)]/h = ab^2/(a^2 + h^2 b^4) \to b^2/a = D_u f(0);$$

if $a = 0$, $D_u f(0) = 0$. Thus **f** has directional derivatives in all directions at **0**. However, **f** is not continuous there, for on the curve $y^2 = x$, f has the value $\frac{1}{2}$.

Differentials

As the above examples indicate, the process of differentiation for functions on **R**^{n} is much more complicated than for functions defined on **R**. For example, in **R**, differentiability at a point implies continuity at the point; but in **R**^{n}, continuity is not implied by the existence of the directional derivatives. To improve our understanding of this problem, let's examine more carefully the implications of differentiability for functions on **R**. One of the motivations for the definition of the derivative was the need to define the tangent line to the graph of $f: [a, b] \to \mathbb{R}$. This leads directly to the definition of the *slope* of the tangent line as the value of the derivative at the point of tangency. (The *tangent line* is well-defined when its slope and the point of tangency are specified.) If we write $y = f(x)$ and $\Delta y = f(x + h) - f(x)$, then

$$\frac{dy}{dx} = f'(x) = \lim_{h \to 0} \frac{\Delta y}{h}. \tag{1}$$

This equation shows that the quantity $f'(x)h = g(x, h)$ approximates the increment Δy, of the function f, at the point x in the sense that

$$\lim_{h \to 0} \frac{\Delta y - g(x, h)}{h} = 0.$$

The function $g(x, \cdot)$ is called the *differential* of the function f at the point x and is usually denoted by dy or $df(x)$. Note that the differential is a function of two variables, x and h. Thus, the differential of f at x is a linear function of h [i.e., $g(x, ay + bz) = ag(x, y) + bg(x, z)$] which approximates the increment, Δy, in the function f. Geometrically, the tangent line is a good approximation to the graph of f in a neighborhood of x. This is illustrated in Figure 10.1.

We formalize and extend this discussion to functions on **R**^{n} with the following definitions.

Definition 2. A function **L**: $\mathbb{R}^n \to \mathbb{R}^m$ is *linear* if $\mathbf{L}(t\mathbf{x} + s\mathbf{y}) = t\mathbf{L}(\mathbf{x}) + s\mathbf{L}(\mathbf{y})$ $\forall s, t \in \mathbb{R}$ and $\forall \mathbf{x}, \mathbf{y} \in \mathbb{R}^n$. When the range of **L** is in **R**, it is often called a *linear functional*.

The following is a nice characterization for linear functionals on **R**^{n}:

Proposition 3. The function $L: \mathbb{R}^n \to \mathbb{R}$ is linear if, and only if, there is a unique $\ell \in \mathbb{R}^n$ such that $L(\mathbf{x}) = \ell \cdot \mathbf{x} \; \forall \mathbf{x} \in \mathbb{R}^n$.

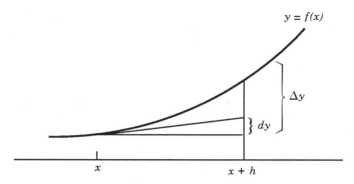

Figure 10.1

Proof. The "only if" part is clear, except perhaps for uniqueness; but $\ell \cdot \mathbf{x} = \ell' \cdot \mathbf{x} \; \forall \mathbf{x}$ implies $(\ell - \ell') \cdot (\ell - \ell') = 0$, and so $\ell - \ell' = 0$.

Next, set $\ell_j = L(\mathbf{e}_j)$ and $\ell = (\ell_1, \ldots, \ell_n)$. Then

$$L(\mathbf{x}) = L\left(\sum_{i=1}^n x_i \mathbf{e}_i \right) = \sum_{i=1}^n x_i L(\mathbf{e}_i) = \sum_{i=1}^n x_i \ell_i = \ell \cdot \mathbf{x}. \qquad \square$$

When L is a linear functional with the representation in Proposition 3, we write $L = \ell$, that is, we identify the linear functional L with its vector representation ℓ.

If $D \subset \mathbb{R}$ and $f: D \to \mathbb{R}$, the differential of f at x, $df(x)(t) = f'(x) \cdot t$, is a linear function that approximates $\Delta f = f(x + t) - f(x)$ in the sense that

$$\lim_{t \to 0} \frac{\Delta f - df(x)(t)}{t} = 0.$$

We now have the necessary tools for defining differentiability and differential for real-valued functions on $\mathbb{R}^n$.

Definition 4. Let D be an open subset of $\mathbb{R}^n$. The function $f: D \to \mathbb{R}$ is *differentiable* at $\mathbf{x} \in D$ if there is a linear function $L_{\mathbf{x}}: \mathbb{R}^n \to \mathbb{R}$ such that

$$\lim_{\mathbf{h} \to 0} \frac{|f(\mathbf{x} + \mathbf{h}) - f(\mathbf{x}) - L_{\mathbf{x}}(\mathbf{h})|}{\|\mathbf{h}\|} = 0.$$

$L_{\mathbf{x}}$ is called the *differential of f at* $\mathbf{x}$; we write $L_{\mathbf{x}} = df(\mathbf{x}): \mathbb{R}^n \to \mathbb{R}$. f is *differentiable* on D if f has a differential at each point of D.

Definition 5. If $f: D \to \mathbb{R}$ has first partial derivatives at $\mathbf{x} \in D$, the *gradient* of f at $\mathbf{x}$ is $\nabla f(\mathbf{x}) = (D_1 f(\mathbf{x}), \dots, D_n f(\mathbf{x}))$.

The gradient gives a useful and descriptive characterization of the differential.

Proposition 6. Suppose f is differentiable at $\mathbf{x} \in D$. Then $D_j f(\mathbf{x})$ exists for $j = 1, \dots, n$ and $df(\mathbf{x})(\mathbf{h}) = \nabla f(\mathbf{x}) \cdot \mathbf{h}\ \forall\, \mathbf{h} \in \mathbb{R}^n$, that is, $df(\mathbf{x}) = \nabla f(\mathbf{x})$. (Thus the differential is unique and our notation is justified.)

Proof. For $t \in \mathbb{R}$, $t \neq 0$,

$$\lim_{t \to 0} \big[f(\mathbf{x} + t\mathbf{e}_j) - f(\mathbf{x}) - df(\mathbf{x})(t\mathbf{e}_j) \big] / t = 0. \tag{2}$$

So $D_j f(\mathbf{x})$ exists and is equal to $df(\mathbf{x})(\mathbf{e}_j)$. It follows from Proposition 3 that $df(\mathbf{x}) = \nabla f(\mathbf{x})$. $\square$

Replacing $\mathbf{e}_j$ in (2) by an arbitrary unit vector $\mathbf{u} \in \mathbb{R}^n$ shows that if f is differentiable, then it has a directional derivative in the direction $\mathbf{u}$ and that $D_{\mathbf{u}} f(\mathbf{x}) = df(\mathbf{x})(\mathbf{u}) = \nabla f(\mathbf{x}) \cdot \mathbf{u}$. This formula is useful in computing directional derivatives. Note also that the Cauchy–Schwarz inequality shows that $\nabla f(\mathbf{x}) / \|\nabla f(\mathbf{x})\|$ solves the optimization problem: Find $\max \{ D_{\mathbf{u}} f(\mathbf{x}) : \|\mathbf{u}\| = 1 \}$. That is, the direction in which the function is increasing (decreasing) most rapidly is given by $\nabla f(\mathbf{x})(-\nabla f(\mathbf{x}))$.

We can finally give an analogue to Proposition 9.2.

Theorem 7. If $f: D \to \mathbb{R}$ is differentiable at $\mathbf{x}$, then it is continuous there.

Proof. For sufficiently small $\|\mathbf{h}\|$,

$$|f(\mathbf{x} + \mathbf{h}) - f(\mathbf{x})| \leq \|\mathbf{h}\| + \|df(\mathbf{x}) \cdot \mathbf{h}\| \leq \|\mathbf{h}\| + \|df(\mathbf{x})\|\, \|\mathbf{h}\|$$

by the Cauchy–Schwarz inequality. Continuity of f at $\mathbf{x}$ follows. $\square$

Proposition 6 and the examples of discontinuous functions with partial derivatives show that the mere existence of the gradient is not adequate to guarantee differentiability of the function. There is, however, the following sufficient condition for differentiability.

Proposition 8. If each $D_j f\ j = 1, \dots, n$ exists in D, and is continuous at $\mathbf{x} \in D$, then f has a differential at $\mathbf{x}$.

Proof. We give the proof for $n = 2$. Let $\mathbf{x} = (x, y) \in D$, $\mathbf{h} = (h, k) \in \mathbb{R}^2$. Application of the mean value theorem yields

$$f(\mathbf{x} + \mathbf{h}) - f(\mathbf{x}) = \{ f(x + h, y + k) - f(x + h, y) \}$$

$$+ \{ f(x + h, y) - f(x, y) \}$$

$$= D_2 f(x + h, y + \theta_1 k)k + D_1 f(x + \theta_2 h, y)h,$$

$$0 < \theta_i < 1.$$

Thus

$$f(x + h, y + k) - f(x, y) - \nabla f(\mathbf{x}) \cdot (h, k)$$

$$= \left[D_1 f(x + \theta_2 h, y) - D_1 f(x, y) \right] h$$

$$+ \left[D_2 f(x + h, y + \theta_1 k) - D_2 f(x, y) \right] k$$

$$= \left(D_1 f(x + \theta_2 h, y) - D_1 f(x, y), D_2 f(x + h, y + \theta_1 k) \right.$$

$$\left. - D_2 f(x, y) \right) \cdot (h, k).$$

The continuity of $D_1 f$ and $D_2 f$ and the Cauchy–Schwarz inequality imply

$$\left[f(x + h, y + k) - f(x, y) - \nabla f(\mathbf{x}) \cdot (h, k) \right] / \| (h, k) \| \to 0 \text{ as } \| \mathbf{h} \| \to 0. \ \square$$

Using differentials, chain rules for functions on $\mathbb{R}^n$ can be obtained. As an example, we establish the following theorem.

Theorem 9 (The Chain Rule). Suppose that f is differentiable at $\mathbf{x} \in D$, g_j: $(a, b) \to \mathbb{R}$ is differentiable at $t_0 \in (a, b)$ for $j = 1, \ldots, n$, the range of $\mathbf{g} = (g_1, \ldots, g_n)$ is contained in D, and $\mathbf{g}(t_0) = \mathbf{x}$. Then $F = f \circ \mathbf{g}$ is differentiable at $t_0 \in (a, b)$ and

$$F'(t_0) = \nabla f(\mathbf{x}) \cdot \left(g_1'(t_0), \ldots, g_n'(t_0) \right). \tag{3}$$

Proof. Define $h: D \to \mathbb{R}$ by

$$h(\mathbf{y}) = \begin{cases} \left[f(\mathbf{y}) - f(\mathbf{x}) - \nabla f(\mathbf{x}) \cdot (\mathbf{y} - \mathbf{x}) \right] / \| \mathbf{y} - \mathbf{x} \|, & \text{if } \mathbf{y} \neq \mathbf{x}, \\ 0 & \text{if } \mathbf{y} = \mathbf{x}. \end{cases}$$

Then h is continuous at $\mathbf{x}$ by the differentiability of f at $\mathbf{x}$. Consequently, $h \circ \mathbf{g}$

is continuous at t_0. It follows that

$$\lim_{\Delta t \to 0} h\big(g(t_0 + \Delta t)\big) \frac{\| g(t_0 + \Delta t) - g(t_0) \|}{\Delta t} = 0$$

$$= \lim_{\Delta t \to 0} \left[\frac{f[g(t_0 + \Delta t)] - f[g(t_0)]}{\Delta t} - \nabla f(\mathbf{x}) \cdot \frac{g(t_0 + \Delta t) - g(t_0)}{\Delta t} \right]$$

$$= F'(t_0) - \nabla f(\mathbf{x}) \cdot (g_1'(t_0), \ldots, g_n'(t_0)). \qquad \square$$

Definition 10. If $g = (g_1, \ldots, g_n)$: $[a, b] \to \mathbb{R}^n$ has the property that each g_j is differentiable at $t_0 \in (a, b)$, then the *derivative* of g at t_0 is $g'(t_0) = (g_1'(t_0), \ldots, g_n'(t_0))$.

With this notation, (3) of Theorem 9 can be written

$$F'(t_0) = \nabla f(\mathbf{x}) \cdot g'(t_0).$$

It's interesting to consider the geometric implications of our development of differentials and differentiability. A *path* in $\mathbb{R}^n$ is a continuous function g, with range in $\mathbb{R}^n$, defined on an interval $[a, b]$ in $\mathbb{R}$. If g is differentiable at t, the *tangent* to g at t is defined to be $g'(t)$. Since $\lim_{h \to 0}[g(t + h) - g(t)]/h = g'(t)$, this agrees with our previous definition of the tangent in $\mathbb{R}$.

If $f: \mathbb{R}^n \to \mathbb{R}$, the set of points $\mathbf{x}$ such that $f(\mathbf{x}) = c$, a constant, is called a *level surface* (in $\mathbb{R}^2$ it's called a *level curve*) of f. For example, the level surfaces (level curves) of the function $f(x, y, z) = x^2 + y^2 + z^2$ ($f(x, y) = x^2 + y^2$) are spheres (circles) centered at the origin.

Let $f: \mathbb{R}^n \to \mathbb{R}$ and let S be the level surface $f = c$. A path $g: [a, b] \to \mathbb{R}^n$ is said to lie in S if $g(t) \in S$ for all $t \in [a, b]$. Suppose $\mathbf{x} \in S$, f is differentiable at $\mathbf{x}$ with $\nabla f(\mathbf{x}) \neq 0$, and $g: [a, b] \to \mathbb{R}^n$ is a path in S with $g(t) = \mathbf{x}$ for some $t \in (a, b)$. If g is differentiable at t, then Theorem 9 implies that $\nabla f(\mathbf{x}) \cdot g'(t) = 0$, that is, $\nabla f(\mathbf{x})$ and the tangent $g'(t)$ are perpendicular (Figure 10.2). In $\mathbb{R}^3$, the collection of all possible tangents at $\mathbf{x}$ to paths lying in S would intuitively, form a plane T and $\nabla f(\mathbf{x})$ would be perpendicular to this plane, that is, $\nabla f(\mathbf{x})$ would be a normal vector to this plane. The plane T is usually referred to as the *tangent plane* to S at $\mathbf{x}$ and the vector $\nabla f(\mathbf{x})$ is called the *normal* to S at $\mathbf{x}$. In $\mathbb{R}^n$, if $\nabla f(\mathbf{x}) \neq 0$, the *tangent plane* to the level surface S at the point $\mathbf{x}$ is defined to be $T = \{\mathbf{h}: \nabla f(\mathbf{x}) \cdot \mathbf{h} = 0\}$. The equation for T is then

$$\sum_{i=1}^{n} D_i f(\mathbf{x}) h_i = 0.$$

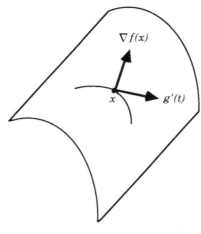

Figure 10.2

If $f\colon \mathbb{R}^2 \to \mathbb{R}$, S is the level curve $f = c$, and if $\Omega = \{\mathbf{x}\colon f(\mathbf{x}) \le c\}$, then whenever $\mathbf{x} \in S$ the gradient $\nabla f(\mathbf{x})$ is an "outer normal" to the region Ω at $\mathbf{x}$ (see Figure 10.3).

An application of Theorem 9 yields a version of the mean value theorem for functions on $\mathbb{R}^n$. Before proceeding, we introduce some terminology. If $\mathbf{x}, \mathbf{y} \in \mathbb{R}^n$, the set of points

$$[\mathbf{x}, \mathbf{y}] = \{t\mathbf{y} + (1 - t)\mathbf{x}\colon 0 \le t \le 1\}$$

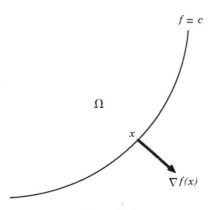

Figure 10.3

is the *line segment* with endpoints **x** and **y**. A subset K of $\mathbb{R}^n$ is called *convex* if $[\mathbf{x}, \mathbf{y}] \subset K$ whenever $\mathbf{x}, \mathbf{y} \in K$. For example, spheres in $\mathbb{R}^n$ are convex (see Exercise 12).

The mean value theorem (**9.5**) extends very nicely to real-valued functions on $\mathbb{R}^n$.

Theorem 11 (Mean Value Theorem). Suppose D is an open subset of $\mathbb{R}^n$ and $f: D \to \mathbb{R}$ is differentiable on D. If $\mathbf{x}, \mathbf{y}$ are points of D for which $[\mathbf{x}, \mathbf{y}] \subseteq D$, then there is $\zeta \in [\mathbf{x}, \mathbf{y}]$ such that

$$f(\mathbf{y}) - f(\mathbf{x}) = \nabla f(\zeta) \cdot (\mathbf{y} - \mathbf{x}).$$

Proof. Define $\psi: [0, 1] \to \mathbb{R}$ by $\psi(t) = f(\mathbf{x} + t(\mathbf{y} - \mathbf{x}))$. Then, by the chain rule, ψ is differentiable on $[0, 1]$ and the mean value theorem for real-valued functions of a real variable implies that there is $t_0 \in (0, 1)$ such that $\psi(1) - \psi(0) = \psi'(t_0) = f(\mathbf{y}) - f(\mathbf{x}) = \nabla f(\mathbf{x} + t_0(\mathbf{y} - \mathbf{x})) \cdot (\mathbf{y} - \mathbf{x})$.

Differentiation, when possible, of a partial derivative of a function defined on $\mathbb{R}^n$ with respect to one of its variables leads to a higher ordered derivative.

Definition 12. Suppose the partial derivatives $D_i f$, $i = 1, \ldots, n$, exist in some neighborhood of $\mathbf{x} \in D$. If the jth partial derivative of $D_i f$ exists at $\mathbf{x}$, it is called a *second order partial derivative* of f and denoted by

$$D_j(D_i f(\mathbf{x})) = D_{ji} f(\mathbf{x}) = f_{x_j x_i}(\mathbf{x}) = \frac{\partial^2 f}{\partial x_j \, \partial x_i}(\mathbf{x}).$$

If $j \neq i$, the $D_{ij} f$ are called *cross partials* or *mixed partial derivatives* of the function f. Third, fourth, and higher ordered derivatives are defined similarly.

In general, the order in which second order partial derivatives are computed is significant. For the function

$$f(x, y) = \begin{cases} xy(x^2 - y^2)/(x^2 + y^2), & (x, y) \neq (0, 0), \\ 0, & (x, y) = (0, 0), \end{cases}$$

$$D_1 f(0, y) = \lim_{h \to 0} \left[y(h^2 - y^2)/(h^2 + y^2) \right] = -y,$$

$$D_2 f(x, 0) = \lim_{h \to 0} \left[x(x^2 - h^2)/(x^2 + h^2) \right] = x,$$

$$D_{21} f(0, 0) = -1, \quad \text{and} \quad D_{12} f(0, 0) = 1,$$

that is, the second order mixed partial derivatives exist at $(0, 0)$ but are not equal there.

Sufficient conditions for the equality of the second order mixed partial derivatives are given by the following:

Theorem 13. If $D_{ij}f$ and $D_{ji}f$ both exist in a neighborhood of a point $\mathbf{x} \in D$ and if both $D_{ij}f$ and $D_{ji}f$ are continuous at $\mathbf{x}$, then $D_{ij}f(\mathbf{x}) = D_{ji}f(\mathbf{x})$.

Proof. By holding all but the ith and jth coordinates fixed, we may assume that we are dealing with functions in $\mathbb{R}^2$. If h, k are small, the rectangle with vertices $(x, y), (x + h, y), (x, y + k), (x + h, y + k)$ lies in D. Applying the mean value theorem to the following functions, we obtain

$$
\begin{aligned}
S_{hk} &= \{f(x + h, y + k) - f(x + h, y)\} - \{f(x, y + k) - f(x, y)\} \\
&= D_1\{f(x', y + k) - f(x', y)\} \cdot h \\
&= D_{21}f(x', y') \cdot hk \\
&= \{f(x + h, y + k) - f(x, y + k)\} - \{f(x + h, y) - f(x, y)\} \\
&= D_2\{f(x + h, y'') - f(x, y'')\} \cdot k \\
&= D_{12}f(x'', y'') \cdot hk,
\end{aligned}
$$

where x' and x'' are between x and $x + h$, and y' and y'' are between y and $y + k$. By the continuity of $D_{12}f$ and $D_{21}f$ at (x, y), we have

$$
\lim_{(h, k) \to (0,0)} S_{hk}/hk = D_{21}f(x, y) = D_{12}f(x, y). \qquad \square
$$

Taylor's theorem (9.11) also extends directly from functions on $\mathbb{R}$ to real-valued functions on $\mathbb{R}^n$. The details become a bit messy as the orders of the partial derivatives become large, but the ideas involved are the same. As was the case in $\mathbb{R}$, this extended result has many important applications in optimization theory.

Theorem 14 (Taylor's Theorem for $N = 1, 2$). Suppose D is an open convex subset of $\mathbb{R}^n$, $f: D \to \mathbb{R}$, and $\mathbf{x}_0, \mathbf{x}_0 + \mathbf{h} \in D$.

(i) If f has continuous second order partial derivatives in D,

$$
f(\mathbf{x}_0 + \mathbf{h}) = f(\mathbf{x}_0) + \nabla f(\mathbf{x}_0) \cdot \mathbf{h} + \frac{1}{2} \sum_{j=1}^{n} \sum_{i=1}^{n} D_{ij}f(\zeta)h_i h_j,
$$

where $\zeta \in [\mathbf{x}_0, \mathbf{x}_0 + \mathbf{h}]$.

(ii) If f has continuous third order partial derivatives in D,

$$f(\mathbf{x}_0 + \mathbf{h}) = f(\mathbf{x}_0) + \nabla f(\mathbf{x}_0) \cdot \mathbf{h} + \frac{1}{2} \sum_{j=1}^{n} \sum_{i=1}^{n} D_{ij}f(\mathbf{x}_0)h_i h_j,$$

$$+ \frac{1}{3!} \sum_{k=1}^{n} \sum_{j=1}^{n} \sum_{i=1}^{n} D_{ijk}f(\boldsymbol{\zeta})h_i h_j h_k, \qquad \text{where } \boldsymbol{\zeta} \in [\mathbf{x}_0, \mathbf{x}_0 + \mathbf{h}].$$

Proof. Define $\psi: [0, 1] \to \mathbb{R}$ by $\psi(t) = f(\mathbf{x}_0 + t\mathbf{h}) - f(\mathbf{x}_0)$. Then, from the chain rule,

$$\psi'(t) = \nabla f(\mathbf{x}_0 + t\mathbf{h}) \cdot \mathbf{h} = \sum_{i=1}^{n} D_i f(\mathbf{x}_0 + t\mathbf{h})h_i,$$

$$\psi''(t) = \sum_{j=1}^{n} \sum_{i=1}^{n} D_{ij}f(\mathbf{x}_0 + t\mathbf{h})h_i h_j,$$

$$\psi'''(t) = \sum_{k=1}^{n} \sum_{j=1}^{n} \sum_{i=1}^{n} D_{ijk}f(\mathbf{x}_0 + t\mathbf{h})h_i h_j h_k.$$

Now the formulas in (i) and (ii) follow from the one-dimensional version of Taylor's theorem. $\qquad\Box$

Let

$$R_1(\mathbf{x}_0, \mathbf{h}) = \frac{1}{2} \sum_{j=1}^{n} \sum_{i=1}^{n} D_{ij}f(\boldsymbol{\zeta})h_{ij} \qquad \text{in (i)}$$

and

$$R_2(\mathbf{x}_0, \mathbf{h}) = \frac{1}{3!} \sum_{k=1}^{n} \sum_{j=1}^{n} \sum_{i=1}^{n} D_{ijk}f(\boldsymbol{\zeta})h_i h_j h_k \qquad \text{in (ii)}.$$

Note that the continuity of the partial derivatives of f implies $|R_1(\mathbf{x}_0, \mathbf{h})| \leq M_1 \|\mathbf{h}\|^2$ and $|R_2(\mathbf{x}_0, \mathbf{h})| \leq M_2 \|\mathbf{h}\|^3$, where M_1 and M_2 are independent of $\mathbf{h}$; consequently,

$$\lim_{\|\mathbf{h}\| \to 0} R_1(\mathbf{x}_0, \mathbf{h}) / \|\mathbf{h}\| = \lim_{\|\mathbf{h}\| \to 0} R_2(\mathbf{x}_0, \mathbf{h}) / \|\mathbf{h}\|^2 = 0.$$

Definition 15. If D is an open subset of $\mathbb{R}^n$ and $f: D \to \mathbb{R}$ has continuous second order partial derivatives at $\mathbf{x}_0 \in D$, the *Hessian* of f at $\mathbf{x}_0$ is the $n \times n$

(symmetric) matrix

$$\left[D_{ij}f(\mathbf{x}_0) \right]_{i,\,j=1,\ldots,\,n} = \nabla^2 f(\mathbf{x}_0) = H(f)(\mathbf{x}_0).$$

Using the Hessian, we can write the formulas in (i) and (ii)

(i) $f(\mathbf{x}_0 + \mathbf{h}) = f(\mathbf{x}_0) + \nabla f(\mathbf{x}_0) \cdot \mathbf{h} + \frac{1}{2}(\nabla^2 f(\boldsymbol{\zeta})\mathbf{h}')^t \cdot \mathbf{h},$

(ii) $f(\mathbf{x}_0 + \mathbf{h}) = f(\mathbf{x}_0) + \nabla f(\mathbf{x}_0) \cdot \mathbf{h} + \frac{1}{2}(\nabla^2 f(\mathbf{x}_0)\mathbf{h}')^t \cdot \mathbf{h} + R_2(\mathbf{x}_0, \mathbf{h}).$

In these expressions, $\mathbf{h}'$ is the transpose of the row matrix $\mathbf{h}$.

As is suggested by the notation, the Hessian is related to the "second derivative" of f; we explore this further in Chapter 29.

Optimization

We have often referred to the optimization of real-valued functions of several real variables. It's now possible to develop results for these functions analogous to those in Chapter 9. An elementary knowledge of linear algebra will help considerably in understanding and appreciating some of the material in this section.

Let $A = [a_{ij}]$ denote an $n \times n$ matrix whose entry in the ith row and jth column is a_{ij}. Then A induces a quadratic form on $\mathbb{R}^n$ by associating with each $\mathbf{h}$ the number

$$\sum_{i,\,j=1}^{n} a_{ij}h_i h_j = (A\mathbf{h}')^t \cdot \mathbf{h} = A[\mathbf{h}].$$

Using this notation, the second order terms in (i) and (ii) above can be written $\frac{1}{2}\nabla^2 f(\boldsymbol{\zeta})[\mathbf{h}]$ and $\frac{1}{2}\nabla^2 f(\mathbf{x}_0)[\mathbf{h}]$, respectively. An $n \times n$ symmetric matrix A is *positive semidefinite* if $A[\mathbf{h}] \geq 0 \; \forall \, \mathbf{h} \in \mathbb{R}^n$; A is *positive definite* if $A[\mathbf{h}] > 0$ for $\mathbf{h} \neq \mathbf{0}$. *Negative semidefinite* and *negative definite* are defined similarly.

For functions of several variables, the following proposition is analogous to **9.14** and **9.15**.

Proposition 16. Suppose D is an open subset of $\mathbb{R}^n$ and $f: D \to \mathbb{R}$ has continuous second order partial derivatives in D. If f has a local maximum at $\mathbf{x}_0 \in D$, then (i) $\nabla f(\mathbf{x}_0) = 0$, and (ii) $\nabla^2 f(\mathbf{x}_0)$ is negative semidefinite.

Proof. (i) is left to Exercise 2. For (ii), fix $\mathbf{h} \in \mathbb{R}^n$. If $\delta > 0$ is sufficiently small, then $\mathbf{x}_0 + t\mathbf{h} \in D$ whenever $|t| \leq \delta$. Define $\psi: [-\delta, \delta] \to \mathbb{R}$ by $\psi(t) = f(\mathbf{x}_0 + t\mathbf{h})$. Then ψ has a local maximum at $t = 0$, and it follows that $\psi''(0) \leq 0$. But $\psi''(0) = \nabla^2 f(\mathbf{x}_0)[\mathbf{h}]$. Thus $\nabla^2 f(\mathbf{x}_0)$ is negative semidefinite. $\square$

The conditions (i) and (ii) are necessary, but not sufficient for the function f to have a local maximum at $\mathbf{x}_0$.

As in **9.16**, we can obtain sufficient conditions for an extreme value by strengthening condition (ii). First, we require a lemma.

Lemma 17. If $A = [a_{ij}]$ is positive definite, then there is $m > 0$ such that $A[\mathbf{h}] \geq m\|\mathbf{h}\|^2 \ \forall \mathbf{h} \in \mathbb{R}^n$.

Proof. The mapping $\bar{A}$: $\mathbf{h} \to A[\mathbf{h}]$ is continuous and positive on $S = \{\mathbf{h}: \ \|\mathbf{h}\| = 1\}$. Therefore $\bar{A}$ attains its minimum, say m, on S. If $\mathbf{h} \in \mathbb{R}^n$, $\mathbf{h} \neq \mathbf{0}$, then $\mathbf{h}/\|\mathbf{h}\| \in S$. Thus $A[\mathbf{h}/\|\mathbf{h}\|] = A[\mathbf{h}]/\|\mathbf{h}\|^2 \geq m$, or $A[\mathbf{h}] \geq m\|\mathbf{h}\|^2$. $\qquad\square$

Proposition 18. Suppose D is an open subset of $\mathbb{R}^n$, $\mathbf{x}_0 \in D$, and $f: D \to \mathbb{R}$ has continuous third order partial derivatives in D. If (1) $\nabla f(\mathbf{x}_0) = \mathbf{0}$ and (2) $\nabla^2 f(\mathbf{x}_0)$ is positive definite, then f has a strict local minimum at $\mathbf{x}_0$.

Proof. By Taylor's theorem, for $\mathbf{x}$ sufficiently close to $\mathbf{x}_0$,

$$f(\mathbf{x}) - f(\mathbf{x}_0) = \nabla f(\mathbf{x}_0) \cdot (\mathbf{x} - \mathbf{x}_0) + \tfrac{1}{2}\nabla^2 f(\mathbf{x}_0)[\mathbf{x} - \mathbf{x}_0] + R_2(\mathbf{x}_0, \mathbf{x} - \mathbf{x}_0),$$

where

$$R_2(\mathbf{x}_0, \mathbf{x} - \mathbf{x}_0)/\|\mathbf{x} - \mathbf{x}_0\|^2 \to 0 \qquad \text{as } \mathbf{x} \to \mathbf{x}_0.$$

By Lemma 17, there is $m > 0$ such that $\nabla^2 f(\mathbf{x}_0)[\mathbf{h}] \geq m\|\mathbf{h}\|^2 \ \forall \mathbf{h} \in \mathbb{R}^n$. There is $\delta > 0$ such that $0 < \|\mathbf{x} - \mathbf{x}_0\| < \delta$ implies

$$|R_2(\mathbf{x}_0, \mathbf{x} - \mathbf{x}_0)|/\|\mathbf{x} - \mathbf{x}_0\|^2 < m/4.$$

Thus, for $0 < \|\mathbf{x} - \mathbf{x}_0\| < \delta$, we have

$$f(\mathbf{x}) - f(\mathbf{x}_0) \geq \|\mathbf{x} - \mathbf{x}_0\|^2(m/4) > 0. \qquad\square$$

The conditions (1) and (2) are sufficient, but not necessary, for f to have a local minimum at $\mathbf{x}_0$.

For a two dimensional example, consider finding local extrema for the function

$$f(x_1, x_2) = 5x_1^2 + 4x_1 x_2 + x_2^2 - 6x_1 - 2x_2 + 6.$$

The gradient of f is

$$\nabla f(x_1, x_2) = (10x_1 + 4x_2 - 6, 4x_1 + 2x_2 - 2)$$

and the gradient vanishes at $(1, -1)$. The Hessian of f at $(1, -1)$ is

$$H = \begin{bmatrix} 10 & 4 \\ 4 & 2 \end{bmatrix}.$$

Now

$$H[h_1, h_2] = 10h_1^2 + 8h_1h_2 + 2h_2^2 = 2h_1^2 + \left(2\sqrt{2}\,h_1 + \sqrt{2}\,h_2\right)^2 \geq 0;$$

so, by Proposition 18, f has a strict local minimum at $(1, -1)$.

Constrained Optimization

We next consider a basic problem in the area of nonlinear programming; namely, that of finding the minimum of a function f over a subset Ω of $\mathbb{R}^n$ that is described by functional constraints. In particular, we develop the basic necessary conditions (usually called *Kuhn–Tucker conditions*) for this problem.

Let $f, g_1, \ldots, g_k, h_1, \ldots, h_m$ be real-valued functions defined on $\mathbb{R}^n$ and having continuous first order partial derivatives. We consider the problem:

minimize $f(\mathbf{x})$

subject to $g_1(\mathbf{x}) \leq 0, \ldots, g_k(\mathbf{x}) \leq 0; \; h_1(\mathbf{x}) = 0, \ldots, h_m(\mathbf{x}) = 0.$ (P)

That is, we seek the minimum of f over the set

$$\Omega = \left\{ \mathbf{x} : g_i(\mathbf{x}) \leq 0, \quad i = 1, \ldots, k; \; h_j(\mathbf{x}) = 0, \quad j = 1, \ldots, m \right\}$$

given by the functional constraints. Note that in general Ω is not open, so the previous necessary conditions are not applicable.

To understand the necessary condition for this problem, consider the situation in $\mathbb{R}^2$ where all the g_i are zero and $m = 1$, that is, where there is only one equality constraint $h_1(x, y) = h(x, y) = 0$. Suppose $\mathbf{p} = (x^*, y^*)$ is a local solution to the problem of minimizing f subject to $h = 0$ and let $f(\mathbf{p}) = c$. Choose a neighborhood N of $\mathbf{p}$ such that $f(\mathbf{p})$ is a minimum of f for all points in N that lie on the level curve $h = 0$. The level curve $f = c$ separates N into two pieces: N_+ where $f > c$ and N_- where $f < c$ (Figure 10.4).

We claim that $\nabla f(\mathbf{p})$ and $\nabla h(\mathbf{p})$ are collinear. If this is not the case, the level curve $h = 0$ crosses the level curve $f = c$ at $\mathbf{p}$ and, therefore, must intersect N_-, which implies that there are points on $h = 0$ where f has values smaller than c (see Figure 10.5). Thus, $\nabla f(\mathbf{p})$ and $\nabla h(\mathbf{p})$ are collinear, that is,

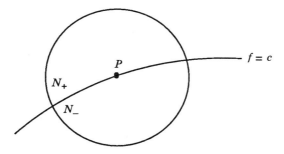

Figure 10.4

there is a $\lambda \neq 0$ such that $\nabla f(\mathbf{p}) + \lambda \nabla h(\mathbf{p}) = 0$. The scalar λ is called a *Lagrange multiplier* for this problem; the function $L(\mathbf{x}, \lambda) = f(\mathbf{x}) + \lambda h(\mathbf{x})$ is called the *Lagrangian* for the problem. The necessary condition for $\mathbf{p}$ to be a local solution to this problem in terms of the Lagrangian is then

$$\nabla L(\mathbf{p}, \lambda) = 0,$$

$$h(\mathbf{p}) = 0.$$

Note that this gives a system of three equations in the three unknowns λ, x^*, y^*.

This necessary condition implies that the level curves $f = c$ and $h = 0$ are tangent at $\mathbf{p}$. This condition is not, in general, a sufficient condition for local extrema (see Figure 10.6, where the two level curves are tangent at $\mathbf{p}$, but $\mathbf{p}$ is not a local maximum or minimum for f subject to $h = 0$).

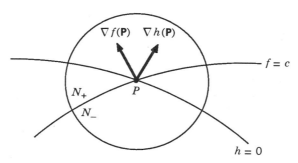

Figure 10.5

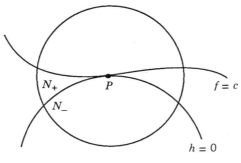

Figure 10.6

A similar situation holds when there is one inequality constraint except that in this case the sign of the Lagrange multiplier becomes an important part of the necessary condition. Suppose **p** is a local minimum for f subject to the constraint $g(x, y) \leq 0$ and let $S = \{\mathbf{x}: g(\mathbf{x}) \leq 0\}$. Then, employing the same notation as before, we indicate the situation as in Figure 10.7.

Now $-\nabla f(\mathbf{p})$ is an outer normal to S at **p** and, similarly, $\nabla g(\mathbf{p})$ is an outer normal to S at **p**. Hence, there is $\lambda > 0$ such that $-\nabla f(\mathbf{p}) = \lambda \nabla g(\mathbf{p})$. Thus, again $\nabla L(\mathbf{p}, \lambda) = 0$, but now the multiplier λ must be positive.

We return to the general problem (P) and develop the multiplier rule for this problem. A point **x** is *feasible* for (P) if it satisfies the constraints for (P). An inequality constraint $g_i(\mathbf{x}) \leq 0$ is *active* at a feasible point **x** if $g_i(\mathbf{x}) = 0$; otherwise, it is said to be *inactive*. Active constraints locally restrict the domain of feasibility; inactive constraints have no local influence (Figure 10.8).

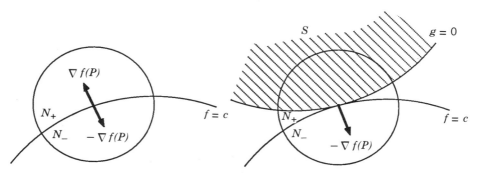

Figure 10.7

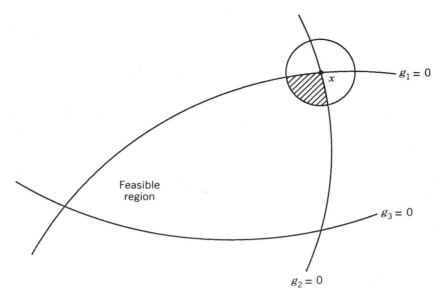

Figure 10.8

The basic first order necessary condition for (P), given below in Theorem 19 is called the *Fritz–John multiplier rule*. If $g: \mathbb{R}^n \to \mathbb{R}$, then we set $g^+(x) = \max\{g(x), 0\}$; recall that $D_j(g^+(x)^2) = 2g^+(x)D_jg(x)$ (Exercise **9.26**).

Theorem 19. Suppose x^* is a local solution to (P). Then there are $\lambda_0^* \in \mathbb{R}$, $\lambda^* \in \mathbb{R}^m$, $\mu^* \in \mathbb{R}^k$, not all zero, such that

(i) $\lambda_0^* \nabla f(x^*) + \sum_{i=1}^m \lambda_i^* \nabla h_i(x^*) + \sum_{i=1}^k \mu_i^* \nabla g_i(x^*) = 0$.

(ii) $\lambda_0^* \geq 0$, $\mu^* \geq 0$.

(iii) $\mu_i^* g_i(x^*) = 0$,   for $i = 1, \ldots, k$.

[Note that (iii) implies that the multiplier μ_j^* associated with any inactive constraint g_j must be zero.]

Proof. We may assume that $x^* = 0$, $f(x^*) = 0$, and $g_1(x^*) = \cdots = g_p(x^*) = 0$, $g_j(x^*) < 0$, for $j = p + 1, \ldots, k$, that is, the first p constraints are the active constraints. Choose $\varepsilon_1 > 0$ such that on $B(\varepsilon_1) = \{x: \|x\| \leq \varepsilon_1\}$ the functions g_j, $j = p + 1, \ldots, k$ are negative and $f(0) \leq f(x)$ for each feasible x in $B(\varepsilon_1)$.

We first claim that for each ε, $0 < \varepsilon \leq \varepsilon_1$, there is N such that

$$f(x) + \|x\|^2 + N\left\{ \sum_{i=1}^p g_i^+(x)^2 + \sum_{i=1}^m h_i(x)^2 \right\} > 0, \qquad (4)$$

for all $\mathbf{x}$ with $\|\mathbf{x}\| = \varepsilon$. If this is not the case, then there are sequences $\{N_j\}$ and $\{\mathbf{x}_j\}$ with $N_j \uparrow \infty$, $\|\mathbf{x}_j\| = \varepsilon$, and

$$f(\mathbf{x}_j) + \|\mathbf{x}_j\|^2 + N_j \left\{ \sum_{i=1}^{p} g_i^+(\mathbf{x}_j)^2 + \sum_{i=1}^{m} h_i(\mathbf{x}_j)^2 \right\} \le 0, \tag{5}$$

A subsequence of $\{\mathbf{x}_j\}$ converges to some point $\mathbf{x}_0$ with $\|\mathbf{x}_0\| = \varepsilon$; for convenience, we assume that $\mathbf{x}_j \to \mathbf{x}_0$. Then $f(\mathbf{x}_j) \to f(\mathbf{x}_0)$. Divide (5) by N_j and let $j \to \infty$ to obtain

$$0 \le \sum_{i=1}^{p} g_i^+(\mathbf{x}_0)^2 + \sum_{i=1}^{m} h_i(\mathbf{x}_0)^2 \le 0.$$

Thus, $\mathbf{x}_0$ is feasible and, from (5), $f(\mathbf{x}_0) \le -\|\mathbf{x}_0\|^2 = -\varepsilon^2$, contradicting the fact that f has a minimum value of zero. Thus (4) holds.

Next, we claim that for each ε, $0 < \varepsilon \le \varepsilon_1$, there is an $\mathbf{x}' \in \mathbb{R}^n$ and a unit vector $(\lambda_0, \lambda_1, \ldots, \lambda_m, \mu_1, \ldots, \mu_k)$ with $\lambda_0 \ge 0$, $\mu_i \ge 0$ such that $\|\mathbf{x}'\| < \varepsilon$ and

$$\lambda_0 \left[D_j f(\mathbf{x}') + 2x_j' \right] + \sum_{i=1}^{k} \mu_i D_j g_i(\mathbf{x}_i) + \sum_{i=1}^{m} \lambda_i D_j h_i(\mathbf{x}) = 0, \tag{6}$$

for $j = 1, \ldots, n$. Let N be as in (4) and define G on $\mathbb{R}^n$ by

$$G(\mathbf{x}) = f(\mathbf{x}) + \|\mathbf{x}\|^2 + N \left\{ \sum_{i=1}^{p} g_i^+(\mathbf{x})^2 + \sum_{i=1}^{m} h_i(\mathbf{x})^2 \right\}.$$

There is a point $\mathbf{x}' \in B(\varepsilon)$ at which G assumes its minimum value. Then $G(\mathbf{x}') \le G(0) = 0$. So, by (4), we cannot have $\|\mathbf{x}'\| = \varepsilon$, that is, $\|\mathbf{x}'\| < \varepsilon$. Thus (Exercise 2),

$$D_j G(\mathbf{x}') = D_j f(\mathbf{x}') + 2x_j' + N \left\{ \sum_{i=1}^{p} 2g_i^+(\mathbf{x}') D_j g_i(\mathbf{x}') \right.$$

$$\left. + \sum_{i=1}^{m} 2h_i(\mathbf{x}') D_j h_i(\mathbf{x}') \right\} = 0. \tag{7}$$

Set $L := [1 + \sum_{i=1}^{p}(2Ng_i^+(\mathbf{x}'))^2 + \sum_{i=1}^{m}(2Nh_i(\mathbf{x}'))^2]^{1/2}$, $\lambda_0 = 1/L$, $\lambda_i = 2Nh_i(\mathbf{x}')/L$, $i = 1, \ldots, m$, $\mu_i = 2Ng_i^+(\mathbf{x}')/L$, $i = 1, \ldots, p$, and $\mu_i = 0$, $i = p + 1, \ldots, k$. Then $(\lambda_0, \lambda_1, \ldots, \lambda_m, \mu_1, \ldots, \mu_k)$ is a unit vector with $\lambda_0 \ge 0$, $\mu_i \ge 0$, $i = 1, \ldots, k$. Dividing (7) by L gives (6).

Now choose a sequence of positive $\varepsilon_\nu \downarrow 0$ and for each ν an $\mathbf{x}_\nu$ with $\|\mathbf{x}_\nu\| < \varepsilon_\nu$ and a unit vector $\lambda_\nu = (\lambda_{0,\nu}, \lambda_{1,\nu}, \ldots, \lambda_{m,\nu}, \mu_{1,\nu}, \ldots, \mu_{k,\nu})$ such that (6) holds. There is a subsequence of $\{\lambda_\nu\}$ that converges to a unit vector $\lambda = (\lambda_0, \lambda_1, \ldots, \lambda_m, \mu_1, \ldots, \mu_k)$ with $\lambda_0 \geq 0$, $\mu_i \geq 0$. Passing to the limit along this subsequence in (6) gives

$$\lambda_0 D_j f(0) + \sum_{i=1}^{k} \mu_i D_j g_i(0) + \sum_{i=1}^{m} \lambda_i D_j h_i(0) = 0, \qquad j = 1, \ldots, n,$$

so that (i) holds, and clearly (ii) and (iii) are satisfied. $\qquad\qquad\qquad\square$

The λ_i and μ_i are called the *Lagrange multipliers* for the problem (P). Conditions (ii) and (iii) are referred to as the *Kuhn–Tucker conditions*. In most introductory courses in real analysis the proof of the Lagrange multiplier rule is based on a deep result, which we will treat later, known as the implicit function theorem. The proof above is due to E. J. McShane, 1973, 922–925, and is elementary (and ingenious) in the sense that only a few elementary properties of compact sets and smooth functions are employed. It should also be noted that the case of inequality constraints is usually not treated, but McShane's proof handles the general case with no difficulty.

It can happen that the multiplier λ_0 in Theorem 19 has the value zero (Exercise 22). When this occurs, the necessary condition above is of little use since it does not involve the minimizing function f. Therefore, it is important to known when the multiplier λ_0 is positive. We give one such simple and useful criterion below.

Corollary 20. If $\{\nabla h_i(\mathbf{x}^*) : i = 1, \ldots, k\} \cup \{\nabla g_i(\mathbf{x}^*) : g_i$ is an active constraint$\}$ is linearly dependent, then $\lambda_0 > 0$.

If $\lambda_0 > 0$, then we can assume that $\lambda_0 = 1$. The *Lagrangian* of (P) is defined to be

$$L(\mathbf{x}, \lambda, \mu) = f(\mathbf{x}) + \sum_{i=1}^{m} \lambda_i h_i(\mathbf{x}) + \sum_{i=1}^{k} \mu_i g_i(\mathbf{x}).$$

If $\lambda_0 > 0$, the necessary condition in Theorem 19 becomes $\nabla_x L(\mathbf{x}^*, \lambda^*, \mu^*) = 0$. This equation, along with $h_i(\mathbf{x}) = 0$, $i = 1, \ldots, m$ and $g_j(\mathbf{x}) = 0$, $j = 1, \ldots, k$ gives a system of $n + m + k$ equations in $n + m + k$ unknowns $\mathbf{x}, \lambda, \mu$.

There are second order sufficient conditions for constrained problems analogous to Proposition 18; see, for example, M. Avriel, 1976 or D. Spring, 1985, 631–643.

Again, it should be emphasized that the conditions in Theorem 19 are necessary conditions; they are not, in general, sufficient.

As an application of Theorem 19, we'll establish a well-known inequality.

EXAMPLE 21. For $a_i \geq 0$, we wish to show that

$$(a_1 a_2 \ldots a_n)^{1/n} \leq \sum_{i=1}^{n} \frac{a_i}{n}, \qquad (8)$$

that is, the geometric mean is less than the arithmetic mean. Let (P) denote the problem

$$\text{minimize } f(x_1, \ldots, x_n) = \sum_{i=1}^{n} x_i/n$$

$$\text{subject to } h(x_1, \ldots, x_n) = 1 - x_1 \cdots \cdot x_n = 0,$$

$$\text{for } \mathbf{x} \in E = \{(x_1, \ldots, x_n): x_i > 0, \, i = 1, \ldots, n\}.$$

First, observe that a minimum for (P) exists since

$$H = \{\mathbf{x}: f(\mathbf{x}) \leq c, \, h(\mathbf{x}) = 0, \, x_i > 0, \, i = 1, \ldots, n\}$$

is compact for all $c > 0$ (see Exercise 23). Suppose the minimum is at $\mathbf{x}$. Then there is $\lambda \in \mathbb{R}$ such that

$$0 = D_j f(\mathbf{x}) + \lambda D_j h(\mathbf{x}) = \frac{1}{n} - \lambda \frac{x_1 \cdots \cdot x_n}{x_j} = \frac{1}{n} - \frac{\lambda}{x_j}$$

for $j = 1, \ldots, n$. So $x_i = x_j = \lambda n$ for $i = 1, \ldots, n$. Therefore $\mathbf{x} = (1, \ldots, 1)$ and $f(\mathbf{x}) = 1$ is the minimum value.

Given $a_i > 0$, for $i = 1, \ldots, n$, set $y_i = a_i/A$ where $A = (a_1 a_2 \ldots a_n)^{1/n}$. Then $h(\mathbf{y}) = 0$ and so

$$f(\mathbf{y}) = \frac{1}{n} \left(\sum_{i=1}^{n} a_i/A \right) \geq 1.$$

This implies (8).   □

EXERCISES 10

1. Let $f(x, y) = x$ for $y = x^2$ and 0 for all other (x, y). Find $D_1 f(0, 0)$ and $D_2 f(0, 0)$.

2. If f has first partial derivatives and a local maximum at a point $\mathbf{x}$ in an open set D, show that $D_j f(\mathbf{x}) = 0$ for $j = 1, \ldots, n$. Can you say more if f has directional derivatives in all directions at $\mathbf{x}$?

3. For the following functions, show that L is linear and give the representation guaranteed by Proposition 3:

  (a) $L(\mathbf{x}) = x_i$ (b) $L(\mathbf{x}) = \sum\limits_{i=1}^{n} x_i$ (c) $L(\mathbf{x}) = 3x_1 - 2x_2$

4. If $L: \mathbb{R}^n \to \mathbb{R}$ is linear, show that L is uniformly continuous on $\mathbb{R}^n$. *Hint:* Use the Cauchy–Schwarz inequality.

5. If $f: [a, b] \to \mathbb{R}$, show that f has a derivative at x iff f is differentiable at x in the sense of Definition 4.

6. Does the converse of Proposition 6 hold?

7. If $f: \mathbb{R}^n \to \mathbb{R}$ has partial derivatives at $\mathbf{x}_0$, is f necessarily differentiable at $\mathbf{x}_0$? What if f has directional derivatives in all directions at $\mathbf{x}_0$? What if f has continuous first partial derivatives at $\mathbf{x}_0$?

8. Let $f(x, y) = x^2 y - 3xy$. Does f have a differential at $(1, 1)$? If so, what is it? Does $D_{\mathbf{u}} f(1, 1)$, where $\mathbf{u} = (1/\sqrt{2}, -1/\sqrt{2})$ exist? If so, what is it?

9. Let $f(x, y) = \sqrt{|xy|}$. Show that $D_1 f(0, 0)$ and $D_2 f(0, 0)$ exist and that f is continuous, but not differentiable at $(0, 0)$.

10. Let f be as in Exercise 1, and $\mathbf{g}(t) = (g_1(t), g_2(t)) = (3t, 9t^2)$. Set $F(t) = f(\mathbf{g}(t))$. Calculate $F'(0)$ and

$$D_1 f(0, 0) \frac{dg_1}{dt}(0) + D_2 f(0, 0) \frac{dg_2}{dt}(0).$$

Compare with (3) of Theorem 9. What happened?

11. Suppose f is differentiable at $\mathbf{x} = (x_1, \ldots, x_n) \in D$, $g_j: \mathbb{R}^2 \to \mathbb{R}$, is differentiable at (u_0, v_0), and $g_j(u_0, v_0) = x_j$ for $j = 1, \ldots, n$. Set $F(u, v) = f(\mathbf{g}(u, v))$. Show that F has first partial derivatives at (u_0, v_0) and give formulas for $D_1 F(u_0, v_0)$ and $D_2 F(u_0, v_0)$.

12. For $\delta > 0$, show that $S(\mathbf{x}_0, \delta)$ and $\{\mathbf{x}: \|\mathbf{x} - \mathbf{x}_0\| \leq \delta\}$ are convex.

13. If $L: \mathbb{R}^n \to \mathbb{R}$ is linear, show that $\{\mathbf{x}: L(\mathbf{x}) \leq c\}$ and $\{\mathbf{x}: L(\mathbf{x}) = c\}$ are closed and convex for every $c \in \mathbb{R}$.

14. If K is an open convex subset of $\mathbb{R}^n$ and $f: K \to \mathbb{R}$ is differentiable on K with $\nabla f(\mathbf{x}) = \mathbf{0}$ $\forall \mathbf{x} \in K$, show that f is a constant function. Can the convexity condition be dropped?

15. Define negative semidefinite and negative definite.

16. Let $A = \begin{bmatrix} a & b \\ b & c \end{bmatrix}$. Show that A is positive definite if and only if $a > 0$ and $\det(A) = ac - b^2 > 0$. *Hint:* Expand $A[\mathbf{h}]$ and complete the square.

17. Formulate Proposition 16 for local minima.

18. Formulate Proposition 18 for local maxima.

19. Examine $f(x, y) = x^2 - 2xy + 2y^2$ for local extrema.

20. Find candidates for solutions to the problem: Minimize $xy + yz + xz$ subject to $x + y + z = 3$.

21. Find candidates for solutions to the problem: Minimize $x + y + z$ subject to $x^2 + y^2 = 2$ and $x + z = 1$.

22. For the problem $\min f(x, y) = -x$ subject to $g_1(x, y) = -(1 - x)^3 + y \leq 0$, $g_2(x, y) = -x \leq 0$, $g_3(x, y) = -y \leq 0$, find a global minimum. (*Hint:* Make a sketch.) Show that the necessary conditions of Theorem 19 are not satisfied with $\lambda_0 > 0$, but are satisfied with $\lambda_0 = 0$.

23. In Example 21, show that H is compact. *Hint:* The set $\{x: h(x) = 0, x_i > 0, i = 1, \ldots, n\}$ is closed and $\{x: f(x) \leq c\}$ is closed and bounded.

24. State and prove a version of Rolle's theorem for functions of two variables.

25. If $f: \mathbb{R}^n \to \mathbb{R}$ is linear, show that f is differentiable and compute $df(x)$.

26. Is

$$f(x, y) = \begin{cases} xy/\sqrt{x^2 + y^2} & \text{if } (x, y) \neq (0, 0), \\ 0 & \text{if } (x, y) = (0, 0), \end{cases}$$

differentiable at $(0, 0)$?

27. Write the equation of the tangent plane to the surface $x^2 + y^2 + z^2 = c$ at any point (x_0, y_0, z_0) on the surface.

28. If f is a smooth function on $\mathbb{R}^2$, $x = r\cos\theta$, $y = r\sin\theta$, and $u(r, \theta) = f(r\cos\theta, r\sin\theta)$ (i.e., polar coordinates), show that

$$\frac{\partial^2 f}{\partial x^2} + \frac{\partial^2 f}{\partial y^2} = \frac{\partial^2 u}{\partial r^2} + \frac{1}{r}\frac{\partial u}{\partial r} + \frac{1}{r^2}\frac{\partial^2 u}{\partial \theta^2}.$$

29. Prove (with appropriate assumptions) that

$$d(af + bg) = a\,df + b\,dg.$$

30. Examine (a) $f(x, y) = x^3 + 6x^2 + 3y^2 - 12xy + 9x$ and (b) $g(x, y) = \sin x + y^2 - 2y + 1$ for local extrema.

31. Write out the first and second order Taylor expansions for the functions $f(x, y) = e^{x+y}$ and $g(x, y) = (x^2 + y^2 + 1)^{-1}$ about the point $(0, 0)$.

32. Let $f(x, y) = x^2 + y^2$ if x and y are both rational and $f(x, y) = 0$ otherwise. Show that f is differentiable at $(0, 0)$. Is f differentiable at other points?

33. If $f(x, y, z) = xyz$, $g(u, v) = (\sin u, e^v, \cos uv)$, and $F(u, v) = f(g(u, v))$, calculate $D_1 F$ and $D_2 F$ directly and by using the chain rule (Exercise 11).

34. Show that $\nabla(fg) = f\nabla g + g\nabla f$ under appropriate assumptions.

35. Let $f: \mathbb{R}^n \to \mathbb{R}$ be differentiable and *positive homogeneous* of degree p, that is, $f(t\mathbf{x}) = t^p f(\mathbf{x}) \; \forall \mathbf{x} \in \mathbb{R}^n$ and $\forall t \in \mathbb{R}$. Prove Euler's relation: $pf(\mathbf{x}) = \sum_{i=1}^{n} x_i D_i f(\mathbf{x})$.

36. If $f: \mathbb{R} \to \mathbb{R}$ is twice differentiable and $u(x, y) = f(x - ay) + f(x + ay)$, show that u satisfies the wave equation

$$a^2 \frac{\partial^2 u}{\partial x^2} = \frac{\partial^2 u}{\partial y^2}.$$

37. Find extrema for the problem: $\min (\max) f(x, y) = x^2 + 2x + y^2$ subject to $|x| \le 1$, $|y| \le 1$.

38. Let $f_1, \ldots, f_n: \mathbb{R} \to \mathbb{R}$ be differentiable. Show that the function $F: \mathbb{R}^n \to \mathbb{R}$ defined by $F(\mathbf{x}) = f_1(x_1) + \cdots + f_n(x_n)$ is differentiable.

39. Show that the continuity of the second order partials in Theorem 13 is not a necessary condition for the equality of the cross partials.
    Hint: Consider $f(x, y) = yg(x)$.

40. If $f: D \to \mathbb{R}$ is differentiable at $\mathbf{x} \in D$, $f(\mathbf{x}) = 0$, and $g: D \to \mathbb{R}$ is continuous at $\mathbf{x}$, show that fg is differentiable at $\mathbf{x}$.

41. If $f: D \to \mathbb{R}$, show that f is differentiable at $\mathbf{x} \in D$ iff there is a function $h: D \to \mathbb{R}^n$ that is continuous at $\mathbf{x}$ and satisfies $f(\mathbf{z}) - f(\mathbf{x}) = h(\mathbf{z}) \cdot (\mathbf{z} - \mathbf{x}) \; \forall \mathbf{z} \in D$.

42. Give an example in which $D_\mathbf{u} f(\mathbf{x})$ is not equal to $\nabla f(\mathbf{x}) \cdot \mathbf{u}$.

43. Let A be an $n \times n$ symmetric matrix. Suppose that the problem $\min \{(A\mathbf{z}^t)^t \cdot \mathbf{z}: \|\mathbf{z}\| = 1\}$ has a minimum at $\mathbf{z}_0$ with $(A\mathbf{z}_0^t)^t \cdot \mathbf{z}_0 = \lambda_0$. Show that λ_0 is the smallest eigenvalue of A with associated eigenvector $\mathbf{z}_0$.

44. (First order sufficient condition for a minimum.) Let $D \subset \mathbb{R}^n$ be open and $f: D \to \mathbb{R}$ be differentiable. Let $\mathbf{x} \in D$ and N be a neighborhood of $\mathbf{x}$ contained in D. If $\nabla f(\mathbf{z}) \cdot (\mathbf{z} - \mathbf{x}) \ge 0$ for all $\mathbf{z} \in N$, show that f has a local minimum at $\mathbf{x}$.

CHAPTER 11

Sequences of Functions

In previous chapters, our investigation of sequences has been restricted to $\mathbb{R}^n$, $n \in \mathbb{N}$. However, the generality of the definition of a sequence in Chapter 3 enables us to consider sequences in any nonempty set and, in particular, sequences of functions. To discuss convergence of such sequences, it will be necessary to give an adequate description for one function to be near another. It turns out that there are a variety of ways to do this, each one leading to a different kind of convergence. In the present chapter we'll consider two kinds of convergence for functions, "pointwise" and "uniform." Some other types of convergence will be discussed in subsequent chapters.

Let D be a subset of $\mathbb{R}^n$ and suppose that for each $k \in \mathbb{N}$ there is defined a function $\mathbf{f}_k \colon D \to \mathbb{R}^m$. If there is a function $\mathbf{f} \colon D \to \mathbb{R}^m$ such that the sequence $\{\mathbf{f}_k(\mathbf{x})\}$ of points in $\mathbb{R}^m$ converges to $\mathbf{f}(\mathbf{x})$ for each $\mathbf{x} \in D$, then the sequence $\{\mathbf{f}_k\}$ is said to *converge pointwise* to $\mathbf{f}$ on D. We write $\mathbf{f}_k \to \mathbf{f}$ pointwise on D, or $\mathbf{f}_k \to_p \mathbf{f}$ on D.

For example, the sequence of functions defined by $f_k(t) = t^k$, $0 \le t \le 1$, $k \in \mathbb{N}$, converges pointwise to the function

$$f(t) = \begin{cases} 0, & \text{if} \quad 0 \le t < 1, \\ 1, & \text{if} \quad t = 1. \end{cases}$$

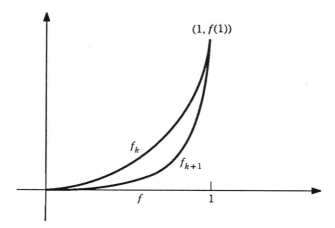

Figure 11.1

Some terms of the sequence and its limit function are shown in Figure 11.1.

If the terms of a sequence of functions $\{f_k\}$ have a common property such as continuity or differentiability, it is natural to inquire about conditions under which a limit function would inherit this property. In general, pointwise convergence does not preserve such properties. The functions in the sequence in the above example are all continuous on $[0, 1]$; however, the limit function fails to be continuous at $x = 1$.

The sequence of differentiable functions $f_k(t) = t^k/k$, $0 \leq t \leq 1$, $k \in \mathbb{N}$, converges to 0, a differentiable function; however, the limit of the sequence of derivatives is not everywhere equal to the derivative of the limit function on $[0, 1]$. An examination of the sequence of derivatives $f_k'(t) = t^{k-1}$ reveals a situation analogous to the first example.

In order to develop sufficient conditions for the preservation in the limit of properties common to the terms of a sequence of functions, we introduce a stronger notion of convergence. For a sequence of functions $\{f_k\}$ to converge pointwise on D to f, it is necessary that for each $\varepsilon > 0$ and for each $x \in D$, there is $N = N(x, \varepsilon)$, generally depending on both x and ε, such that $k \geq N$ implies that $|f_k(x) - f(x)| < \varepsilon$. For example, if $f_k(x) = x^k$, $0 < x < 1$, then $N(x, \varepsilon) > \ln \varepsilon / \ln x$ guarantees that $|f_k(x)| < \varepsilon$ whenever $k \geq N$. The dependence of N on x in this example is quite clear. An examination of the sequence of differentiable functions given above reveals similar behavior. It is more than coincidence that the dependence of N on x occurs near the same point at which the limit function fails to inherit the property common to all the terms of the sequence. If it is possible to find an N that depends only on ε, and not on x, then the convergence will be stronger in the sense that for each

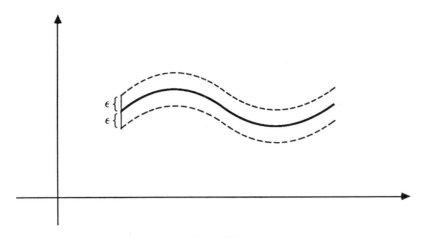

Figure 11.2

ε, a single N will work uniformly for all x under consideration. We formalize this with a definition.

Definition 1. The sequence $\{\mathbf{f}_k\}$ of functions *converges uniformly* to $\mathbf{f}$ on D if for each $\varepsilon > 0$, there is N such that $\|\mathbf{f}_k(\mathbf{x}) - \mathbf{f}(\mathbf{x})\| < \varepsilon$, whenever $k \geq N$ and $\mathbf{x} \in D$. We write $\mathbf{f}_k \to \mathbf{f}$ uniformly, or $\mathbf{f}_k \to_u \mathbf{f}$ on D.

We reemphasize that the term "uniform" refers to the existence of an N that, for a given ε, works uniformly for all $\mathbf{x}$ in D. Unless it is specified otherwise, we assume throughout the current chapter that we are dealing with functions $\mathbf{f}, \mathbf{f}_k \colon D \to \mathbb{R}^m$, where D is a subset of $\mathbb{R}^n$. It is apparent that if a sequence of functions converges uniformly on D, then it converges uniformly on every subset of D. The geometric interpretation of uniform convergence is illustrated in Figure 11.2. In order for a sequence to converge uniformly to the function f, whose graph is pictured in Figure 11.2, for each $\varepsilon > 0$, the graph of each f_k for sufficiently large k must lie in the "ε-strip", indicated by the dotted curves, about the graph of f.

Figure 11.3 indicates a sequence of functions $[f_k(t) = kt(1 - t)^k$ from Exercise 2(a)] which converges pointwise, but not uniformly, to zero.

Proposition 2. The sequence $\mathbf{f}_k \to \mathbf{f}$ uniformly on D if, and only if,

$$\sup_{\mathbf{x} \in D} \|\mathbf{f}_k(\mathbf{x}) - \mathbf{f}(\mathbf{x})\| \to 0 \text{ as } k \to \infty.$$

Proof. (Exercise 1).   □

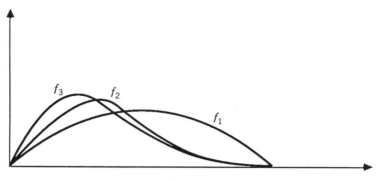

Figure 11.3

In the first example above, $\sup\{|f_k(t) - f(t)| : 0 \le t \le 1\} = \sup\{t^k : 0 \le t \le 1\} = 1$. Therefore f_k does not converge uniformly to f on $[0, 1]$. The reason for the failure of uniform convergence becomes apparent when one notes as above that $|t^k| < \varepsilon < 1$ iff $k \ge n_{t,\varepsilon} = \ln \varepsilon / \ln t$, and so, $n_{t,\varepsilon} \to \infty$ as $t \to 1^-$. However, if $0 < \delta < 1$, then $f_k \to_u f$ on $[0, 1 - \delta]$, showing that the troublesome point is indeed at $t = 1$.

Two further examples will be useful for our continued discussion of uniform convergence.

Let g_k be defined by $g_k(t) = t/(kt + 1)$, $0 < t < 1$, and $k \in \mathbb{N}$. Then $g_k \to_p 0$ on $(0, 1)$, and, since

$$\sup\{|g_k(t)| : 0 < t < 1\} = (k + 1)^{-1},$$

we also have $g_k \to_u g$ on $(0, 1)$.

If h_k is defined by $h_k(t) = 1/(kt + 1)$, $0 < t < 1$, and $k \in \mathbb{N}$, then $h_k \to_p 0$ on $(0, 1)$; however,

$$\sup\{|h_k(t)| : 0 < t < 1\} = 1.$$

Consequently, $\{h_k\}$ does not converge uniformly on $(0, 1)$. A sketch of the graphs for a few of the h_k should illustrate to the reader exactly where the difficulties with uniform convergence arise.

Another problem with the sequence $\{h_k\}$ is that the two iterated limits, $\lim_{t \to 0} \lim_{k \to \infty} h_k(t) = 0$ and $\lim_{k \to \infty} \lim_{t \to 0} h_k(t) = 1$, exist but are not equal. There are conditions under which the existence and equality of the iterated limits can be assured:

Theorem 3. Suppose $\mathbf{f}, \mathbf{f}_k : D \to \mathbb{R}^m$, and that $\mathbf{x}_0$ is an accumulation point of D. If $\{\mathbf{f}_k\}$ converges uniformly to $\mathbf{f}$ on D and if $\lim_{\mathbf{x} \to \mathbf{x}_0} \mathbf{f}_k(\mathbf{x}) = \mathbf{A}_k$ exists for each $k \in \mathbb{N}$, then $\lim_{k \to \infty} A_k = \lim_{k \to \infty} \lim_{x \to x_0} f_k(x)$ exists, and

$$\lim_{x \to x_0} f(x) = \lim_{x \to x_0} \lim_{k \to \infty} f_k(x) = \lim_{k \to \infty} \lim_{x \to x_0} f_k(x).$$

Proof. Since $f_k \to_u f$ on D, for each $\varepsilon > 0$ there is $N > 0$ such that $\|f_k(x) = f(x)\| < \varepsilon/6$ for $k \ge N$ and $x \in D$. Thus

$$\|f_k(x) - f_j(x)\| < \varepsilon/3 \tag{1}$$

whenever $x \in D$ and $j, k \ge N$. If $x \to x_0$, this becomes

$$\|A_k - A_j\| \le \varepsilon/3 \tag{2}$$

for $j, k \ge N$. Thus $\{A_k\}$ is a Cauchy sequence in $\mathbb{R}^m$ and converges to some point $A \in \mathbb{R}^m$. Letting $j \to \infty$ in (2) yields $\|A_k - A\| \le \varepsilon/3$, for $k \ge N$. There is $\delta > 0$ such that $\|f_N(x) - A_N\| < \varepsilon/3$, whenever $x \in D$ and $0 < \|x - x_0\| < \delta$. For such x,

$$\|f(x) - A\| \le \|f(x) - f_N(x)\| + \|f_N(x) - A_N\| + \|A_N - A\|$$
$$< \varepsilon/6 + \varepsilon/3 + \varepsilon/3 < \varepsilon.$$

So, $\lim_{x \to x_0} f(x) = A$. $\square$

As a corollary to Theorem 3, we obtain conditions sufficient for the continuity of the limit of a sequence of continuous functions.

Corollary 4. If $f_k: D \to \mathbb{R}^m$ is continuous at $x_0 \in D$ for each $k \in \mathbb{N}$ and $\{f_k\}$ converges uniformly to f on D, then f is continuous at $x_0 \in D$.

Proof. Let $\{x_k\}$ be a sequence in D converging to $x_0 \in D$. If we take $A_k = f(x_k)$ in Theorem 3, the result is then immediate. $\square$

The sequence $\{h_k\}$ above shows that uniform convergence is a sufficient, but not necessary, condition for the conclusion of Corollary 4. In fact, there is a canard to the effect that "uniform convergence is sufficient for just about anything, but necessary for nothing."

As another application of Theorem 3, we establish a useful corollary on limits of double sequences. Recall that a double sequence is a mapping $a: \mathbb{N} \times \mathbb{N} \to \mathbb{R}^m$; we write $a_{ij} = a(i, j)$.

Corollary 5. Let a_{ij} be such that

(i) $\lim_j a_{ij} = r_i$ exists for each i, and
(ii) $\lim_i a_{ij} = c_j$ converges uniformly for $j \in \mathbb{N}$.

Then $\lim_i r_i$ exists and $\lim_i r_i = \lim_j c_j = \lim_i \lim_j a_{ij} = \lim_j \lim_i a_{ij}$. (A sketch of an infinite matrix with convergent rows and columns is useful for understanding the content of the corollary.)

Proof. Let $D = \{1/j: j \in \mathbb{N}\}$ and note that 0 is an accumulation point of D. Define $\mathbf{f}_i: D \to \mathbb{R}^m$ by $\mathbf{f}_i(1/j) = \mathbf{a}_{ij}$. Now apply Theorem 3.   □

There is a Cauchy-type criterion for uniform convergence analogous to that for convergence of a sequence in $\mathbb{R}^n$.

Proposition 6. Suppose $\mathbf{f}_k: D \to \mathbb{R}^m$ for $k \in \mathbb{N}$. The sequence $\{\mathbf{f}_k\}$ is uniformly convergent (to some $\mathbf{f}: D \to \mathbb{R}^m$) if, and only if, it is uniformly Cauchy, that is, for each $\varepsilon > 0$, there is $N > 0$ such that

$$\|\mathbf{f}_k(\mathbf{x}) - \mathbf{f}_j(\mathbf{x})\| < \varepsilon \tag{3}$$

whenever $j, k \geq N$ and $\mathbf{x} \in D$.

Proof. If $\mathbf{f} \to_u \mathbf{f}$, then a uniform Cauchy condition follows from the inequality

$$\|\mathbf{f}_k(\mathbf{x}) - \mathbf{f}_j(\mathbf{x})\| \leq \|\mathbf{f}_k(\mathbf{x}) - \mathbf{f}(\mathbf{x})\| + \|\mathbf{f}(\mathbf{x}) - \mathbf{f}_j(\mathbf{x})\|.$$

If $\{\mathbf{f}_k\}$ satisfies the uniform Cauchy condition, then it is a Cauchy sequence in the "pointwise" sense; thus, there is $\mathbf{f}: D \to \mathbb{R}^m$ such that $\mathbf{f}_k \to_p \mathbf{f}$ on D. Let k be fixed and $j \to \infty$ in (3). It follows that $\|\mathbf{f}_k(\mathbf{x}) - \mathbf{f}(\mathbf{x})\| \leq \varepsilon$ whenever $k \geq N$ and $\mathbf{x} \in D$, that is, $\mathbf{f}_k \to_u \mathbf{f}$.   □

Differentiation of Limit Functions

If

$$f_k(t) = \begin{cases} 1/k & \text{for} \quad |t| \leq 1/k, \\ |t| & \text{for} \quad |t| \geq 1/k, \end{cases}$$

then $f_k \to_u f$, on $\mathbb{R}$, where $f(t) = |t|$. Each f_k is differentiable at $t = 0$; however, this is not true of the limit function f. The situation can be even worse than that in this last example. If $g_k(t) = (\sin kt)/\sqrt{k}$, then $g_k \to_u 0$ while the derived sequence $\{g_k'\}$ does not even converge pointwise.

Conditions under which the limit function will inherit differentiability from the terms of the sequence are given in the following.

Theorem 7. Suppose $f_k: [a, b] \to \mathbb{R}$ is a sequence of functions that are differentiable on $[a, b]$ and $f_k \to_p f$ on $[a, b]$. If the sequence $\{f_k'\}$ of derivatives converges uniformly to h on $[a, b]$, then f is differentiable on $[a, b]$ and $f' = h$.

Proof. For fixed $x \in [a, b]$, define $\phi, \phi_n: [a, b] \setminus \{x\} \to \mathbb{R}$ by $\phi_n(t) = (f_n(t) - f_n(x))/(t - x)$, $\phi(t) = (f(t) - f(x))/(t - x)$. Then $\lim_{t \to x} \phi_n(t) =$

$f_n'(x)$, and $\lim_{n \to \infty} \phi_n(t) = \phi(t)$. Application of the mean value theorem yields

$$|\phi_n(t) - \phi_m(t)| = |[f_n(t) - f_m(t)] - [f_n(x) - f_m(x)]|/|t - x|$$

$$= |(f_n' - f_m')(\zeta_{n,m,t,x})|,$$

where $\zeta_{n,m,t,x}$ lies between t and x. Since $\{f_k'\}$ is uniformly convergent, Proposition 6 can be applied to show that $\{\phi_k\}$ is uniformly convergent. Theorem 3 then implies

$$h(x) = \lim_{n \to \infty} f_n'(x) = \lim_{n \to \infty} \lim_{t \to x} \phi_n(t)$$

$$= \lim_{t \to x} \lim_{n \to \infty} \phi_n(t) = \lim_{t \to x} \phi(t) = f'(x). \qquad \square$$

For the sequence of functions defined by $f_k(t) = k^2 t e^{-kt}$, for $t \in [0, 1]$, we have $f_k \to_p 0$ and $f_k' \to_p 0$; however, the convergence of $\{f_k'\}$ is not uniform on $[0, 1]$. Thus the conditions given in Theorem 7 are sufficient, but not necessary for differentiability of the limit function.

Infinite Series of Functions

Definition 8. Let $f, f_k: D \to \mathbb{R}^m$, $k \in \mathbb{N}$. The infinite series $\sum_{i=1}^{\infty} f_i$ *converges pointwise (uniformly)* to f on D if the sequence of partial sums $s_k = \sum_{i=1}^{k} f_i$ converges pointwise (uniformly) to f on D. With this definition, Corollary 4 and Theorem 7 extend immediately to infinite series.

Theorem 9. Let $f_i: D \to \mathbb{R}^m$, $i \in \mathbb{N}$, be functions that are continuous at $x_0 \in D$. If $\sum_{i=1}^{\infty} f_i$ converges uniformly to f on D, then f is continuous at x_0.

Theorem 10. Let $f_k: [a, b] \to \mathbb{R}$, $k \in \mathbb{N}$ be differentiable functions for which the series $\sum_{k=1}^{\infty} f_k$ converges pointwise to f on $[a, b]$. If $\sum_{k=1}^{\infty} f_k'$ converges uniformly to h on $[a, b]$, then f is differentiable on $[a, b]$ and $f' = h$.

The following theorem, proved by the German mathematician Karl Weierstrass (1815–1897), gives sufficient conditions for uniform convergence of an infinite series of functions.

Theorem 11 (Weierstrass M-test). Let $f_k: D \to \mathbb{R}^m$, $k \in \mathbb{N}$. If for each $k \in \mathbb{N}$ there is $M_k \geq 0$ such that $\|f_k(x)\| \leq M_k$ for all $x \in D$ and if $\sum_{k=1}^{\infty} M_k$ converges, then $\sum_{k=1}^{\infty} f_k$ converges uniformly on D.

Proof. For $k > j$, $\|s_k(x) - s_j(x)\| = \|\sum_{i=j+1}^{k} f_i(x)\| \leq \sum_{i=j+1}^{k} M_i$. Now apply Proposition 6. $\qquad \square$

Power Series

An infinite series of the form

$$\sum_{k=0}^{\infty} a_k(x - x_0)^k, \qquad a_k, x, x_0 \in \mathbb{R}$$

is called a *power series* about x_0. Since a simple translation transforms this series to a series of the form

$$\sum_{k=0}^{\infty} a_k x^k, \qquad a_k, x, \in \mathbb{R} \tag{4}$$

we'll generally consider power series about the origin. The convergence properties that we develop for such series can then be easily rephrased to apply to power series about x_0. We sometimes write $\Sigma a_k x^k$ instead of (4), understanding that the index of summation runs from 0 to ∞. Clearly the series (4) converges for $x = 0$. The following theorem shows that if (4) converges for some other value, say $x = c$, then it converges in the symmetric interval $(-|c|, |c|)$ about $x = 0$.

Proposition 12. If $\sum_{k=0}^{\infty} a_k x^k$ converges for some $x = c \neq 0$, then

(i) $\sum_{k=0}^{\infty} a_k x^k$ converges absolutely for $|x| < |c|$.

(ii) $\sum_{k=0}^{\infty} a_k x^k$ converges uniformly on $[-b, b]$ for $0 < b < |c|$.

Proof. To establish (i), let $M > 0$. There is N such that $|a_k c^k| \leq M$ for all $k \geq N$. If $|x| < |c|$, then

$$\sum_{k=N}^{\infty} |a_k| \, |x|^k = \sum_{k=N}^{\infty} |a_k| \, |c|^k \left|\frac{x}{c}\right|^k \leq M \sum_{k=N}^{\infty} \left|\frac{x}{c}\right|^k.$$

The last series above is a geometric series with common ratio $|x/c| < 1$ and therefore converges.

For (ii), if $b < |c|$, then (i) implies that $\sum_{k=0}^{\infty} |a_k b^k|$ converges, and so, (ii) follows from the Weierstrass M-test.   □

The possible convergence behavior of a power series is described more explicitly by Theorem 13.

Theorem 13. Let $S = \{x: \sum_{k=0}^{\infty} a_k x^k$ converges$\}$. Then, exactly one of the following holds:

(i) $S = \{0\}$.

(ii) $S = \mathbb{R}$.

(iii) There is r, $0 < r < \infty$, such that the series (4) converges for $|x| < r$ and diverges for $|x| > r$.

Proof. If S is not bounded above, then Proposition 12(i) implies that $S = \mathbb{R}$. If S is bounded above, let $r = \sup \{|x|: x \in S\}$. If $r = 0$, then (i) holds. If $0 < r < \infty$ and $x \in (-r, r)$, then there is x_0 such that $|x| < |x_0| < r$ and $\sum_{k=0}^{\infty} a_k x_0^k$ converges. It follows from 12(i) that $\sum_{k=0}^{\infty} a_k x^k$ converges. □

The *radius of convergence* of the power series (4) is $r = 0$ if (i) holds in Theorem 13, $r = \infty$ if (ii) holds, and r in case (iii). In this last case, the series may or may not converge at the endpoints of the interval $(-r, r)$. Examination of the series

$$\Sigma x^n, \quad \Sigma x^n/n, \quad \Sigma x^n/n^2.$$

shows that anything can happen at these endpoints. The examples in Exercise 11.12 show that all of the situations described in Theorem 13 can occur.

If $\lim \sqrt[k]{|a_k|} = R$, then the root test (4.15) implies that the series $\Sigma a_k x^k$ converges for $R|x| < 1$ and diverges for $R|x| > 1$, that is, the radius of convergence is given by

$$r = 1/R = 1/\lim \sqrt[k]{|a_k|}$$

We've used the convention $1/0 = \infty$, $1/\infty = 0$. A similar formula for the radius of convergence can be based on the ratio test (see Exercise 18).

A power series and its series of derivatives, called its *derived series*, have the same radius of convergence. So the derived series is uniformly convergent on every closed subinterval of the interval of convergence with the result that the original series is termwise differentiable and the derived series converges to the derivative of the limit function. This is made explicit in the following theorem.

Theorem 14. Suppose (4) has radius of convergence $r > 0$, and write $f(x) = \sum_{k=0}^{\infty} a_k x^k$. Then f is differentiable on $(-r, r)$ and

$$f'(x) = \sum_{k=1}^{\infty} ka_k x^{k-1}.$$

Proof. If $|x| < r$, choose c so that $|x| < c < r$. Then $\Sigma a_k c^k$ converges absolutely. Since $\lim_{k \to \infty} k^{1/k} = 1$ (3.5), we have for k sufficiently large, $\sqrt[k]{k} |x| < c$. Hence, $k|a_k| |x^k| \leq |a_k|c^k$ for large enough k. Consequently, $\sum_{k=1}^{\infty} ka_k x^{k-1}$ converges absolutely and uniformly for $|x| \leq c$. The result now follows from Theorem 10. □

Corollary 15. The function in Theorem 14 has derivatives of all orders in $(-r, r)$. Moreover, the coefficients in the series satisfy

$$a_k = f^{(k)}(0)/k!.$$

The formula given for the a_k in Corollary 15 shows first that if $\Sigma a_k x^k = \Sigma b_k x^k$, then $a_k = b_k$ and, second, that the power series in (4) is the Taylor series for f about the point $x = 0$.

As an example of a power series, we give the following generalization of the classical binomial expansion of $(1 + x)^n$ to the case where n need not be a positive integer.

EXAMPLE 16. The series $\Sigma_{k=0}^{\infty} \binom{\alpha}{k} x^k$, where

$$\binom{\alpha}{k} = \frac{\alpha(\alpha - 1)\ldots(\alpha - k + 1)}{k!}$$

for $k \in \mathbb{N}$ and $\binom{\alpha}{0} = 1$, is called the *binomial series* and converges to $(1 + x)^\alpha$ for $|x| < 1$. The numbers $\binom{\alpha}{k}$ are called the *binomial coefficients*. First, the radius of convergence r of the series $f(x) = \Sigma_{k=0}^{\infty} \binom{\alpha}{k} x^k$ is given by

$$\frac{1}{r} = \lim_{k \to \infty} \frac{|\alpha(\alpha - 1)\ldots(\alpha - k)|}{(k + 1)!} \cdot \frac{k!}{|\alpha(\alpha - 1)\ldots(\alpha - k + 1)|}$$

$$= \lim_{k \to \infty} \frac{|\alpha - k|}{k + 1} = 1$$

(see Exercise 18). So the series can be differentiated termwise and we find that

$$(1 + x)f'(x) = \sum_{k=1}^{\infty} \frac{\alpha(\alpha - 1)\ldots(\alpha - k + 1)}{(k - 1)!} \cdot x^{k-1}$$

$$+ \sum_{k=1}^{\infty} \frac{\alpha(\alpha - 1)\ldots(\alpha - k + 1)}{(k - 1)!} \cdot x^k$$

$$= \alpha + \sum_{k=1}^{\infty} \left\{ \frac{\alpha(\alpha - 1)\ldots(\alpha - k)}{k!} + \frac{\alpha(\alpha - 1)\ldots(\alpha - k + 1)}{(k - 1)!} \right\} x^k$$

$$= \alpha + \sum_{k=1}^{\infty} \frac{\alpha(\alpha - 1)\ldots(\alpha - k + 1)}{(k - 1)!} \left\{ \frac{\alpha - k}{k} + 1 \right\} x^k$$

$$= \alpha \left\{ 1 + \sum_{k=1}^{\infty} \frac{\alpha(\alpha - 1)\ldots(\alpha - k + 1)}{k!} x^k \right\}$$

$$= \alpha f(x).$$

Thus, for $|x| < 1$,

$$\frac{d}{dx}\{(1 + x)^{-\alpha}f(x)\} = -\alpha(1 + x)^{-\alpha-1}f(x) + (1 + x)^{-\alpha}f'(x) = 0.$$

Therefore, $(1 + x)^{-\alpha}f(x) = C$ (constant). Finally, $f(0) = 1$ implies that $C = 1$. □

The study of the relationship between possible convergence of a power series at the endpoints of the interval of convergence to convergence at interior points usually takes place in the more general setting of functions of a complex variable. However, a result of the Norwegian mathematician, Niels Abel (1802–1829), shows that if a power series converges at an endpoint of the interval of convergence, then the function defined by the series has one-sided continuity from within the interval of convergence. As we'll see later, this theorem can sometimes be used for the evaluation of series, that is, to actually find the sum; for example, in Chapter 15 we'll show that

$$\sum_{k=0}^{\infty} (-1)^k/k = -\ln 2.$$

Theorem 17 (Abel). Suppose $\sum_{k=0}^{\infty}a_k$ converges. Set $f(x) = \sum_{k=0}^{\infty}a_k x^k$, $|x| < 1$. Then $\lim_{x \to 1^-} f(x) = \sum_{k=0}^{\infty}a_k$, that is, f is continuous at $x = 1$ if it is defined to have the value $\sum a_k$ there.

Proof. Let $s_{-1} = 0$, $s_n = \sum_{k=0}^{n}a_k$, $n \geq 0$. Then

$$\sum_{k=0}^{m} a_k x^k = \sum_{k=0}^{m} (s_k - s_{k-1})x^k = (1 - x)\sum_{k=0}^{m-1} s_k x^k + s_m x^m.$$

For $|x| < 1$, let $m \to \infty$ to obtain $f(x) = (1 - x)\sum_{k=0}^{\infty}s_k x^k$. Let $s = \lim_{n \to \infty} s_n$ and $\varepsilon > 0$. Choose N so that $|s - s_n| < \varepsilon/2$ whenever $n \geq N$. Since $(1 - x)\sum_{k=0}^{\infty}x^k = 1$ for $|x| < 1$, we have,

$$|f(x) - s| = \left|(1 - x)\sum_{k=0}^{\infty} (s_k - s)x^k\right|$$

$$\leq (1 - x)\sum_{k=0}^{N} |s_k - s|\,|x|^k + (1 - x)\frac{\varepsilon}{2}\sum_{k=N+1}^{\infty} |x|^k$$

$$\leq (1 - x)\sum_{k=0}^{N} |s_k - s| + \frac{\varepsilon}{2} < \varepsilon$$

when $x > 1 - \delta$, provided $\delta > 0$ is chosen sufficiently small.   □

The examples presented at the beginning of this chapter show that, in general, pointwise convergence does not imply uniform convergence. However, the following result of the Italian mathematician Ulisse Dini (1845–1918) gives useful criteria under which pointwise convergence yields uniform convergence.

Theorem 18 (Dini). Let K be a compact subset of $\mathbb{R}^n$ and $f, f_k \colon K \to \mathbb{R}$ be continuous on K. If $f_k(\mathbf{x})$ decreases (increases) to $f(\mathbf{x})$ for all $\mathbf{x} \in K$, then $f_k \to_u f$ uniformly on K.

Proof. By replacing f_k by $f_k - f$, we may assume $f = 0$. If the conclusion fails, then there is $\varepsilon > 0$ such that for all k there is $n_k \in \mathbb{N}$ and $\mathbf{x}_k \in K$ with $f_{n_k}(\mathbf{x}_k) \geq \varepsilon$. We may assume $n_{k+1} > n_k$ for all k. Since K is compact $\{\mathbf{x}_k\}$ has a subsequence that converges to $\mathbf{x} \in K$. To avoid messy subscripts, assume that $\mathbf{x}_k \to \mathbf{x}$. Fix N. If $n_k \geq N$, then $\varepsilon \leq f_{n_k}(\mathbf{x}_k) \leq f_N(\mathbf{x}_k)$. Let $k \to \infty$ to obtain $\varepsilon \leq f_N(\mathbf{x})$. Hence, $\{\mathbf{f}_N(\mathbf{x})\}$ doesn't decrease to zero. This is a contradiction. $\square$

In Exercise 14, the reader is asked to examine the importance of the various hypotheses in Theorem 18.

As an application of our previous results, we show how to construct an example of a continuous function that is nowhere differentiable. For another interesting example of a continuous function defined on an interval whose range is the unit square in $\mathbb{R}^2$ (i.e., a so-called *space filling curve*), see Wen, 1983, 283.

Lemma 19. Let $f \colon [a, b] \to \mathbb{R}$ be differentiable at $t \in (a, b)$. If $a < \alpha_n < t < \beta_n < b$ and $\alpha_n \to t$, $\beta_n \to t$, then

$$\frac{\lim [f(\beta_n) - f(\alpha_n)]}{(\beta_n - \alpha_n)} = f'(t).$$

Proof. Let $\lambda_n = (\beta_n - t)/(\beta_n - \alpha_n)$. Then $0 < \lambda_n < 1$ and

$$[f(\beta_n) - f(\alpha_n)]/(\beta_n - \alpha_n) - f'(t)$$

$$= \lambda_n\{[f(\beta_n) - f(t)]/(\beta_n - t) - f'(t)\}$$

$$+ (1 - \lambda_n)\{[f(\alpha_n) - f(t)]/(\alpha_n - t) - f'(t)\} \to 0. \qquad \square$$

EXAMPLE 20 (A Nondifferentiable Continuous Function). Let

$$\phi(t) = \begin{cases} t & \text{if } 0 \leq t \leq 1, \\ 2 - t & \text{if } 1 \leq t \leq 2, \end{cases}$$

and extend ϕ to a periodic function on $\mathbb{R}$ by defining $\phi(t) = \phi(t + 2)$. Define $f: \mathbb{R} \to \mathbb{R}$ by $f(t) = \sum_{k=0}^{\infty}(3/4)^k\phi(4^k t)$. (Sketch the graphs of the first few terms of the series to see how the construction goes.) The Weierstrass M-test shows that this series is uniformly convergent; thus, f is continuous on $\mathbb{R}$.

We claim that f fails to have a derivative at every point $x \in \mathbb{R}$. Let $m \in \mathbb{N}$. There is an integer $k = k(m, x)$ such that $k < 4^m x \le k + 1$. Set $\alpha_m = 4^{-m}k$, $\beta_m = 4^{-m}(k + 1)$ and consider $4^n\alpha_m$, $4^n\beta_m$: if $n > m$, $|4^n\beta_m - 4^n\alpha_m|$ is an even integer; if $n = m$, $|4^n\beta_m - 4^n\alpha_m| = 1$; if $n < m$, no integer lies between $4^n\alpha_m$ and $4^n\beta_m$. Hence,

$$|\phi(4^n\beta_m) - \phi(4^n\alpha_m)| = \begin{cases} 0 & \text{if} \quad n > m, \\ 4^{n-m} & \text{if} \quad n \le m. \end{cases}$$

Therefore,

$$|f(\beta_m) - f(\alpha_m)| = \left| \sum_{k=0}^{m} (3/4)^k \{ \phi(4^k\beta_m) - \phi(4^k\alpha_m) \} \right|$$

$$\ge (3/4)^m - \sum_{k=0}^{m-1} (3/4)^k 4^{k-m}$$

$$= (3/4)^m - (3^{m-1} - 1)/(4^m 2) > (3/4)^m/2.$$

So, $|f(\beta_m) - f(\alpha_m)|/(\beta_m - \alpha_m) > 3^m/2$, and, by Lemma 19, f is not differentiable at x. $\square$

We will show in Chapter 24 that in a certain sense, "most" continuous functions are nowhere differentiable.

EXERCISES 11

1. Prove Proposition 2.
2. In each of the following, find a function f such that $f_k \to_p f$. Is the convergence uniform?
   (a) $f_k(t) = kt(1 - t)^k$, $t \in [0, 1]$
   (b) $f_k(t) = t^{2k}/(1 + t^{2k})$, $t \in \mathbb{R}$
   (c) $f_k(t) = (\sin kt)/k\sqrt{t}$, $t > 0$
3. Let $D \subseteq \mathbb{R}^n$ and suppose $f: [a, \infty) \times D \to \mathbb{R}$. Define $\lim_{t \to \infty} f(t, x) = F(x)$ uniformly for $x \in D$.

4. Let $g\colon [a, b] \to \mathbb{R}$ and $\mathbf{f}, \mathbf{f}_k\colon [a, b] \to \mathbb{R}^m$. If $\mathbf{f}_k \to_u \mathbf{f}$ on $[a, b]$, does $g\mathbf{f}_k \to_p g\mathbf{f}$ on $[a, b]$? Does $g\mathbf{f} \to_u g\mathbf{f}$ on $[a, b]$? If not, what condition on g ensures uniform convergence?

5. If $\mathbf{f}_k, \mathbf{g}_k\colon E \to \mathbb{R}^n$ and $\mathbf{f}_k \to_p 0$, $\mathbf{g}_k \to_p 0$, does $\mathbf{f}_k \cdot \mathbf{g}_k \to_p 0$? What about uniform convergence?

6. Let $f_k(t) = t^k$, $t \in [0, 1]$. Calculate $\lim_{t \to 1} \lim_{k \to \infty} f_k(t)$ and $\lim_{k \to \infty} \lim_{t \to 1} f_k(t)$. What do you conclude?

7. Let $f_k(t) = k^2 t e^{-kt}$, $t \in (0, 1)$. Show that $f_k \to_p 0$ and $f_k' \to_p 0$, but not uniformly. What does this imply concerning Theorem 7?

8. Show that $\sum_{k=1}^{\infty} (\sin \sqrt{k}\, t)/k^2$ converges uniformly on $\mathbb{R}$ to a function that is differentiable on $\mathbb{R}$.

9. Show that $\sum_{k=1}^{\infty} t^k/k!$ converges uniformly on $[-r, r]\ \forall\, r > 0$. What is its limit function? Does the series converge uniformly on $\mathbb{R}$?

10. Show that $\zeta(x) = \sum_{k=1}^{\infty} k^{-x}$ converges for $x > 1$ and defines a continuous function ζ on $(1, \infty)$. [ζ is called the Riemann-zeta function.]

11. Let $\{b_k\}$ be a bounded sequence and define $F(t) = \sum_{k=1}^{\infty} b_k e^{-k^2 t}$. Show that F is a continuous and differentiable function on $(0, \infty)$. Calculate $F'(t)$.

12. Find the radius of convergence r of each of the following series and check the convergence at $\pm r$ when applicable:
    (a) Σx^k (b) $\Sigma x^k/k$ (c) $\Sigma x^k/k!$ (d) $\Sigma k! x^k$
    (e) Σx^{2k} (f) $\Sigma x^{k!}$ (g) $\Sigma \dfrac{(-1)^k}{(2k)!} x^{2k}$

13. Show that the series $\sum_{k=0}^{\infty} k^2 x^k$ converges for $|x| < 1$. Determine its sum. Hint: $k^2 = k(k-1) + k$.

14. In Theorem 18, is the condition of compactness important? Is monotonicity important? Is continuity of f important?

15. Use the difference quotient to give a direct proof of Theorem 14.

16. Find the convergence set for each of the following sequences of functions:
    (a) $\{(1 + x/n)^n\}$ (b) $\{nx/(1 + n^2 x^2)\}$
    (c) $\{n^x\}$ (d) $\{(1/x)\sin nx\}$
    (e) $\{xe^{-nx}\}$,   (f) $x^{2n}/(1 + x^{2n})$
    Find sets on which these sequences converge uniformly.

17. Examine each of the following series for uniform convergence:
    (a) $\displaystyle\sum_{k=1}^{\infty} \frac{1}{x^2 + k^2}$ (b) $\displaystyle\sum_{k=1}^{\infty} \left(\frac{x+1}{x-1}\right)^k$ (c) $\displaystyle\sum_{k=1}^{\infty} \frac{1}{(x-k)^2}$

18. Prove that the radius of convergence of $\Sigma a_k x^k$ is given by $r = \lim_{k \to \infty} |a_k/a_{k+1}|$, provided this limit exists.

19. State Theorem 16 for an arbitrary interval $|x| < \rho$.

20. Show by example that the converse of Abel's theorem is false.

21. Let $g: [0, 1] \to \mathbb{R}$ be continuous with $g(1) = 0$. Show that $\{g(x)x^n\}$ converges uniformly for $x \in [0, 1]$.

22. Let $f_k \to f$ pointwise on D and set $u_k = \sup \{f_j: j \geq k\}$. Show that $u_k \downarrow f$ pointwise on D.

23. For the double sequences below, compute $\lim_i \lim_j a_{ij}$ and $\lim_j \lim_i a_{ij}$.
    (a) $ij/(i^2 + j^2)$ (b) $i/(i + j)$
    (c) $(-1)^i i/(i + j)$ (d) $(-1)^i/j$
    (e) $(-1)^{i+j}$ (f) $(-1)^{i+j}(1/i + 1/j)$

24. Under the hypotheses of Corollary 5, show that $\lim_i a_{ii}$ exists.

25. If $f: [0, \infty) \to \mathbb{R}$ is differentiable and $f'(x) + f(x) > 0$ for $x \geq 0$, show that $f(x) > f(0)e^{-x}$ for $x \geq 0$.

26. Let $f: \mathbb{R} \to \mathbb{R}$ be uniformly continuous. For $k \in \mathbb{N}$ set $f_k(t) = f(t + 1/k)$. Show that $\{f_k\}$ converges uniformly on $\mathbb{R}$.

27. Let $\{f_k\}$ be a sequence of continuous functions that converges uniformly on $[0, 1]$. Show that there is M such that $|f_k(t)| \leq M \; \forall k \in \mathbb{N}, t \in [0, 1]$. Can "uniform convergence" be replaced by "pointwise convergence"?

28. If $\{f_k\}$ is a sequence of functions converging uniformly on $\mathbb{R}$ to a function f that is continuous on $\mathbb{R}$, show that $\lim_{k \to \infty} f_k(x + 1/k) = f(x)$ for $x \in \mathbb{R}$.

29. Does the series $\Sigma_{k=0}^{\infty} x^2/(1 + x^2)^k$ converge pointwise on $\mathbb{R}$? Does it converge uniformly on $\mathbb{R}$?

30. Let $f_k: [a, b] \to \mathbb{R}$ be continuous and such that $f_k \to f$ uniformly on (a, b). Show that $f_k \to f$ uniformly on $[a, b]$.

31. Show that $s(x) = \Sigma_{k=1}^{\infty} 1/(k^2 + k^4 x^2)$ converges uniformly on $\mathbb{R}$. Show that s is differentiable for $x \neq 0$, but $s'(0)$ does not exist.

32. Let $f: [\frac{1}{2}, 1] \to \mathbb{R}$ be continuous at $x = 1$. Show that $\{x^k f(x)\}$ converges $\forall x \in [\frac{1}{2}, 1]$ and that the convergence is uniform iff f is bounded and $f(1) = 0$.

33. Let $g: D \to \mathbb{R}$ and $a > 0$ be such that $|g(x)| \geq a$ for $x \in D$. Does the sequence $g_k(x) = kg(x)/(1 + kg(x))$ converge pointwise? Does it converge uniformly?

34. If $f_k: [a, b] \to \mathbb{R}$ is (strictly) increasing and $f_k \to f$ pointwise on $[a, b]$, is f (strictly) increasing?

35. If $\{f_k\}$ converges to f uniformly on every closed subinterval of $(0, 1)$, does it follow that $\{f_k\}$ converges uniformly to f on $(0,1)$? Support your statement with either a proof or an example.

36. (Dirichlet Test) Let $f_k, g_k: D \to \mathbb{R}$. If the partial sums of the series Σf_k are uniformly bounded on D and if $g_k \to 0$ uniformly on D with $g_k(x) \downarrow$ $\forall\, x \in D$, show that the series $\Sigma f_k g_k$ is uniformly convergent on D. *Hint:* See Exercise **4**.22.

37. (Abel's Test) Let $f_k, g_k: D \to \mathbb{R}$. If the series Σf_k converges uniformly on D and $\{g_k\}$ is uniformly bounded with $g_k(x) \downarrow \forall\, x \in D$, show that the series $\Sigma f_k g_k$ converges uniformly on D. *Hint:* Use the Dirichlet test.

38. If $\mathbf{f}_k, \mathbf{f}: D \to \mathbb{R}^m$, each $\mathbf{f}_k$ is uniformly continuous, and $\mathbf{f}_k \to \mathbf{f}$ uniformly on D, show that $\mathbf{f}$ is uniformly continuous.

39. Let $\mathbf{f}_k, \mathbf{f}: D \to \mathbb{R}^m$ be continuous and $\mathbf{f}_k \to \mathbf{f}$ uniformly on D. Suppose that $\mathbf{x}_k \to \mathbf{x}$ in D. Show that $\mathbf{f}_k(\mathbf{x}_k) \to \mathbf{f}(\mathbf{x})$.

40. Show that every derivative is the pointwise limit of a sequence of continuous functions.

41. Let $f_k \to f$ pointwise on S and $|f(t)| \le M$ for all $t \in S$. Show that the truncated sequence

$$g_k(t) = \begin{cases} M, & M < f_k(t), \\ f_k(t), & -M \le f_k(t) \le M, \\ -M, & -M < f_k(t) \end{cases}$$

converges pointwise to f.

42. Suppose that $\{P_n\}$ is a sequence of polynomials that converges uniformly on $\mathbb{R}$ to a function $f: \mathbb{R} \to \mathbb{R}$. Show that f is a polynomial.

CHAPTER 12

The Riemann
Integral Reviewed

In the calculus, one of the basic problems considered is that of defining the area under a given curve. That is, given a nonnegative function $f: [a, b] \to \mathbb{R}$, how does one first define the area of the region $R = \{(x, y): a \leq x \leq b, 0 \leq y \leq f(x)\}$ under the graph of f and then how does one devise techniques for evaluating the area? The basic idea, adopted from one used by Archimedes to find the area enclosed by a circle, is to approximate the area of the region R by simple geometric figures whose areas are easily defined and evaluated and then use a limiting technique to define the area. Perhaps the simplest approximation is by means of rectangles and yields the so-called Riemann sum.

We begin with the introduction of some useful terminology. A *partition* $\mathscr{P} = \{x_0, x_1, \ldots, x_n\}$ of a closed interval $I = [a, b]$ is a finite ordered set of points with $a = x_0 < x_1 < \cdots < x_n = b$. The points $\{x_i\}$ divide or partition the interval I into subintervals $[x_{-1}, x_i]$. The length of the largest of the subintervals induced by P is called the *mesh* of the partition. We write

$$\|\mathscr{P}\| = \max \{x_i - x_{i-1}: 1 \leq i \leq n\}.$$

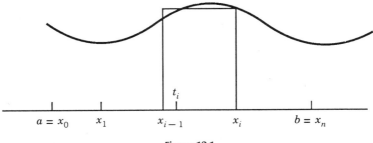

Figure 12.1

In each subinterval $[x_{i-1}, x_i]$, choose a point t_i. The rectangle with base $[x_{i-1}, x_i]$ and height $f(t_i)$ is used to approximate the area under the curve $y = f(x)$ over the interval $[x_{i-1}, x_i]$ (Figure 12.1). The area of the region R is then approximated by the *Riemann sum*

$$S(f, \mathscr{P}, \{t_i\}) = \sum_{i=1}^{n} f(t_i)(x_i - x_{i-1}).$$

Intuitively, if the interval is subdivided more finely so that the subintervals in the partition have shorter lengths, then the resulting Riemann sum should give a better approximation to what we think the area of the region R ought to be. This suggests that we should examine the behavior of the Riemann sums as the mesh of the partition approaches zero. Since the Riemann sums depend on both the partition $\mathscr{P}$ and the choice of the $\{t_i\}$, it's clear that this is not a straightforward limit in the sense of Definition **6**.1. One of the ways of dealing with this limiting process leads to the classical Riemann integral (Definition 1).

Definition 1. The function $f: [a, b] \to \mathbb{R}$ is *Riemann integrable* over $[a, b]$ if there is a number $A \in \mathbb{R}$ with the property that for each $\varepsilon > 0$, there is $\delta > 0$ such that if $\mathscr{P} = \{a = x_0 < x_1 < \cdots < x_n = b\}$ is any partition of $[a, b]$ with mesh $\|\mathscr{P}\| < \delta$ and if $t_i \in [x_{i-1}, x_i]$ $(1 \le i \le n)$, then

$$|S(f, \mathscr{P}, \{t_i\}) - A| < \varepsilon.$$

The number A is unique (Exercise 1) and is called the *Riemann integral* of f over $[a, b]$. We write $A = \int_I f = \int_a^b f$.

This definition is essentially that originally given by the German mathematician, Bernhard Riemann (1826–1866). The integral defined above is usually first encountered in the elementary calculus; it enjoys a large number of useful properties and has many applications. Not long after its introduction, it was noted that this integral has some serious shortcomings and this motivated a search for a more general theory of integration. Among the many notions of integral that evolved from this search, the most prominent was that of the Frenchman Henri Lebesgue (1875–1941). The Lebesgue integral is much more general that the Riemann integral, has many more desirable properties, and over the years has become a powerful tool in the mathematical analysis and applications to scientific problems. While the advantages of the Lebesgue integral over that of Riemann are generally acknowledged, the increased complexity of the traditional treatments of the Lebesgue integral has often resulted in the exclusive use of the Riemann integral (or its close relative, the Riemann–Stieltjes integral) in elementary mathematical analysis texts. In Chapter 13, we will present an integral, called the gauge integral, which is very much in the spirit of the Riemann integral defined above and which not only gives a generalization of the Riemann integral but actually generalizes the classical Lebesgue integral. Since the gauge integral is a natural and straightforward generalization of the classical Riemann integral, we will not develop any of the basic properties of the Riemann integral at this point; for the most part, these properties will be special cases of results to be established for the gauge integral.

Before proceeding to the gauge integral, it is desirable to justify its introduction by considering some of the deficiencies of the Riemann integral. This will help to later develop an appreciation of the advantages of the gauge integral. We begin with an example:

EXAMPLE 2. Let $f: [0,1] \to \mathbb{R}$ be defined by $f(t) = 1$ if t is irrational and $f(t) = 0$ if t is rational. Then f is not Riemann integrable on $[0,1]$, for if $\mathscr{P} = \{0 = x_0 < x_1 < \cdots < x_n = 1\}$ is any partition of $[0,1]$, then pick $t_i \in [x_{i-1}, x_i]$ so that $t_i \in \mathbb{Q}$, and $s_i \in [x_{i-1}, x_i]$ so that $s_i \notin \mathbb{Q}$. Then $-S(f, \mathscr{P}, \{t_i\}) + S(f, \mathscr{P}, \{s_i\}) = 1$, and it follows that the approximation condition in Definition 1 cannot be satisfied by any real number A.

It is left to Exercise 2 to establish the following proposition.

Proposition 3.

  (i) If $f(x) = c$, a constant, on $[a, b]$, then f is Riemann integrable on $[a, b,]$ and $\int_a^b f = c(b - a)$.

  (ii) The conclusion of (i) also holds if f is constant on $[a, b]$ except for at most a finite number of points in $[a, b]$.

Using Example 2 and Proposition 3, we can observe one of the defects of the Riemann integral.

EXAMPLE 4. Let $\{r_i\}$ be an enumeration of the rational numbers in $[0, 1]$. Define $f_n\colon [0, 1] \to \mathbb{R}$ by $f_n(t) = 1$ if $t \neq r_1, \ldots, r_n$, and $f_n(r_i) = 0$ if $1 \leq i \leq n$. Then each f_n is Riemann integrable and $\int_a^b f = 1$ (Exercise 2). Note that the sequence $\{f_n\}$ converges pointwise to the function f defined in Example 2 and f is not Riemann integrable. Thus, we have the existence of a sequence of uniformly bounded Riemann integrable functions that converges pointwise to a function which is not Riemann integrable. We will discover that this situation does not occur with the gauge integral.   □

Another problem with the Riemann integral concerns the fundamental theorem of calculus, which relates the basic concepts of derivative and integral. This theorem, discovered independently by the Englishman, Isaac Newton (1642–1727) and the German, Gottfried Wilhelm von Leibniz (1646–1716), is the principal reason for the success of the calculus. Essentially, the fundamental theorem states that differentiation and integration are inverse operations of each other; for the Riemann integral, it can be stated as follows.

Theorem 5. Let $f\colon [a, b] \to \mathbb{R}$ be differentiable on $[a, b]$. If f' is Riemann integrable on $[a, b]$, then $\int_a^b f' = f(b) - f(a)$.

The primary shortcoming in this theorem is the need for the assumption that the derivative f' is Riemann integrable. Indeed, there are examples of functions f with (bounded) derivatives f' that are not Riemann integrable. There are also examples which show that the Lebesgue integral exhibits similar behavior. We'll see that this is not a problem with the gauge integral, that is, any derivative f' is gauge integrable and its gauge integral is $\int_a^b f' = f(b) - f(a)$.

A third problem with the class of Riemann integrable functions is that it exhibits behavior analogous to that of the rational numbers as an incomplete ordered field. This will be discussed later in greater depth in the context of metric spaces.

A version of the Riemann integral, given by the French mathematician Gaston Darboux (1842–1917), is often found in elementary texts. We review it briefly before proceeding to the gauge integral.

Let $f\colon [a, b] \to \mathbb{R}$ be bounded and $\mathscr{P} = \{a = x_0 < x_1 < \cdots < x_n = b\}$ be a partition of $[a, b]$. Again for the case when f is nonnegative, we consider the problem of defining the area of the region R under the curve $y = f(x)$. For each subinterval $[x_{i-1}, x_i]$,

$$\inf \{ f(t)\colon x_{i-1} \leq t \leq x_i \}(x_i - x_{i-1}),$$

$$\left(\sup \{ f(t)\colon x_{i-1} \leq t \leq x_i \}(x_i - x_{i-1})\right),$$

will give a lower (upper) approximation to the area of the region under the

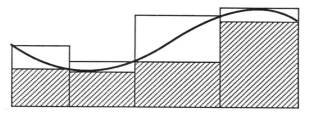

Figure 12.2

curve $y = f(x)$ and over the interval $[x_{i-1}, x_i]$ and the sum

$$L(f, \mathscr{P}) = \sum_{i=1}^{n} \inf \{ f(t): x_{i-1} \le t \le x_i \}(x_i - x_{i-1}),$$

$$\left(U(f, \mathscr{P}) = \sum_{i=1}^{n} \sup \{ f(t): x_{i-1} \le t \le x_i \}(x_i - x_{i-1}) \right),$$

will give a lower (upper) approximation to the area of the region R. $L(f, \mathscr{P})(U(f, \mathscr{P}))$ is called a *lower sum* (*upper sum*) relative to f and $\mathscr{P}$; these quantities are finite since f is bounded. The sketch in Figure 12.2 indicates $L(f, \mathscr{P})$, the area of the shaded region, and U(f, $\mathscr{P}$), the area of the union of the shaded and the unshaded rectangles.

The number

$$\underline{\int_a^b} f = \sup \{ L(f, \mathscr{P}): \mathscr{P} \text{ is a partition of } [a, b] \},$$

$$\left(\overline{\int_a^b} f = \inf \{ U(f, \mathscr{P}): \mathscr{P} \text{ is a partition of } [a, b] \} \right)$$

should then be a lower(upper) estimate of what the area of the region R ought to be. The region R is said to have *area* if $\underline{\int_a^b} f = \overline{\int_a^b} f$ and, in this case, the *area* is defined to be this common value; it is denoted by $\int_a^b f$.

For an arbitrary (not necessarily nonnegative) bounded function, we define $L(f, \mathscr{P})$, $U(f, \mathscr{P})$, $\underline{\int_a^b} f$, $\overline{\int_a^b} f$ as above. We then say that the function f is *Riemann integrable* on $[a, b]$ if $\underline{\int_a^b} f = \overline{\int_a^b} f$, and define the integral, $\int_a^b f$, of f to be this common value. The Darboux and Riemann definitions of the integral are equivalent; a proof of this equivalence is given in M. H. Protter and C. B. Morrey, 1977.

EXERCISES 12

1. Show that the number A in Definition 1 is unique.

2. Prove Proposition 3.

3. Compute $\underline{\int_a^b} f$ and $\overline{\int_a^b} f$ for the function in Example 2.

4. Repeat Exercise 2 using the Darboux definition of the Riemann integral.

5. Show that if f is integrable in the sense of Definition 1, then f must be bounded. Thus, a necessary condition for a function to be Riemann integrable by either Definition 1 or the Darboux definition is that the function is bounded.

CHAPTER 13

The Gauge Integral

We begin by introducing some terminology that will help to describe the essential ingredients involved in defining the integral. If $I = [a, b]$ is an interval in $\mathbb{R}$, then a *division* of I is a finite set of closed subintervals $\{J_1, \ldots, J_n\}$ of I such that $J_i^0 \cap J_j^0 = \emptyset$ for $i \neq j$ and $\bigcup_{i=1}^n J_i = I$. (J_i^0 denotes the ith subinterval with endpoints deleted.) A *tagged division* of I is a finite set $\mathcal{D} = \{(z_i, J_i): 1 \leq i \leq n\}$ of ordered pairs such that $\{J_i: 1 \leq i \leq n\}$ is a division of I and $z_i \in J_i$ for $i = 1, \ldots, n$. The elements J_i of $\mathcal{D}$ are called *subintervals* of $\mathcal{D}$ and each z_i is called the *tag* of J_i.

In a division of I, it's not necessary that J_i and J_{i+1} be contiguous. In the event that the $\{J_i\}$ are arranged so that $\max J_{i-1} = \min J_i$, and $J_i = [x_{i-1}, x_i]$, then $\{a = x_0 < x_1 < \cdots < x_n = b\}$ is a partition of I with $x_{i-1} \leq z_i \leq x_i$. Conversely, if

$$\mathcal{P} = \{x_0 < x_1 < \cdots < x_n\}$$

is a partition of I and z_i is chosen in $[x_{i-1}, x_i]$, then

$$\mathcal{D} = \{(z_i, [x_{i-1}, x_i]): 1 \leq i \leq n\}$$

is a tagged division of I.

Definition 1. If $f: [a, b] \to \mathbb{R}$ and

$$\mathscr{D} = \{(z_i, J_i): 1 \leq i \leq n\}$$

is a tagged division of $[a, b]$, then

$$S(f, \mathscr{D}) = \sum_{i=1}^{n} f(z_i)\ell(J_i),$$

where $\ell(J)$ is the length of an interval J, is called the *Riemann sum* of f with respect to $\mathscr{D}$.

Using the above terminology, we can restate the definition of the Riemann integral given in Chapter 12 as follows.

Definition 2. A function $f: [a, b] \to \mathbb{R}$ is *Riemann integrable* on $[a, b]$ if there is $A \in \mathbb{R}$ with the property that for each $\varepsilon > 0$, there is $\delta > 0$ such that if

$$\mathscr{D} = \{(z_i, [x_{i-1}, x_i]): 1 \leq i \leq n\}$$

is any tagged division of $[a, b]$ satisfying

$$[x_{i-1}, x_i] \subset (z_i - \delta, z_i + \delta), \qquad \text{for } i = 1, \ldots, n, \tag{1}$$

then

$$|S(f, \mathscr{D}) - A| < \varepsilon. \tag{2}$$

Note that (1) guarantees that any tagged division $\mathscr{D} = \{(z_i, J_i): 1 \leq i \leq n\}$ that satisfies this condition is such that the associated partition of $[a, b]$ induced by the J_i as discussed above has mesh smaller than 2δ. Conversely, if $\mathscr{P} = \{x_0 < x_1 < \cdots < x_n\}$ is a partition of $[a, b]$ with mesh smaller than δ and if $z_i \in [x_{i-1}, x_i]$, then the tagged division $\mathscr{D} = \{(z_i, [x_{i-1}, x_i]): 1 \leq i \leq n\}$ satisfies (1). Thus, our earlier definition of Riemann integrability of a function and Definition 2 are equivalent.

An important characteristic of the Riemann integral is that the division of the interval $[a, b]$ is independent of the function f and therefore fails to take into account any special properties that might be possessed by the function. This leads to a uniform and relatively simple development of the calculus of the integral; however, by failing to take into account any singular properties of the functions involved, the integral is restricted to a smaller class of functions than is desirable.

For example, to approximate the area between the curve

$$y = 1 + \sin\frac{1}{x}$$

and the x-axis on $(0, 10]$, it seems natural to use a partition that is fine near 0 where the function oscillates wildly and more coarse near 10 where the function is "better behaved." If a point x is the tag for a subinterval J in some tagged division $\mathcal{D}$ and the function behaves badly near x, then the interval J should be small enough so that the contribution of $f(x)\ell(J)$ to the sum $S(f, \mathcal{D})$ is not disproportionate. On the other hand, if the function is relatively "flat" near x, it would not be necessary for J to be small in order to obtain a good approximation to that part of the area under the curve. Similar remarks would apply to a non-Riemann integrable function such as $f(t) = 1/\sqrt{t}$, $0 < t \leq 10$, which has a "singularity" at $t = 0$. The reader should make sketches of these two examples to better understand the motivation for the varying lengths of the subintervals in a division.

This "variable length" partitioning of an interval can be achieved by slightly altering the definition given above for the Riemann integral. For the intervals defined in (1), we adopt the notation $\gamma(t) = (t - \delta, t + \delta)$. Condition (1) can then be written as

$$[x_{i-1}, x_i] \subseteq \gamma(z_i). \tag{1'}$$

Now, instead of taking the intervals $\gamma(t)$ of constant length 2δ, we allow the lengths to depend on $t \in [a, b]$, that is, $\gamma(t) = (t - \delta(t), t + \delta(t))$, where δ is now a function of t rather than a constant. For such intervals, condition (1') would imply that the lengths of the subintervals in a tagged division $\mathcal{D}$ can vary with the location of z_i in $[a, b]$.

We formalize these ideas with the following definition.

Definition 3. A *gauge* on a subset E of $\mathbb{R}$ is a function γ that associates with each point $t \in E$ an open interval $\gamma(t)$ that contains t.

If $\delta > 0$, then $\gamma(t) = (t - \delta, t + \delta)$ defines a gauge of constant length on E. If δ is any positive valued function $\delta \colon E \to \mathbb{R}^+$, then δ induces a gauge γ on E by

$$\gamma(t) = (t - \delta(t), t + \delta(t)).$$

Definition 4. If γ is a gauge on $I = [a, b]$ and

$$\mathcal{D} = \{(z_i, J_i) \colon 1 \leq i \leq n\}$$

is a tagged division of I, then $\mathcal{D}$ is said to be γ-*fine* if $J_i \subset \gamma(z_i)$ for $1 \leq i \leq n$. $\mathcal{D}$ is also said to be *compatible* with γ.

Theorem 5. Let γ be a gauge on $I = [a, b]$. Then there exists a γ-fine tagged division of I.

Proof. Define a subset E of I by $E = \{t \in (a, b]: \exists$ a tagged division of $[a, t]$ which is γ-fine$\}$. First, $E \neq \varnothing$, for we can choose $x \in \gamma(a)$ such that $a < x < b$. Then $\{(a, [a, x])\}$ is a γ-fine tagged division of $[a, x]$.

Let $y = \sup E$. We claim that $y \in E$. Choose $x \in E$ such that $x \in \gamma(y)$ and $x < y$. There is a γ-fine tagged division $\mathscr{D}$ of $[a, x]$. The set $\mathscr{D} \cup \{(y, [x, y])\}$ is a γ-fine tagged division of $[a, y]$, that is, $y \in E$.

Next, we wish to show that $y = b$. Suppose $y < b$, and choose $w \in \gamma(y) \cap (y, b)$. Let $\mathscr{D}$ be a γ-fine tagged division of $[a, y]$. Then $\mathscr{D}' = \mathscr{D} \cup \{(y, [y, w])\}$ is a γ-fine tagged division of $[a, w]$. This contradicts the definition of y, and so, $y = b$. $\qquad\square$

The tools necessary for the definition of the gauge integral are now available.

Definition 6. The function $f: [a, b] \to \mathbb{R}$ is *gauge integrable* on $I = [a, b]$ if there is $A \in \mathbb{R}$ with the property that for each $\varepsilon > 0$, there is a gauge γ on I such that for every γ-fine tagged division $\mathscr{D}$ of I, $|S(f, \mathscr{D}) - A| < \varepsilon$.

It is shown below that the number A, if it exists, is unique; it is called the *gauge integral* of f over I, and is denoted by $\int_I f$, $\int_a^b f$, or $\int_a^b f(t)\, dt$. Other names that have been used for the gauge integral are generalized Riemann integral, and Riemann-complete integral. Since we will deal almost exclusively with the gauge integral, we shall drop the adjective "gauge" and speak simply of the integral and functions being integrable. In order to distinguish between the gauge and the Riemann integrals, we will always use the terms "Riemann integral" or "Riemann integrable" for the latter and write $R\int_a^b f$ for the Riemann integral of a function f over $[a, b]$.

From Definitions 2 and 4 it follows that a function is Riemann integrable if, and only if, it is gauge integrable with respect to gauges of constant length. In particular, every Riemann integrable function is gauge integrable and the two integrals are the same. Note also that it follows from Theorem 5 that the definition of the gauge integral is meaningful, since every gauge on an interval I has at least one tagged division which is γ-fine. That is, we'll never have the case in which there is a gauge γ for which there is no γ-fine tagged division.

We next establish the uniqueness of the gauge integral.

Theorem 7. A function $f: [a, b] \to \mathbb{R}$ can have at most one integral.

Proof. Let $\varepsilon > 0$. Suppose A_1 and A_2 satisfy the conditions of Definition 6, with respect to the gauges γ_1 and γ_2, respectively. Let $\gamma(t) = \gamma_1(t) \cap \gamma_2(t)$. If $\mathscr{D}$ is any tagged division which is γ-fine, then $|A_1 - A_2| \leq |-S(f, \mathscr{D}) + A_1| + |S(f, \mathscr{D}) - A_2| < 2\varepsilon$, by Exercise 4. Thus $|A_1 - A_2|$, which is a fixed nonnegative number, is smaller than every positive number. The only possibility is that $A_1 - A_2 = 0$, and so, $A_1 = A_2$. $\qquad\square$

At this point, it's not clear that we have done anything beyond the introduction of a more complicated version of the Riemann integral. It's not obvious that there are functions that are gauge integrable, but not Riemann integrable. The following example shows that such functions exist.

EXAMPLE 8. Let $f: [a, b] \to \mathbb{R}$ have the constant value c except possibly at a countable number of points $C = \{c_i: i \in \mathbb{N}\}$. Then f is integrable over $[a, b]$ with integral $\int_a^b f = c(b - a)$.
    Let $\varepsilon > 0$ be given. If $\mathcal{D} = \{(z_j, J_j): 1 \leq j \leq n\}$ is a tagged division of $[a, b]$, consider

$$\left| \sum_{j=1}^n f(z_j)\ell(J_j) - c(b - a) \right| = \left| \sum_{j=1}^n [f(z_j) - c]\ell(J_j) \right|. \tag{3}$$

If $z_j \notin C$, the term $[f(z_j) - c]\ell(J_j)$ in the right-hand side of (3) is 0, so we may set $\gamma(z) = (z - 1, z + 1)$ whenever $z \notin C$. If $z_j = c_k$ for some k, and if $\mathcal{D}$ is γ-fine with respect to some gauge, then the term $|[f(z_j) - c]\ell(J_j)|$ is not greater than $|[f(z_j) - c]|\ell\gamma(c_k)$. If we choose $\delta_k = \varepsilon/|f(c_k) - c|2^{k+2}$ and set $\gamma(c_k) = (c_k - \delta_k, c_k + \delta_k)$, then we have defined a gauge γ on $[a, b]$ such that

$$\left|[f(z_j) - c]\right|\ell(J_j) < \varepsilon/2^{k+1}$$

whenever $z_j = c_k$ and $\mathcal{D}$ is γ-fine. In particular, if $\mathcal{D}$ is γ-fine, we have from (3) that

$$\left| \sum_{j=1}^n f(z_j)\ell(J_j) - c(b - a) \right| \leq 2 \sum_{k=1}^\infty \varepsilon/2^{k+1} = \varepsilon,$$

since each c_k is the tag for at most two subintervals in D. This establishes the integrability and the formula above.   □

Example 8 is applicable to the function f defined by

$$f(t) = \begin{cases} 0 & \text{if } t \in [0, 1] \text{ and rational,} \\ 1 & \text{if } t \in [0, 1] \text{ and irrational,} \end{cases}$$

and so provides an example of a function that is gauge integrable, but not Riemann integrable. In contrast to the Riemann integral, the gauge γ in Example 8 is not a gauge of constant length.

The Fundamental Theorem of Calculus

The most beautiful result in elementary calculus is the Fundamental Theorem, which relates the basic concepts of differentiation and integration and shows that they are essentially inverse operations of each other. To give partial consideration of this in the context of the gauge integral we need the following lemma.

Lemma 9 (Straddle Lemma). Let $F: [a, b] \to \mathbb{R}$ be differentiable at $z \in [a, b]$. Then for each $\varepsilon > 0$, there is $\delta > 0$ such that

$$|F(v) - F(u) - F'(z)(v - u)| \le \varepsilon(v - u)$$

whenever $u \le z \le v$ and $[u, v] \subseteq [a, b] \cap (z - \delta, z + \delta)$.

Proof. Since F is differentiable at z, there is $\delta > 0$ such that

$$\left| \frac{F(x) - F(z)}{x - z} - F'(z) \right| < \varepsilon$$

for $0 < |x - z| < \delta$, $x \in [a, b]$. If $z = u$ or $z = v$, the conclusion of the lemma is immediate; so suppose $u < z < v$. Then

$$|F(v) - F(u) - F'(z)(v - u)|$$
$$\le |F(v) - F(z) - F'(z)(v - z)| + |F(z) - F(u) - F'(z)(z - u)|$$
$$< \varepsilon(v - z) + \varepsilon(z - u) = \varepsilon(v - u). \qquad \square$$

The geometrical interpretation of Lemma 9 can be seen in Figure 13.1. If u and v "straddle" z, then for u and v near z, the slope of the chord between $(u, F(u))$ and $(v, F(v))$ is close to the slope of the tangent line at $(z, F(z))$. The conclusion of the lemma is false if u and v do not straddle z. Visualize a curve, such as $f(t) = t \sin(1/t)$ near $t = 0$, which oscillates near $z = 0$ and has arbitrarily large slopes in every neighborhood of z.

The Fundamental Theorem of calculus can be separated into two parts: the integration of a derivative, and the differentiation of an integral. The first part is considered below in Theorem 10. The second part will be dealt with later in Theorem 28.

Theorem 10 (The Fundamental Theorem of Calculus—Part 1). If $F: [a, b] \to \mathbb{R}$ is differentiable on $[a, b]$, then F' is integrable on $[a, b]$ with

$$\int_a^b F' = F(b) - F(a).$$

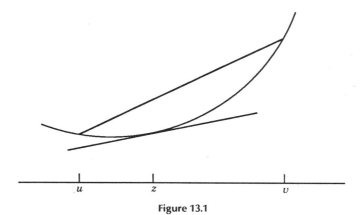

Figure 13.1

Proof. Let $\varepsilon > 0$ be given. For $z \in [a, b]$ let $\delta_z > 0$ be the δ given by the straddle lemma and define a gauge on $[a, b]$ by $\gamma(z) = (z - \delta_z, z + \delta_z)$. Suppose $\mathscr{D} = \{(z_i, J_i): 1 \le i \le n\}$ is a γ-fine tagged division of $[a, b]$. We arrange the subintervals of $\mathscr{D}$ so that $\max J_{i-1} = \min J_i$ and $J_i = [x_{i-1}, x_i]$, $i = 1, \ldots, n$. Note that $F(b) - F(a) = \sum_{i=1}^{n}[F(x_i) - F(x_{i-1})]$. Hence

$$|S(F', \mathscr{D}) - [F(b) - F(a)]|$$

$$= \left| \sum_{i=1}^{n} \{F'(z_i)(x_i - x_{i-1}) - [F(x_i) - F(x_{i-1})]\} \right|$$

$$< \sum_{i=1}^{n} \varepsilon(x_i - x_{i-1}) = \varepsilon(b - a),$$

by the straddle lemma.   □

Note that the gauge of Theorem 10 will generally vary with the behavior of F' near z and so is not of constant length.

Recall that the version of the fundamental theorem for the Riemann integral requires the additional hypothesis that the derivative F' be Riemann integrable. This is not necessary for the gauge integral, for all derivatives are gauge integrable. The reader familiar with the Lebesgue integral should note that it has this same "defect." Consider

EXAMPLE 11. Let

$$F(t) = \begin{cases} t^2 \cos\left(\pi/t^2\right), & \text{if } 0 < t \le 1, \\ 0 & \text{if } t = 0. \end{cases}$$

Then

$$F'(t) = \begin{cases} 2t\cos(\pi/t^2) + (2\pi/t)\sin(\pi/t^2), & \text{if } 0 < t \le 1, \\ 0 & \text{if } t = 0. \end{cases}$$

F' is integrable on $[0, 1]$ and

$$\int_0^1 F' = F(1) - F(0) = -1.$$

However, F' is not Riemann integrable because it is not bounded on $[0, 1]$. In fact, F' is also not Lebesgue integrable. There are functions F with bounded derivatives F' that are not Riemann integrable, but these are much more difficult to construct.   □

It follows from Theorem 10 that the rules developed in elementary calculus for calculating integrals using antiderivatives hold for the gauge integral. One of the common methods used to evaluate integrals in elementary calculus is integration by substitution. A version of this for the gauge integral follows easily from Theorem 10.

$\int_b^a f$ and $\int_a^b f$ are calculated using opposite orderings of the interval $[a, b]$ and so should be defined, when one of them exists, as negatives of each other, that is, we define $\int_b^a f = -\int_a^b f$; we also define $\int_a^a f = 0$.

Theorem 12 (Integration by Substitution). Let $f: [a, b] \to \mathbb{R}$ and $\varphi: [\alpha, \beta] \to [a, b]$ both be differentiable functions. Then $(f' \circ \varphi)\varphi'$ is integrable on $[\alpha, \beta]$ and f' is integrable on the interval $[\min\{\varphi(\alpha), \varphi(\beta)\}, \max\{\varphi(\alpha), \varphi(\beta)\}]$ with

$$\int_{\varphi(\alpha)}^{\varphi(\beta)} f' = \int_\alpha^\beta (f' \circ \varphi)\varphi'.$$

Proof. By the chain rule, $(f \circ \varphi)' = (f' \circ \varphi)\varphi'$. Theorem 10 implies

$$\int_\alpha^\beta (f' \circ \varphi)\varphi' = f \circ \varphi(\beta) - f \circ \varphi(\alpha) = \int_{\varphi(\alpha)}^{\varphi(\beta)} f'.$$ □

Basic Properties of the Integral

Among the properties of the integral $\int_I f$, there are basically two quantities that determine the value of the integral; the function f, called the *integrand*, and the interval I over which the integration is performed. We'll work first with f and consider the integral as a function of the integrand.

Proposition 13 (Linearity). Let $f, g \colon I = [a, b] \to \mathbb{R}$ be integrable functions over I. Then

    (i) $f + g$ is integrable over I and $\int_I (f + g) = \int_I f + \int_I g$.

    (ii) for $c \in \mathbb{R}$, cf is integrable over I and $\int_I cf = c\int_I f$.

Proof. Let $\varepsilon > 0$. For $i = 1, 2$, there are gauges γ_i on I such that if $\mathscr{D}_i$ is a γ_i-fine tagged division of I, then

$$\left| S(f, \mathscr{D}_1) - \int_I f \right| < \frac{\varepsilon}{2} \quad \text{and} \quad \left| S(g, \mathscr{D}_2) - \int_I g \right| < \frac{\varepsilon}{2}.$$

Let $\gamma(z) = \gamma_1(z) \cap \gamma_2(z)$. If $\mathscr{D}$ is a γ-fine tagged division of I, then

$$\left| S(f + g, \mathscr{D}) - \left(\int_I f + \int_I g \right) \right| = \left| S(f, \mathscr{D}) - \int_I f + S(g, \mathscr{D}) - \int_I g \right| < \varepsilon,$$

by Exercise 4. Part (ii) is left to Exercise 8.   □

Proposition 14 (Positivity). If $f \colon [a, b] \to \mathbb{R}$ is integrable over $I = [a, b]$ and $f(t) \geq 0$ for $t \in I$, then $\int_I f \geq 0$.

Proof. For $\varepsilon > 0$, there is a gauge γ on I such that

$$\left| S(f, \mathscr{D}) - \int_I f \right| < \varepsilon$$

for every γ-fine division $\mathscr{D}$ of I. Since $f \geq 0$,

$$0 \leq S(f, \mathscr{D}) \leq \int_I f + \varepsilon \quad \text{for every } \varepsilon > 0.$$

Hence $0 \leq \int_I f$.   □

Corollary 15. If $f, g \colon I \to \mathbb{R}$ are both integrable over I and $f(t) \leq g(t)$ for all $t \in I$, then $\int_I f \leq \int_I g$.

Proof. Propositions 13 and 14 imply that $\int_I g - \int_I f = \int_I (g - f) \geq 0$. The result follows.   □

A function f is called *absolutely integrable* over I if both f and $|f|$ are integrable over I.

Corollary 16. If $f \colon I \to \mathbb{R}$ is absolutely integrable over I, then

$$\left| \int_I f \right| \leq \int_I |f|.$$

Proof. Since $-f \le |f|$, Corollary 15 implies that $\int_I(-f) = -\int_I f \le \int_I |f|$. It follows that $|\int_I f| \le \int_I |f|$. □

For both the Riemann and Lebesgue integrals, the integrability of f implies that f is absolutely integrable. This is not the case for the gauge integral over I (see Example 32).

An interesting application of some of the above results yields the well-known integration by parts formula for the gauge integral.

Proposition 17. Let $f, g: I \to \mathbb{R}$ be differentiable on I. Then $f'g$ is integrable over I if and only if fg' is integrable over I and in this case, $\int_a^b f'g = f(b)g(b) - f(a)g(a) - \int_a^b fg'$.

Proof. By the product rule for differentiation, $(fg)' = fg' + f'g$. The result now follows from Proposition 13 and Theorem 10. □

The Integral as a Set Function.

We next direct our attention to the interval of integration and consider the integral $\int_I f$ as a function of the set I.

Theorem 18. Let $f: [a, b] = I \to \mathbb{R}$ and $\{a = x_0 < x_1 < \cdots < x_n = b\}$ be a partition of I. If f is integrable on $I_i = [x_{i-1}, x_i]$ for $i = 1, \ldots, n$, then f is integrable on I and

$$\int_I f = \sum_{i=1}^n \int_{I_i} f.$$

Proof. We give the proof for $n = 2$. Let $\varepsilon > 0$ and $x_1 = c$ where $a < c < b$. There are gauges γ_1 and γ_2 on $[a, c]$ and $[c, b]$, respectively, such that if $\mathscr{D}_1(\mathscr{D}_2)$ is a γ_1-fine (γ_2-fine) tagged division of $[a, c]([c, b])$ then

$$\left| S(f, \mathscr{D}_1) - \int_a^c f \right| < \varepsilon/2 \left(\left| S(f, \mathscr{D}_2) - \int_c^b f \right| < \varepsilon/2 \right).$$

Define a gauge γ on $[a, b]$ by

$$\begin{aligned}
\gamma(z) &= (a, c) \cap \gamma_1(z) & &\text{if } z \in (a, c), \\
&= (c, b) \cap \gamma_2(z) & &\text{if } z \in (c, b), \\
&= \gamma_1(c) \cap \gamma_2(c) & &\text{if } z = c, \\
&= \gamma_1(a) \cap (-\infty, c) & &\text{if } z = a, \\
&= \gamma_2(b) \cap (c, \infty) & &\text{if } z = b.
\end{aligned}$$

If $\mathcal{D}$ is a γ-fine tagged division of $[a, b]$, then $\mathcal{D}$ contains either one subinterval J with c as a tag or $\mathcal{D}$ contains two subintervals with c as a tag. (Note that c must be a tag.) In the former case, we can "divide" J at c without changing the sum $S(f, \mathcal{D})$ and, therefore, obtain the latter case. In the latter case

$$\mathcal{D}_1 = \{(z, J) \in \mathcal{D}: J \subset [a, c]\} \quad \text{and} \quad \mathcal{D}_2 = \{(z, J) \in \mathcal{D}: J \subset [c, b]\}$$

are tagged divisions of $[a, c]$ and $[c, b]$, respectively, where $\mathcal{D}_i$ is γ_i-fine. Then

$$\left| S(f, \mathcal{D}) - \left(\int_a^c f + \int_c^b f \right) \right| = \left| S(f, \mathcal{D}_1) + S(f, \mathcal{D}_2) - \left(\int_a^c f + \int_c^b f \right) \right| < \varepsilon.$$

The general case for $n \in \mathbb{N}$ can be established by mathematical induction. The details are left to the reader. □

Remark 19. It follows from Theorem 18 and Example 8 that if J is a subinterval of $[a, b]$, then the characteristic function f of J is integrable with $\int_a^b f = \ell(J)$. Thus, if g is a step function, that is,

$$g = \sum_{i=1}^{n} a_i C_{I_i}$$

where $a_i \in \mathbb{R}$, I_i is a bounded interval, and C_{I_i} is the characteristic function of I_i, then g is integrable over I with

$$\int_I g = \sum_{i=1}^{n} a_i \ell(I_i \cap I).$$

To establish the converse of Theorem 18, we need the following Cauchy criterion for the integral. This condition has the same advantage as the Cauchy criterion for sequences, that is, it relieves us of the necessity of having in hand a value for the integral.

Theorem 20 (Cauchy Criterion). Let $f: I \to \mathbb{R}$. Then f is integrable over I if, and only if, for every $\varepsilon > 0$, there is a gauge γ on I such that if $\mathcal{D}_1, \mathcal{D}_2$ are γ-fine tagged divisions of I, then

$$|S(f, \mathcal{D}_1) - S(f, \mathcal{D}_2)| < \varepsilon.$$

Proof. By now the reader can surely supply the proof that the condition is necessary. To prove that it is also sufficient, note that for each $n \in \mathbb{N}$, there is a gauge γ_n on I such that if $\mathcal{D}_1, \mathcal{D}_2$ are γ_n fine tagged divisions of I, then $|S(f, \mathcal{D}_1) - S(f, \mathcal{D}_2)| < 1/n$. We may arrange that $\gamma_1(z) \supset \gamma_2(z) \supset \cdots \supset \gamma_n(z)$ for all z and n.

For each n, let $\mathcal{D}_n$ be a γ_n-fine tagged division of I. Then for $m > n$, $|S(f, \mathcal{D}_n) - S(f, \mathcal{D}_m)| < 1/n$. Thus, $\{S(f, \mathcal{D}_n)\}$ is a Cauchy sequence. Let $A = \lim S(f, \mathcal{D}_n)$. Then $|S(f, \mathcal{D}_n) - A| \le 1/n$ for each n.

Let $\varepsilon > 0$ be given and pick n_0 such that $1/n_0 < \varepsilon/2$. Suppose $\mathcal{D}$ is a γ_{n_0}-fine tagged division. Then

$$|A - S(f, \mathcal{D})| < |A - S(f, \mathcal{D}_{n_0})|$$

$$+ |S(f, \mathcal{D}_{n_0}) - S(f, \mathcal{D})| < \frac{1}{n_0} + \frac{1}{n_0} < \varepsilon. \qquad \square$$

Using Theorem 20, we can now establish the converse of Theorem 18.

Theorem 21. Let $f: I \to \mathbb{R}$ be integrable over I. If J is a closed subinterval of I, then f is integrable over J.

Proof. Since we have no obvious candidate for the value of $\int_J f$, it is natural to use the Cauchy criterion. For $\varepsilon > 0$, there is a gauge γ on I such that if $\mathcal{D}_1, \mathcal{D}_2$ are γ-fine tagged divisions of I, then $|S(f, \mathcal{D}_1) - S(f, \mathcal{D}_2)| < \varepsilon$.

Consider the case where $a < c < d < b$ and $J = [c, d]$; the other cases are similar. Let $\bar{\gamma}$ be the restriction of γ to J. Suppose $\mathcal{D}$ and $\mathcal{E}$ are $\bar{\gamma}$-fine tagged divisions of J. Let $\gamma_1(\gamma_2)$ be the restriction of γ to $[a, c]([d, b])$ and let $\mathcal{D}_1(\mathcal{D}_2)$ be a γ_1-fine (γ_2-fine) tagged division of $[a, c]([d, b])$. Then $\mathcal{D}' = \mathcal{D} \cup \mathcal{D}_1 \cup \mathcal{D}_2$, $\mathcal{E}' = \mathcal{E} \cup \mathcal{D}_1 \cup \mathcal{D}_2$ are γ-fine tagged divisions of I; thus $|S(f, \mathcal{D}') - S(f, \mathcal{E}')| = |S(f, \mathcal{D}) - S(f, \mathcal{E})| < \varepsilon$. Hence, f is integrable over J by Theorem 20. $\qquad \square$

It follows from Theorems 18 and 21 that a function f is integrable over an interval I iff if it is integrable on every subinterval of I. Moreover, if $\mathcal{P} = \{J_1, \ldots, J_n\}$ is a division of I, then

$$\int_I f = \sum_{i=1}^{n} \int_{J_i} f.$$

This is called the *additivity property* of the integral and in the case where the function is positive, refers to a desirable additivity property of areas (see Figure 13.2).

Existence of the Integral

The Riemann integrability of the continuous functions was first brought to the reader's attention in the calculus. It's been observed that every Riemann integrable function is integrable; thus, the continuous functions are also gauge

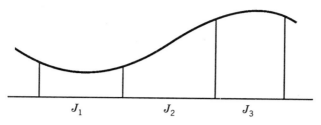

J_1 J_2 J_3

Figure 13.2

integrable. We shall establish the integrability of the continuous functions directly by appealing to the Cauchy criterion. To do this, we introduce the following lemma.

Lemma 22. Let $f\colon I \to \mathbb{R}$ and suppose that for each $\varepsilon > 0$ there are integrable functions $g_1, g_2\colon I \to \mathbb{R}$ such that $g_1(t) \le f(t) \le g_2(t)$ for all $t \in I$ and

$$\int_I g_2 \le \int_I g_1 + \varepsilon.$$

Then f is integrable on I.

Proof. Let $\varepsilon > 0$. There are gauges γ_1, γ_2 on I such that for each γ_i-fine tagged division $\mathscr{D}_i$ of I

$$\left| S(g_i, \mathscr{D}_i) - \int_I g_i \right| < \varepsilon.$$

Set $\gamma = \gamma_1 \cap \gamma_2$ and suppose that $\mathscr{D}$ is a γ-fine tagged division of I. Then

$$\int_I g_1 - \varepsilon < S(g_i, \mathscr{D}) \le S(f, \mathscr{D}) \le S(g_2, \mathscr{D}) < \int_I g_2 + \varepsilon \le \int_I g_1 + 2\varepsilon.$$

Thus any Riemann sum for f with respect to a γ-fine tagged division lies in the interval with endpoints $\int_I g_1 - \varepsilon$ and $\int_I g_1 + 2\varepsilon$ and so any two such sums differ by at most 3ε. The result now follows from Theorem 20. $\square$

Theorem 23. If $f\colon I \to \mathbb{R}$ is continuous, then f is integrable over I.

Proof. Let $\varepsilon > 0$. Since f is uniformly continuous on $[a, b]$, there is $\delta > 0$ such that $|f(x) - f(y)| < \varepsilon/(b - a)$ for $|x - y| < \delta$. Let $\mathscr{P} = \{x_0 < x_1 < \cdots < x_n\}$ be a partition of $[a, b]$ with mesh $< \delta$. For $i = 1, \ldots, n$, set

$$m_i = \inf \{ f(t)\colon x_{i-1} \le t \le x_i \}, \qquad M_i = \sup \{ f(t)\colon x_{i-1} \le t \le x_i \}$$

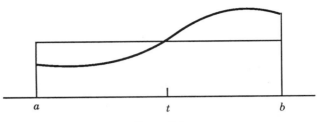

Figure 13.3

and define step functions g_1, g_2 by

$$g_1 = m_1 C_{[x_0, x_1]} + \sum_{j=2}^{n} m_j C_{(x_{j-1}, x_j]},$$

$$g_2 = M_1 C_{[x_0, x_1]} + \sum_{j=2}^{n} M_j C_{(x_{j-1}, x_j]}.$$

Then $g_1 \leq f \leq g_2$ and

$$\int_I (g_2 - g_1) \leq [\varepsilon/(b - a)](b - a).$$

Thus f is integrable by Lemma 22.   □

We can now use Theorem **8**.15 to establish a mean value theorem for integrals.

Theorem 24. Let $f: [a, b] \to \mathbb{R}$ be continuous. Then there is $t \in [a, b]$ such that $(b - a)f(t) = \int_a^b f$. (See Figure 13.3).

Proof. Let $M = \max \{ f(t): a \leq t \leq b \}$, $m = \min \{ f(t): a \leq t \leq b \}$. Since $m \leq f(t) \leq M$, we have $m(b - a) \leq \int_a^b f \leq M(b - a)$, or

$$m \leq \frac{1}{b - a} \int_a^b f \leq M.$$

By the intermediate value property (**8**.15), there is $t \in [a, b]$ such that

$$f(t) = \frac{1}{b - a} \int_a^b f.$$ □

Henstock's Lemma

Up to this point, we have considered the basic elementary properties of the integral. We now proceed to develop the deeper properties of the integral including the monotone and dominated convergence theorems; almost all of these results depend on the following result of R. Henstock.

Lemma 25 (Henstock's Lemma). Let $f: [a, b] = I \to \mathbb{R}$ be integrable over I. For $\varepsilon > 0$, suppose γ is a gauge on I such that if $\mathcal{D}$ is a γ-fine tagged division of I, then

$$\left| S(f, \mathcal{D}) - \int_I f \right| < \varepsilon.$$

If $\mathcal{J} = \{(x_1, J_1), \ldots, (x_n, J_n)\}$ is such that each J_i is a closed subinterval of I, $x_i \in J_i \subset \gamma(x_i)$, and $J_i^0 \cap J_j^0 = \varnothing$, $i \neq j$ (it is possible that $\bigcup_{i=1}^n J_i \neq I$), then

$$\left| \sum_{i=1}^n \left\{ f(x_i)\ell(J_i) - \int_{J_i} f \right\} \right| \leq \varepsilon \qquad \text{and} \qquad \sum_{i=1}^n \left| f(x_i)\ell(J_i) - \int_{J_i} f \right| \leq 2\varepsilon.$$

Proof. The set $I \setminus \bigcup_{i=1}^n J_i$ consists of a finite number of disjoint intervals. Let $K_1, \ldots, K_m$ be these intervals with their endpoints adjoined to them. For $\eta > 0$, the integrability of f over each K_j implies that there is a γ-fine tagged division $\mathcal{K}_j$ of K_j such that

$$\left| S(f, \mathcal{K}_j) - \int_{K_j} f \right| < \frac{\eta}{m}.$$

Then $\mathcal{D} = \mathcal{J} \cup \mathcal{K}_1 \cup \cdots \cup \mathcal{K}_m$ is a γ-fine tagged division of I. If $S(f, \mathcal{J}) = \sum_{i=1}^n f(x_i)\ell(J_i)$, then

$$\left| S(f, \mathcal{D}) - \int_I f \right| = \left| S(f, \mathcal{J}) - \sum_{i=1}^n \int_{J_i} f + \sum_{i=1}^m \left\{ S(f, \mathcal{K}_i) - \int_{K_i} f \right\} \right| < \varepsilon$$

and

$$\left| S(f, \mathcal{J}) - \sum_{i=1}^n \int_{J_i} f \right| < \varepsilon + \sum_{i=1}^m \left| S(f, \mathcal{K}_i) - \int_{K_i} f \right| < \varepsilon + m \cdot \eta/m = \varepsilon + \eta$$

for all $\eta > 0$. It follows that $|S(f, \mathcal{J}) - \sum_{i=1}^n \int_{J_i} f| \leq \varepsilon$.

For the second inequality, let $\mathscr{J}^+ (\mathscr{J}^-)$ be those $(x_i, J_i) \in \mathscr{J}$ such that $f(x_i)\ell(J_i) - \int_{J_i} f \geq 0 \ (\leq 0)$. Then, by the first inequality,

$$0 \leq \sum_{\mathscr{J}^+} \left\{ f(x_i)\ell(J_i) - \int_{J_i} f \right\} = \sum_{\mathscr{J}^+} \left| f(x_i)\ell(J_i) - \int_{J_i} f \right| \leq \varepsilon$$

and

$$- \sum_{\mathscr{J}^-} \left\{ f(x_i)\ell(J_i) - \int_{J_i} f \right\} = \sum_{\mathscr{J}^-} \left| f(x_i)\ell(J_i) - \int_{J_i} f \right| \leq \varepsilon.$$

Therefore,

$$\sum_{i=1}^{m} \left| f(x_i)\ell(J_i) - \int_{J_i} f \right| \leq 2\varepsilon. \qquad \square$$

A set $\mathscr{J}$ satisfying the hypotheses of Lemma 25 is called a *partial tagged division* of I. If γ is a gauge on I and $\mathscr{J}$ is a partial tagged division of I, then $\mathscr{J}$ is said to be γ-*fine* if $\gamma(x) \supset J$ for all $(x, J) \in \mathscr{J}$. If $\mathscr{J} = \{(x_1, J_1), \ldots, (x_n, J_n)\}$ is a partial tagged division of I and $J = \bigcup_{i=1}^{n} J_i$, we write $S(f, \mathscr{J}) = \sum_{i=1}^{n} f(x_i)\ell(J_i)$ and

$$\int_J f = \sum_{i=1}^{n} \int_{J_i} f.$$

$S(f, \mathscr{J})$ is called a *Riemann sum induced by* $\mathscr{J}$. If $\mathscr{J}$ is actually a tagged division of I, then Theorem 18 shows that this agrees with our previous notation and so should cause no confusion. Using this terminology and notation, Henstock's lemma states that if γ is a gauge on I such that γ-fine tagged divisions of I induce Riemann sums that give good approximations to the integral of f over I, then likewise any γ-fine partial tagged division induces Riemann sums that give good approximations to the integral of f over the union of the subintervals making up the partial tagged division. With this notation, the first conclusion of Lemma 25 becomes $|S(f, \mathscr{J}) - \int_J f| \leq \varepsilon$ for any partial tagged division $\mathscr{J}$ of I.

With the help of Henstock's lemma, we can develop some of the results on integration mentioned earlier.

Corollary 26. Let $f: [a, b] \to \mathbb{R}$ be integrable over $[a, b]$ and such that $\int_a^c f = 0$ for all $c \in (a, b)$. Then $|f|$ is integrable on $[a, b]$ and $\int_a^b |f| = 0$.

Proof. First, note that if $a < c < d \leq b$, then $\int_c^d f = 0$ since $\int_a^d f - \int_a^c f = \int_c^d f$. Let $\varepsilon > 0$ and let γ be a gauge on $[a, b]$ such that if $\mathscr{D}$ is a γ-fine tagged division of $[a, b]$, then $|S(f, \mathscr{D}) - \int_a^b f| < \varepsilon$. Then for every $(x, J) \in \mathscr{D}$,

$$\left| f(x)\ell(J) - \int_J f \right| = |f(x)\ell(J)|.$$

By Henstock's lemma,

$$\sum_{(x, J) \in \mathscr{D}} |f(x)|\ell(J) = S(|f|, \mathscr{D}) \le 2\varepsilon.$$

This implies $\int_a^b |f| = 0$. $\qquad\qquad\qquad\qquad\qquad\qquad\qquad\qquad\qquad\qquad\quad\square$

The Fundamental Theorem of Calculus (Second Part)

In Theorem 10, we considered the first half of the fundamental theorem of calculus: integrating derivatives. We now consider the second half: differentiating integrals.

Let $f: [a, b] \to \mathbb{R}$ be integrable over $[a, b]$. For $a \le x \le b$, we set $F(x) = \int_a^x f$. The function F is called the *indefinite integral* of f; it is clearly a function of its upper limit, that is, $F: [a, b] \to \mathbb{R}$. First we show that indefinite integrals are always continuous.

Theorem 27. If $f: [a, b] \to \mathbb{R}$ is integrable over $[a, b]$ and F is the indefinite integral of f, then F is continuous on $[a, b]$.

Proof. Let $x \in I$ and $\varepsilon > 0$. There is a gauge γ on $[a, b]$ such that if $\mathscr{D}$ is a γ-fine tagged division of $[a, b]$, then $|S(f, \mathscr{D}) - \int_a^b f| < \varepsilon/2$. Suppose $\gamma(x) = (\alpha, \beta)$. Set

$$\delta = \min \{ \beta - x, x - \alpha, \varepsilon/2(1 + |f(x)|) \}.$$

Suppose further that $y \in [a, b]$, $|y - x| < \delta$. Then from Henstock's lemma applied to $\{(x, [x, y])\}$ or $\{(x, [y, x])\}$, we have

$$\left| f(x)(y - x) - \int_x^y f \right| \le \varepsilon/2$$

or

$$\left| f(x)(x - y) - \int_y^x f \right| \le \varepsilon/2.$$

In either case,

$$\left| \int_x^y f \right| = |F(y) - F(x)| \le \varepsilon/2 + |f(x)| \, |y - x| < \varepsilon/2 + \varepsilon/2 = \varepsilon,$$

so that F is continuous at x. $\qquad\qquad\qquad\qquad\qquad\qquad\qquad\qquad\qquad\quad\square$

The other half of the fundamental theorem of calculus on the differentiability of the integral is now easy to establish.

Theorem 28 (The Fundamental Theorem of Calculus—Part 2). Let $f: [a, b] \to \mathbb{R}$. If f and $|f|$ are integrable over $[a, b]$, continuous at $x \in [a, b]$, and if F is the indefinite integral of f, then F is differentiable at x and its derivative is given by $F'(x) = f(x)$.

Proof. First consider the case where $h > 0$ and $x + h \in [a, b]$. Then

$$\left| \frac{[F(x + h) - F(x)]}{h} - f(x) \right| = \left| \frac{1}{h} \int_x^{x+h} f - f(x) \right|$$

$$= \left| \frac{1}{h} \int_x^{x+h} \{f - f(x)\} \right| \leq \frac{1}{h} \int_x^{x+h} |f - f(x)|$$

$$\leq \sup \{|f(t + x) - f(x)| : 0 \leq t \leq h\}. \quad (4)$$

A similar set of inequalities hold for $h < 0$. The continuity of f at x and the inequalities (4), (for arbitrary h), then yield the conclusion of the theorem. $\square$

Without the continuity assumption, the conclusion of Theorem 28 may fail (see Exercise 22). However, it is true that the indefinite integral F is always differentiable at "most" points of $[a, b]$.

Absolute Integrability

A function f is *absolutely integrable* if both f and $|f|$ are integrable. In this section we give conditions for the absolute integrability of a function and, in particular, show that an integrable function need not be absolutely integrable. For a function that is integrable, but not absolutely integrable, one visualizes that there is sufficient cancellation of terms in the Riemann sums for f so that convergence to a limit (the integral of f) occurs; however, when absolute values are taken, this cancellation effect does not happen for $|f|$ and the Riemann sums fail to have a limit. The key to the solution of this problem lies in controlling the oscillation or variation in the values of the function.

Definition 29. Let $\varphi: [a, b] \to \mathbb{R}$. The *variation* of φ over $[a, b]$, denoted by $\mathrm{Var}(\varphi: [a, b])$, is defined by $\mathrm{Var}(\varphi: [a, b]) = \sup \{\sum_{i=1}^n |\varphi(x_i) - \varphi(x_{i-1})| : \{x_0 < x_1 < \cdots < x_n\}$ is a partition of $[a, b]\}$. φ is of *bounded variation* over $[a, b]$ if $\mathrm{Var}(\varphi: [a, b]) < \infty$. The set of all functions that are of bounded variation on $[a, b]$ is denoted by $\mathscr{BV}[a, b]$.

Geometrically, the variation of a function is a measure of how much the function oscillates over an interval. The next example shows that even a continuous function can have infinite variation.

EXAMPLE 30. Let

$$f(t) = \begin{cases} 0 & \text{if} \quad t = 0, \\ t \sin(1/t) & \text{if} \quad 0 < t \le 1. \end{cases}$$

Set $x_n = 1/(n + 1/2)\pi$. Then

$$f(x_n) = \begin{cases} 1/(n + \frac{1}{2})\pi, & n \text{ even,} \\ -1/(n + \frac{1}{2})\pi, & n \text{ odd.} \end{cases}$$

If $\mathscr{P}_n$ is the partition $\{0 < x_n < x_{n-1} < \cdots < x_1 < 1\}$, then

$$\sum_{i=1}^{n-1} |f(x_i) - f(x_{i+1})| \ge \frac{2}{\pi} \sum_{i=1}^{n-1} \frac{1}{i+1}.$$

But, $\sum_{i=1}^{\infty} 1/(i+1)$ diverges. Thus $\text{Var}(f: [0,1]) = \infty$. $\square$

The introduction of the property of bounded variation enables us to give a necessary and sufficient condition for the absolute integrability of an integrable function f:

Theorem 31. Let $f: I \to \mathbb{R}$ be integrable over $I = [a, b]$. Then $|f|$ is integrable over I if, and only if, the indefinite integral of f, $F(x) = \int_a^x f$, is of bounded variation over I. In this case,

$$\text{Var}(F: [a, b]) = \int_a^b |f|.$$

Proof. Let $V = \text{Var}(F: [a, b])$. If $|f|$ is integrable and $\{x_0 < x_1 < \cdots < x_n\}$ is a partition of I, then

$$\sum_{i=1}^n \left| \int_{x_{i-1}}^{x_i} f \right| \le \sum_{i=1}^n |f| = \int_I |f|,$$

and so, $V \le \int_I |f| < \infty$.

Next, let $\varepsilon > 0$. There is a partition $\mathscr{P} = \{x_0 < x_1 < \cdots < x_n\}$ of I such that

$$V - \varepsilon \le \sum_{i=1}^n \left| \int_{K_i} f \right| \le V,$$

where $K_i = [x_{i-1}, x_i]$. Note that if $\mathscr{P}_1 = \{ y_0 < y_1 < \cdots < y_m \}$ is a refinement of the partition $\mathscr{P}$, (i.e., $\mathscr{P}_1 \supset \mathscr{P}$) and $L_i = [y_{i-1}, y_i]$, then

$$V - \varepsilon \leq \sum_{i=1}^{n} \left| \int_{K_i} f \right| \leq \sum_{i=1}^{m} \left| \int_{L_i} f \right| \leq V, \tag{5}$$

by the additivity of the integral over subintervals (See Lemma 35).

Let γ_1 be a gauge on I such that $|S(f, \mathscr{D}) - \int_I f| < \varepsilon$ for every γ_1-fine tagged division $\mathscr{D}$ of I. By Exercise 20, if $\mathscr{D} = \{(z_i, J_i): 1 \leq i \leq p\}$ is γ_1-fine, then

$$\left| \sum_{i=1}^{p} \left\{ |f(z_i)| \ell(J_i) - \left| \int_{J_i} f \right| \right\} \right| \leq 2\varepsilon. \tag{6}$$

Let γ be a gauge on I such that $\gamma(x) \subset \gamma_1(x)$, for all $x \in I$; $\gamma(x) \subset K_i$ when $x \in K_i^0$; and $\gamma(x_i) \subset (x_{i-1}, x_{i+1})$, where $x_{-1} = -\infty$, $x_{n+1} = \infty$. If $\mathscr{D} = \{(z_i, I_i): 1 \leq i \leq q\}$ is a γ-fine tagged division of I, then there is a γ-fine tagged division $\mathscr{D}' = \{(z_i', I_i'): 1 \leq i \leq r\}$ such that the division $\{I_1', \ldots, I_r'\}$ is a refinement of $\{K_1, \ldots, K_n\}$ and such that $S(|f|, \mathscr{D}) = S(|f|, \mathscr{D}')$. In fact, by the definition of γ, we may take $\mathscr{D}' = \{(z_i, I_i \cap K_j): 1 \leq i \leq q, 1 \leq j \leq n, I_i^0 \cap K_j^0 \neq \varnothing\}$ (Figure 13.4). Then (5) and (6) imply

$$V - \varepsilon \leq \sum_{i=1}^{r} \left| \int_{I_i'} f \right| \leq V \quad \text{and} \quad \left| \sum_{i=1}^{r} \left\{ |f(z_i')| \ell(I_i') - \left| \int_{I_i'} f \right| \right\} \right| \leq 2\varepsilon.$$

Consequently,

$$|S(|f|, \mathscr{D}) - V| = \left| S(|f|, \mathscr{D}') \pm \sum_{i=1}^{r} \int_{I_i'} |f| - V \right| \leq 2\varepsilon + \varepsilon,$$

and the conclusion of the theorem follows.   □

Using Theorem 31, we present next an example of an integrable function that is not absolutely integrable. Following the usual terminology of the calculus such functions are said to be *conditionally integrable*.

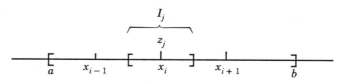

Figure 13.4

EXAMPLE 32. The function

$$f(t) = \begin{cases} t^2 \cos\left(\pi/t^2\right), & 0 < t \le 1 \\ 0, & t = 0 \end{cases}$$

has derivative

$$f'(t) = \begin{cases} 2t \cos\left(\pi/t^2\right) + \dfrac{2\pi}{t} \sin\left(\pi/t^2\right), & 0 < t \le 1 \\ 0, & t = 0. \end{cases}$$

Thus, f' is integrable by Theorem 10. An argument much like that in Example 30 shows, however, that $|f'|$ is not integrable. To demonstrate this, let $\beta_n = 1/\sqrt{2n}$ and $\alpha_n = \sqrt{2/(4n+1)}$. Then

$$\int_{\alpha_n}^{\beta_n} f' = \frac{1}{2n}.$$

The intervals $\{[\alpha_n, \beta_n]\colon n \in \mathbb{N}\}$ are pairwise disjoint. It follows that

$$\mathrm{Var}\,(f\colon [0,1]) \ge \sum_{n=1}^{N} \left| \int_{\alpha_n}^{\beta_n} f' \right| = \sum_{n=1}^{N} \frac{1}{2n},$$

for all N, and $f \notin \mathscr{BV}[0,1]$.   □

From Theorem 31 we obtain the following comparison test for absolute integrability.

Corollary 33. Let $f, g\colon I \to \mathbb{R}$ be integrable over I with $|f(t)| \le g(t)$ for all $t \in I$. Then f is absolutely integrable over I and

$$\left| \int_I f \right| \le \int_I |f| \le \int_I g.$$

Proof. For any partition $\mathscr{P} = \{x_0 < x_1 < \cdots < x_n\}$ of I,

$$\sum_{i=1}^{n} \left| \int_{x_{i-1}}^{x_i} f \right| \le \sum_{i=1}^{n} \int_{x_{i-1}}^{x_i} g = \int_I g.$$

It follows from Theorem 31 that f is absolutely integrable with

$$\int_I |f| \le \int_I g.$$ □

Functions of Bounded Variation

The functions of bounded variation are an important class of functions that arise in many areas of analysis. We pause in our study of integration to list at least a few of the basic properties of these functions. We conclude the section with an application to arclength.

If $\phi: I \to \mathbb{R}$ and $\mathscr{P} = \{x_0 < x_1 < \cdots < x_n\}$ is a partition of I, we set

$$v(\phi: \mathscr{P}) = \sum_{i=1}^{n} |\phi(x_i) - \phi(x_{i-1})|,$$

and call this the *variation of ϕ with respect to $\mathscr{P}$*. Thus,

$$\text{Var}(\phi: I) = \sup\{v(\phi: \mathscr{P}): \mathscr{P} \text{ is a partition of } I\}.$$

We show first that functions of bounded variation are bounded.

Proposition 34. If $\phi: [a, b] \to \mathbb{R}$ is of bounded variation over $[a, b]$, then ϕ is bounded on $[a, b]$.

Proof. Let $x \in (a, b)$. Then

$$|\phi(x) - \phi(a)| + |\phi(b) - \phi(x)| \le \text{Var}(\phi: [a, b]).$$

This implies that

$$2|\phi(x)| \le |\phi(a)| + |\phi(b)| + \text{Var}(\phi: [a, b]). \qquad \square$$

Next we consider $\text{Var}(\phi: I)$ as a set function of the interval I. Some of the elementary properties of $\text{Var}(\phi: I)$ as a function of ϕ are given in the exercises at the end of this section.

Lemma 35. Let $\phi: [a, b] \to \mathbb{R}$. If $\mathscr{P}$ and $\mathscr{P}'$ are partitions of $[a, b]$ such that $\mathscr{P} \subseteq \mathscr{P}'$, then $v(\phi: \mathscr{P}) \le v(\phi: \mathscr{P}')$.

The proof is an easy consequence of the triangle inequality.

Proposition 36. Let $\phi: [a, b] \to \mathbb{R}$ and $a < c < b$. Then

$$\text{Var}(\phi: [a, b]) = \text{Var}(\phi: [a, c]) + \text{Var}(\phi: [c, b]).$$

Proof. Let $\mathscr{P}$ be a partition of $[a, b]$ and $\mathscr{P}'$ the partition obtained by adding the point c to $\mathscr{P}$. Let $\mathscr{P}_1$ and $\mathscr{P}_2$ be the partitions of $[a, c]$ and $[c, b]$, respectively, induced by $\mathscr{P}'$. Then $v(\phi: \mathscr{P}) \le v(\phi: \mathscr{P}') \le v(\phi: \mathscr{P}_1) + v(\phi:$

$\mathscr{P}_2) \leq \text{Var}(\phi: [a, c]) + \text{Var}(\phi: [c, b])$. Taking the supremum over all partitions $\mathscr{P}$ yields $\text{Var}(\phi: [a, b]) \leq \text{Var}(\phi: [a, c]) + \text{Var}(\phi: [c, b])$.

If $\mathscr{P}_1$ and $\mathscr{P}_2$ are partitions of $[a, c]$ and $[c, b]$, respectively, then $\mathscr{P} = \mathscr{P}_1 \cup \mathscr{P}_2$ is a partition of $[a, b]$. Consequently, $v(\phi: \mathscr{P}) = v(\phi: \mathscr{P}_1) + v(\phi: \mathscr{P}_2) \leq \text{Var}(\phi: [a, b])$; thus, $\text{Var}(\phi: [a, c]) + \text{Var}(\phi: [c, b]) \leq \text{Var}(\phi: [a, b])$. The equality in Proposition 36 now follows. $\qquad\square$

Using the terminology that was employed when we considered the integral as a set function, we say that for fixed ϕ, the variation function $\text{Var}(\phi: \cdot)$ is *additive*. This additive property of the variation function enables us to give a characterization of functions of bounded variation.

Definition 37. Let $\phi \in \mathscr{BV}[a, b]$. The function v_ϕ, defined by

$$v_\phi(x) = \begin{cases} 0, & x = a, \\ \text{Var}(\phi: [a, x]), & a < x \leq b, \end{cases}$$

is called the *total variation of* ϕ.

From Proposition 36 it follows that v_ϕ is increasing. It follows from Exercise 24 that the sum and difference of functions of bounded variation are of bounded variation. Exercise 23 shows that the difference of two increasing functions is of bounded variation. Next we show that the converse of this is also true, that is, that a function of bounded variation can always be written as the difference of two increasing functions.

Theorem 38. Let $\phi: [a, b] \to \mathbb{R}$. Then $\phi \in \mathscr{BV}[a, b]$ if, and only if, there are increasing functions p and q on $[a, b]$ such that $\phi = p - q$.

Proof. As already mentioned, Exercises 23 and 24 show that the condition is sufficient.

To establish necessity, suppose that $\phi \in \mathscr{BV}[a, b]$. Set $p = v_\phi$ and $q = p - \phi$. It suffices to show that q is increasing. Let $a \leq x < y \leq b$. Then $q(y) = p(y) - \phi(y) = p(x) + \text{Var}(\phi: [x, y]) - \phi(y)$ implies $q(y) - q(x) = p(x) - \phi(y) + \text{Var}(\phi:[x, y]) - p(x) + \phi(x) = \text{Var}(\phi:[x, y]) - (\phi(y) - \phi(x)) \geq 0$. $\qquad\square$

It is also the case that if ϕ is a continuous function of bounded variation, then ϕ can be written as the difference of two continuous functions of bounded variation. From the proof of Theorem 38, this statement would follow immediately if the total variation function v_ϕ is continuous when ϕ is continuous. This is precisely the content of the next proposition.

Proposition 39. If $\phi \in \mathscr{BV}[a, b]$ and is continuous at $x_0 \in [a, b]$, then v_ϕ is continuous at x_0.

Proof. Let $\varepsilon > 0$ and suppose $x_0 < b$. There is a partition $\mathscr{P}$ of $[x_0, b]$ such that $v(\phi: \mathscr{P}) > \mathrm{Var}\,(\phi: [x_0, b]) - \varepsilon/2$. Since ϕ is continuous at x_0, we may add a point x_1 to $\mathscr{P}$ to obtain a partition $\mathscr{P}' = \{x_0 < x_1 < \cdots < x_n\}$ of $[x_0, b]$ such that $|\phi(x_0) - \phi(x_1)| < \varepsilon/2$. Then

$$\frac{\varepsilon}{2} + v(\phi: \mathscr{P}') = |\phi(x_0) - \phi(x_1)| + \sum_{j=1}^{n-1} |\phi(x_{j+1}) - \phi(x_j)| + \varepsilon/2$$

$$< \varepsilon + \mathrm{Var}\,(\phi: [x_1, b])$$

implies

$$\mathrm{Var}\,(\phi: [x_0, b]) < v(\phi: \mathscr{P}) + \varepsilon/2$$

$$\le v(\phi: \mathscr{P}') + \varepsilon/2 < \varepsilon + \mathrm{Var}\,(\phi: [x_1, b]),$$

and this last inequality implies

$$0 \le \mathrm{Var}\,(\phi: [x_0, b]) - \mathrm{Var}\,(\phi: [x_1, b]) < \varepsilon.$$

This is equivalent to $0 \le v_\phi(x_1) - v_\phi(x_0) < \varepsilon$. Since $v_\phi \uparrow$,

$$\lim_{x \to x_0^+} v_\phi(x)$$

exists and by the inequality above, must be $v_\phi(x_0)$.

Similarly, if $a < x_0$,

$$\lim_{x \to x_0^-} v_\phi(x) = v_\phi(x_0);$$

so v_ϕ is continuous at x_0. $\square$

We conclude with a brief discussion of arclength (in $\mathbb{R}^2$). A *path* in $\mathbb{R}^2$ is a continuous function ψ from an interval $[a, b]$ into $\mathbb{R}^2$. The points $\psi(a)$ and $\psi(b)$ are called the *initial* and *terminal* points of the path, and the path is said to join $\psi(a)$ to $\psi(b)$. Intuitively, the graph of ψ, $\{\psi(t): a \le t \le b\}$ is a curve joining $\psi(a)$ to $\psi(b)$. For example, $\psi(t) = (\cos t, \sin t)$, $0 \le t \le 2\pi$, has for its graph the unit circle centered at the origin. Consider the problem of defining the length of such a path. We proceed as we did in defining the area under a curve and try to approximate the path, in some sense, by a "simple" path whose length is easily defined. A natural candidate for an approximating path is a polygonal path, constructed by selecting a finite number of points, including the endpoints, on the path, order them using the natural ordering induced by the domain (interval) of definition of ϕ, and connect consecutive points by line segments. More precisely, these polygonal approximations can

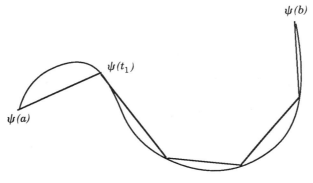

Figure 13.5

be generated in the following way: let $\mathscr{P} = \{t_0 < t_1 < \cdots < t_n\}$ be a partition of $[a, b]$ and consider the points $\{\psi(t_i): 0 \leq i \leq n\}$ on the graph of ψ. Connect the points $\psi(t_{i-1})$ and $\psi(t_i)$ by straight line segments for $i = 1, \ldots, n$ (see Figure 13.5). The *length* of the polygonal path is defined to be

$$\ell(\mathscr{P}) = \sum_{i=1}^{n} \left\{ (x(t_i) - x(t_{i-1}))^2 + (y(t_i) - y(t_{i-1}))^2 \right\}^{1/2},$$

where $\psi = (x, y)$. This quantity obviously underestimates what we would want to call the length of the path ψ, but as the partition becomes finer, we should obtain better approximations. This suggests that a natural definition for the *arclength of the path* ψ would be

$$L(\psi) = \sup \{\ell(\mathscr{P}): \mathscr{P} \text{ is a partition of } [a, b]\}.$$

Note that

$$\sum_{i=1}^{n} |x(t_i) - x(t_{i-1})| \leq \ell(\mathscr{P})$$

$$\leq \sum_{i=1}^{n} \left[|x(t_i) - x(t_{i-1})| + |y(t_i) - y(t_{i-1})| \right],$$

with a similar inequality holding for $\sum_{i=1}^{n} |y(t_i) - y(t_{i-1})|$. Thus, ψ has finite arclength, that is, $L(\psi) < \infty$, if and only if both the coordinate functions x and y are of bounded variation over $[a, b]$. In particular, Example 30 furnishes an example of a path with infinite arclength.

Lattice Properties of Integrable Functions

In this section we give some of the order, or lattice, properties of the set of integrable functions. If f and g are real-valued functions on a set S, we write $f \vee g = \max\{f, g\}$, $f \wedge g = \min\{f, g\}$, $f^+ = f \vee 0 = (f + |f|)/2$, $f^- = (-f) \vee 0 = (-f + |f|)/2$. Consequently, $|f| = f^+ + f^-$ and $f = f^+ - f^-$.

Proposition 40. Let $f, g: [a, b] \to \mathbb{R}$.

  (i) f is absolutely integrable over $[a, b]$ if, and only if, both f^+ and f^- are integrable over $[a, b]$.
  (ii) If f and g are absolutely integrable over I, then $f \vee g$ and $f \wedge g$ are integrable over I.

  Proof. (i) follows from the formulas immediately preceding Proposition 40.
  For (ii), we have the formulas $f \vee g = \frac{1}{2}[f + g + |f - g|]$ and $f \wedge g = \frac{1}{2}[f + g - |f - g|]$. Since $|f - g| \le |f| + |g|$, (ii) now follows from Corollary 33.   □

Despite the fact that Exercise 31 indicates that, in general, the maximum and minimum of integrable functions need not be integrable, the following result gives a sufficient condition, not involving absolute integrability, for this to hold,

Proposition 41. Let f, g, h be integrable on $I = [a, b]$.

  (i) If $f \le h$ and $g \le h$, then $f \vee g$ and $f \wedge g$ are integrable over I.
  (ii) If $h \le f$ and $h \le g$, then $f \vee g$ and $f \wedge g$ are integrable over I.

  Proof. To prove (i), observe that $f \vee g \le h$, and therefore $|f - g| = 2f \vee g - f - g \le 2h - f - g$. Thus $|f - g|$ is integrable by Corollary 33. It follows from the formulas in the proof of Proposition 40 that $f \vee g$ and $f \wedge g$ are integrable.
  (ii) is established in the same way by showing that

$$|f - g| \le f + g - 2h.$$ □

Improper Integrals

We finally consider the place of "improper integrals" in the context of the gauge integral. In the calculus of the Riemann integral, functions such as $f(t) = 1/\sqrt{t}$, with singularities, must be given special treatment. It requires an extension of the Riemann integral, sometimes called the Cauchy–Riemann integral, to properly handle such functions. Functions with several singularities on an interval become cumbersome to treat. Theorem 42 will show that it is not necessary to give such functions special treatment in the theory of the gauge integral.

Recall that if $f: [a, b] \to \mathbb{R}$ has a singularity at a in the sense that $\lim_{x \to a} |f(x)| = \infty$ and if f is R-integrable on $[c, b]$ for each $c > a$, then f is said to be *Cauchy–Riemann integrable* over $[a, b]$ iff

$$\lim_{c \to a^+} R\int_c^b f$$

exists and, in this case, the *Cauchy–Riemann integral* of f over $[a, b]$ is defined to be this limit. The following result shows that such a special procedure is unnecessary for the gauge integral, that is, for the gauge integral, there are no "improper integrals" over bounded intervals.

Theorem 42. Let $f: [a, b] \to \mathbb{R}$ be integrable over $[c, b]$ for all $a < c < b$. Then f is integrable over $[a, b]$ if, and only if,

$$\lim_{c \to a^+} \int_c^b f$$

exists. In this case

$$\int_a^b f = \lim_{c \to a^+} \int_c^b f.$$

Proof. To show the condition necessary, let $\varepsilon > 0$. There is a gauge γ on $[a, b]$ such that for each γ-fine tagged division $\mathcal{D}$ of $[a, b]$,

$$\left| S(f, \mathcal{D}) - \int_a^b f \right| < \varepsilon/3.$$

For each $c \in (a, b)$, there is a gauge γ_c on $[c, b]$ such that

$$\left| S(f, \mathcal{E}) - \int_c^b f \right| < \varepsilon/3$$

for each γ_c-fine tagged division $\mathcal{E}$ of $[c, b]$. We may assume that $\gamma_c(z) \subset \gamma(z)$ for all $z \in [c, b]$. Choose c such that $c \in \gamma(a)$ and $(c - a)|f(a)| < \varepsilon/3$, and let $s \in (a, c)$. Let $\mathcal{E}$ be a γ_s-fine tagged division of $[s, b]$ and set $\mathcal{D} = \{(a, [a, s])\} \cup \mathcal{E}$. Then $\mathcal{D}$ is a γ-fine tagged division of $[a, b]$, and

$$\left| \int_a^b f - \int_s^b f \right| \leq \left| \int_a^b f - S(f, \mathcal{D}) \right| + \left| S(f, \mathcal{E}) - \int_s^b f \right| + |f(a)|(s - a) < \varepsilon.$$

Thus,

$$\int_a^b f = \lim_{c \to a^+} \int_c^b f.$$

To show the condition sufficient, choose a sequence $\{c_n\}_{n=0}^\infty$ from $[a, b]$ such that $c_0 = b$, $c_n > c_{n+1}$, and $c_n \to a$. Pick a gauge γ_1 on $[c_1, c_0]$ such that

$$\left| S(f, \mathcal{D}) - \int_{c_1}^{c_0} f \right| < \varepsilon/2$$

whenever $\mathcal{D}$ is a γ_1-fine tagged division of $[c_1, c_0]$. For each $n \geq 2$, pick a gauge γ_n on $[c_n, c_{n-2}]$ such that

$$\left| S(f, \mathcal{D}) - \int_{c_n}^{c_{n-2}} f \right| < \varepsilon/2^n$$

whenever $\mathcal{D}$ is a γ_n-fine tagged division of $[c_n, c_{n-2}]$.
    Put

$$A = \lim_{c \to a^+} \int_c^b f$$

and choose N such that

$$\left| \int_s^{c_0} f - A \right| < \varepsilon$$

for $a < s \leq c_N$ and $|f(a)|(c_N - a) < \varepsilon$. Now define a gauge γ on $[a, b]$ as follows:

$$\gamma(z) = \begin{cases} (-\infty, c_N), & z = a, \\ \gamma_1(z) \cap (c_1, \infty), & c_1 < z \leq c_0, \\ \gamma_n(z) \cap (c_n, c_{n-2}), & c_n < z \leq c_{n-1}, \quad (n = 2, 3, \dots). \end{cases}$$

Now suppose that $\mathcal{D} = \{(x_i, I_i): 1 \leq i \leq p\}$ is a γ-fine tagged division of $[a, b]$. For each $n \geq 1$ let $\mathcal{D}_n$ be the subset of $\mathcal{D}$ whose tags are in $(c_n, c_{n-1}]$. Only a finite number of the $\mathcal{D}_n$ are nonvoid and no two have common elements. Let J_n be the union of the subintervals belonging to $\mathcal{D}_n$. By the definition of γ, each $\mathcal{D}_n$ is γ_n-fine on J_n, $J_1 \subset (c_1, c_0]$, and $J_n \subset (c_n, c_{n-2}]$ for $n \geq 2$. By Henstock's lemma,

$$\left| \int_{J_n} f - S(f, \mathcal{D}_n) \right| \leq \varepsilon/2^n.$$

Note that if $(x, K) \in \mathcal{D}$ is such that the subinterval $K = [a, d]$ is the subinterval in $\mathcal{D}$ containing a, then the tag associated with K must be a. [Assume that $a < x$. Then there is n such that $a < c_n < x$ and $x \in K \subset \gamma(x) \subset (c_n, c_{n-2})$. This is impossible, since $a \in K$.] Thus

$$|A - S(f, \mathcal{D})| \leq |f(a)|\ell(K) + \left| \sum_{n=1}^{\infty} \left\{ \int_{J_n} f - S(f, \mathcal{D}_n) \right\} \right| + \left| A - \int_d^b f \right|$$

$$< \varepsilon + \sum_{n=1}^{\infty} \varepsilon/2^n + \varepsilon = 3\varepsilon. \qquad \square$$

For an interesting discussion comparing the gauge integral with the improper Riemann integral, see J. T. Lewis and O. Shisha, 1983, 192–199.

Theorem 42 yields the following conditions for the existence of integrals.

Corollary 43. Let $f: [a, b] \to \mathbb{R}$ be such that f is integrable over $[c, b]$ for all $a < c \leq b$.

(i) If $f \geq 0$, then f is integrable over $[a, b]$ if, and only if, $\{ \int_c^b f: a < c \leq b \}$ is bounded.

(ii) (Comparison.) If $g: [a, b] \to \mathbb{R}$ is integrable over $[a, b]$ and $|f(t)| \leq g(t)$ for all $t \in [a, b]$, then f is (absolutely) integrable over $[a, b]$.

Proof. (i) follows by noting that the function $c \to \int_c^b f$ is decreasing on $(a, b]$.

For (ii), we need only show that the function F defined by $F(s) = \int_s^b f$ has a limit at a. For this it suffices to check the Cauchy criterion. For $a < s < t < b$,

$$|F(s) - F(t)| = \left| \int_s^t f \right| \leq \int_s^t g. \tag{7}$$

But the function g is integrable over $[a, b]$ and so the function $G(s) = \int_s^b g$ has a limit at a. Hence, (7) implies the Cauchy criterion for the function F at a.

The statement about absolute integrability follows from Corollary 33. $\square$

We conclude this section with two examples.

EXAMPLE 44. Let $f(t) = 1/t^k$ for $0 < t \leq 1$, and $f(0) = 0$, where $k \in \mathbb{R}$. If $0 < c < 1$, then $\int_c^1 f = (1 - c^{1-k})/(1 - k)$ for $k \neq 1$ and $\int_c^1 f = -\ln c$ for $k = 1$. Thus $|\int_c^1 f| \to \infty$ as $c \to 0^+$ if $k \geq 1$; so f is integrable over $[0, 1]$ iff $k < 1$ and, in this case, $\int_0^1 f = 1/(1 - k)$. $\square$

The beta function, which is useful in certain areas of probability and statistics, is considered in the next example.

EXAMPLE 45. The *beta function* is defined by

$$B(x, y) = \int_0^1 t^{x-1}(1 - t)^{y-1}\, dt.$$

If $x \geq 1$ and $y \geq 1$, the integrand is continuous and so the integral exists; however, for $x < 1$ ($y < 1$) the integrand has a singularity at $t = 0$ ($t = 1$). By splitting the integral into two parts,

$$\int_0^{1/2} + \int_{1/2}^1 t^{x-1}(1 - t)^{y-1}\, dt,$$

we may consider these cases separately. First, consider

$$\int_0^{1/2} t^{x-1}(1 - t)^{y-1}\, dt.$$

Since $(1 - t)^{y-1}$ is bounded on the interval $[0, \frac{1}{2}]$, the comparison result in Corollary 43(ii) and Example 44 show that the integral converges for $x > 0$. The change of variable $s = 1 - t$ transforms the second integral into an integral of the same form as the first integral. So the analysis above implies that the second integral converges for $y > 0$. Thus, B is defined for $x, y > 0$. $\qquad\qquad\square$

Historical Remarks

The gauge integral was originally introduced by the Czech mathematician J. Kurzweil in 1957. Although he did not give a detailed study of the properties of the integral he used it in his work on differential equations (see *Nichtabsolut Konvergente Integrale*, Tuebner, 1980). The integral was rediscovered independently by R. Henstock in 1961 who did develop the basic properties of the integral and established the convergence theorems for the integral. (See *Canadian Journal of Mathematics 20*, 1968, 79–87 for Henstock's expository account of the integral.) The books *The Generalized Riemann Integral* by R. M. McLeod and *Introduction A L'Analyse* by Jean Mawhin contain expository accounts of the gauge integral.

As we noted earlier, the general version of the fundamental theorem of calculus as given in Theorem 10 is false for both the Riemann and the Lebesgue integrals. Both integrals require the assumption that the derivative F' be integrable in order to obtain the basic formula $\int_a^b F' = F(b) - F(a)$.

This motivated mathematicians in the early 1900s to seek an integration theory for which the fundamental theorem of calculus is valid in its full generality as given in Theorem 10. This problem was solved by A. Denjoy and O. Perron (the gauge integral is equivalent to the Perron integral); however, the Denjoy and Perron integrals are technically much more difficult to describe than the gauge integral.

The Perron (and, therefore, the gauge) integral is more general than the Lebesgue integral and, in fact, a function f is Lebesgue integrable iff it is absolutely gauge integrable. (Thus, the function in Example 32 furnishes an example of a function that is gauge integrable, but not Lebesgue integrable.) E. J. McShane noted that a simple alteration in the definition of the gauge integral produces exactly the classical Lebesgue integral. In our definition of a tagged division $\mathscr{D} = \{(z_i, J_i): 1 \le i \le n\}$, McShane simply drops the requirement that the tag z_i must belong to the interval J_i. An expository version of McShane's integral is given in "A Unified Theory of Integration," *American Mathematical Monthly, 80* (1973), 349–359. E. J. McShane's book *Unified Integration* (Academic Press, 1983) contains an exhaustive description of the integral.

EXERCISES 13

1. Give an example of a gauge which is not of constant length.

2. Let $\mathscr{P} = \{I_1, \ldots, I_k\}$ be a division of I and γ a gauge on I. Suppose that for $i = 1, \ldots, k$, $\mathscr{D}_i$ is a γ-fine tagged division of I_i. Show that $\mathscr{D} = \bigcup_{i=1}^{k} \mathscr{D}_i$ is a γ-fine tagged division of I.

3. Suppose that γ_1, γ_2 are gauges on I such that $\gamma_1(t) \subset \gamma_2(t)$, for all $t \in I$. Show that any tagged division of I that is γ_1-fine is also γ_2-fine.

4. Suppose that γ_1, γ_2 are gauges on I. Set $\gamma(t) = \gamma_1(t) \cap \gamma_2(t)$. Show that γ is a gauge on I and that any tagged division of I that is γ-fine is also γ_i-fine.

5. Let $f: I \to \mathbb{R}$. Suppose there is $A \in \mathbb{R}$ such that for each $\varepsilon > 0$ there are integrable functions g and h with $g \le f \le h$ and $A - \varepsilon < \int_I g \le \int_I h < A + \varepsilon$. Show that f is integrable with $\int_I f = A$.

6. Let $f: [a, b] \to \mathbb{R}$ be integrable over $[a, b]$ and let $\alpha \in \mathbb{R}$. Define f_α: $[a + \alpha, b + \alpha] \to \mathbb{R}$ by $f_\alpha(t) = f(t - \alpha)$. Show that f_α is integrable with $\int_{a+\alpha}^{b+\alpha} f_\alpha = \int_a^b f$ (linear change of variable).

7. Evaluate $\int_1^2 t^{-2} e^{-1/t}\, dt$.

8. Prove part (ii) of Proposition 13.

9. Let $f: I \to \mathbb{R}$ be integrable on I and suppose $g: I \to \mathbb{R}$ is equal to f except possibly at a countable number of points of I. Show that g is integrable on I and that $\int_I f = \int_I g$.

10. Let

$$f(t) = \begin{cases} \sin t & \text{if } t \in [0,1] \setminus \mathbb{Q}, \\ t & \text{if } t \in \mathbb{Q} \cap [0,1]. \end{cases}$$

Show that f is integrable on $[0,1]$ and calculate $\int_0^1 f$.

11. Can the condition $f(t) \le g(t) \; \forall \, t \in I$ in Corollary 15 be relaxed?

12. Suppose $g: I \to \mathbb{R}$ is nonnegative and integrable and $f: I \to \mathbb{R}$ satisfies $|f(t)| \le g(t) \; \forall \, t \in I$. If $\int_I g = 0$, show that f is integrable over I and $\int_I f = 0$.

13. Suppose that $f: I \to \mathbb{R}$ has the property that $|f|$ is integrable over I with $\int_I |f| = 0$. Show that f is integrable over I.

14. Evaluate $\int_0^1 x e^{-x} \, dx$.

15. Show that f is R-integrable if, and only if, there is a gauge of constant length satisfying the Cauchy criterion of Theorem 20.

16. Suppose that $\int_I |f| = 0$. Show that f is integrable over every closed subinterval $J \subset I$ and $\int_J f = 0$.

17. Suppose that $\int_I |f - g| = 0$. Show that f is integrable over I if, and only if, g is integrable over I and in this case, $\int_J f = \int_J g$ for each closed subinterval $J \subset I$.

18. Suppose that $f: I \to \mathbb{R}$ is nonnegative and continuous. Show that if $\int_I f = 0$, then $f \equiv 0$ on I. Can the condition that f be nonnegative be dropped? Can the continuity condition be dropped?

19. Let $f, g: I \to \mathbb{R}$ be continuous functions with $g \ge 0$. Show that there is $t \in I$ such that $\int_a^b fg = f(t) \int_a^b g$.

20. Using the notation of Lemma 25, show that

$$\left| \sum_{i=1}^n \left\{ |f(x_i)| |\ell(J_i)| - \left| \int_{J_i} f \right| \right\} \right| \le 2\varepsilon.$$

21. Can the condition $\int_a^c f = 0$ for all $c \in (a, b]$, in Corollary 26, be replaced by $\int_a^b f = 0$?

22. Let

$$f(t) = \begin{cases} 1 & \text{if } \quad 0 \le t \le 1, \\ -1 & \text{if } -1 \le t < 0. \end{cases}$$

Is the indefinite integral F, of f, differentiable at $t = 0$? Why doesn't this violate Theorem 28?

23. If $f \uparrow (\downarrow)$ on $[a, b]$, show that $f \in \mathcal{BV}[a, b]$ and compute $\mathrm{Var}(f: [a, b])$.

24. If $\phi, \varphi \in \mathcal{BV}[a, b]$, then

(a) $\phi + \varphi \in \mathcal{BV}[a, b]$ and $\mathrm{Var}(\phi + \varphi: [a, b]) \leq \mathrm{Var}(\phi: [a, b]) + \mathrm{Var}(\varphi: [a, b])$.

(b) for each $t \in \mathbb{R}$, $t\phi \in \mathcal{BV}[a, b]$ and $\mathrm{Var}(t\phi: [a, b]) = |t|\mathrm{Var}(\phi: [a, b])$.

25. Is the decomposition in Theorem 38 unique?

26. Prove the converse of Proposition 39.

27. If $\phi, \varphi \in \mathcal{BV}[a, b]$, show that $\phi\varphi \in \mathcal{BV}[a, b]$.

28. If $f, g: I \to \mathbb{R}$ are absolutely integrable over I, show that $f \pm g$ and tf, for $t \in \mathbb{R}$, are absolutely integrable over I.

29. Give necessary and sufficient conditions on a function ϕ in order that $\mathrm{Var}(\phi: [a, b]) = 0$.

30. Let $f: [a, b] \to \mathbb{R}$ have a continuous derivative f'. Show that f has finite arclength [i.e., the path $t \to (t, f(t))$ has finite arclength] with

$$L(f) = \int_a^b \{1 + [f'(t)]^2\}^{1/2} \, dt.$$

Hint: Use the MVT to rewrite $\ell(\mathcal{P})$.

31. Show that (ii) in Proposition 40 is false if absolute integrability is replaced by integrability.
Hint: Consider Example 32 and $g = 0$.

32. Show that $B(\frac{1}{2}, \frac{1}{2}) = \pi$.
Hint: Try $t = \cos^2 u$.

33. Show that $B(x, y) = B(y, x)$.

34. Show that the integral $\int_0^1 (\ln x)/(x^r) \, dx$ exists for $0 < r < 1$.

35. If $f: [a, b] \to \mathbb{R}$ is continuous, $|f| \leq 1$, and $\int_a^b f = b - a$, what is f?

36. Let $f: [a, b] \to \mathbb{R}$ be bounded on $[a, b]$ and integrable over $[c, b]$ for all $c \in (a, b)$. Show that f is integrable over $[a, b]$ and $\lim_{c \to a^+} \int_c^b f = \int_a^b f$.

37. Let $f: [a, b] \to \mathbb{R}$ be continuously differentiable with $f(a) = f(b) = 0$. Show that if $\int_a^b f^2 = 1$, then $\int_a^b tf(t)f'(t) \, dt = -\frac{1}{2}$.

38. Let $f: [a, b] \to \mathbb{R}$ be continuous and differentiable on (a, b). If $f(a) = 0$, $f(b) = -1$, and $\int_a^b f = 0$, show that there is $c \in (a, b)$ such that $f'(c) = 0$.

39. Let f be positive and integrable on $[0, b]$. If f is decreasing on $(0, b]$, show that $\lim_{x \to 0^+} xf(x) = 0$.

40. Let $f: \mathbb{R} \to \mathbb{R}$ be continuous and set $f_n(x) = n\int_x^{x+1/n} f$. Show that $f_n \to f$ pointwise and that the convergence is uniform if f is uniformly continuous.

41. Let $a \leq x_0 \leq b$. Show that there is a gauge γ on $[a, b]$ such that if $\mathcal{D}$ is a γ-fine tagged division and J is the subinterval of $\mathcal{D}$ containing x_0, then x_0 must be the tag for J. Generalize this result to a finite number of points.

42. Give an example of two functions f and g that are absolutely integrable and whose product fg is not integrable.

43. Let $f: [0, 1] \to \mathbb{R}$ be continuous. Let $f_0 = f$ and $f_{k+1}(t) = \int_0^t f_k$, for $k \geq 0$. Show that if $f_k = 0$ for some k, then $f = 0$.

44. Let $f: [a, b] \to \mathbb{R}$ be integrable. For $\alpha \neq 0$, let g be defined by $g(s) = f(\alpha s)$. Show that g is integrable and $\int_a^b f = \alpha \int_{a/\alpha}^{b/\alpha} g$.

45. Show that the continuity assumption cannot be eliminated from the mean value theorem 24.
    Hint: Consider $f(t) = 0, 0 \leq t \leq 1$, $f(t) = 1, 1 < t \leq 2$.

46. Let $f: [0, 1] \to \mathbb{R}$ be continuous and increasing on $[0, 1]$. If $F(x) = (1/x)\int_0^x f$ for $0 < x < 1$, show that F is increasing.

47. Let $f: [a, b] \to \mathbb{R}$ be positive and continuous. If $M = \sup\{f(t): a \leq t \leq b\}$, show that

$$M = \lim \left\{ \int_a^b f^k \right\}^{1/k}.$$

48. Determine which of the following integrals exist:

(a) $\int_0^1 \sqrt{x}/\ln x \, dx$ (b) $\int_0^1 \frac{\sin x}{x^{3/2}} \, dx$

(c) $\int_2^3 \frac{x^2 + 1}{x^2 - 4} \, dx$ (d) $\int_0^1 \frac{dx}{\sqrt{x + x^2}}$

(e) $\int_0^1 \frac{\ln x}{1 - x^2} \, dx$

49. Let $f, g: [a, b] \to \mathbb{R}$ be integrable over $[c, b]$ for every $a < c < b$, and g be nonnegative. If $\lim_{x \to a} f(x)/g(x) = L \neq 0$, show that either both f and g are absolutely integrable over $[a, b]$ or both fail to be integrable.

50. Let $f_k: [a, b] \to \mathbb{R}$ be of bounded variation with $\mathrm{Var}(f_k: [a, b]) \leq M \; \forall k$. If f_k converges pointwise to a function f, show that f has bounded variation with $\mathrm{Var}(f: [a, b]) \leq M$.

CHAPTER 14

The Gauge Integral over Unbounded Intervals

In the previous chapter, the definition of the gauge integral was restricted to bounded intervals. In this chapter, this restriction is removed and the gauge integral is defined over unbounded intervals. If f is a function defined on some interval I (bounded or unbounded), then f can be extended to a function on $\mathbb{R}$ by setting $f(t) = 0$ for $t \notin I$. A reasonable definition of the integral of the extended function over $\mathbb{R}$ would then give a definition of the integral of the original function over I. So, we consider the problem of defining the integral of a function $f \colon \mathbb{R} \to \mathbb{R}$ over the entire line $\mathbb{R}$.

To give a definition analogous to that of the gauge interval over bounded intervals, it is first necessary to extend the definitions of partition and tagged division to the set $\mathbb{R}$. If a partition of $\mathbb{R}$ is to be a finite collection of closed intervals whose union is $\mathbb{R}$, then at least one of the subintervals of the partition must be unbounded. Such an unbounded interval would obviously have infinite length. If the function f was strictly positive on $\mathbb{R}$, then every tag associated with this interval would contribute an infinite term to the corresponding Riemann sum. Thus, every Riemann sum would be infinite and no positive function could possibly be integrable. If the integral of a positive function is to be associated with the area under its graph, this is clearly an undesirable situation.

The problem is easily solved. We simply add the points at infinity, $\pm\infty$, to $\mathbb{R}$, call the new set $\overline{\mathbb{R}}$, and then extend the function f to $\overline{\mathbb{R}}$ by defining $f(\pm\infty) = 0$. Intervals in $\overline{\mathbb{R}}$ of the form

$$[a, \infty] = \{t \in \overline{\mathbb{R}} : a \le t \le \infty\}, \qquad -\infty \le a < \infty,$$

and

$$[-\infty, a] = \{t \in \overline{\mathbb{R}} : -\infty \le t \le a\}, \qquad -\infty < a \le \infty$$

are called *closed intervals* containing ∞ and $-\infty$, respectively. Similarly, intervals of the form

$$(a, \infty] = \{t \in \overline{\mathbb{R}} : a < t \le \infty\}, \qquad -\infty \le a < \infty,$$

and

$$[-\infty, a) = \{t \in \overline{\mathbb{R}} : -\infty \le t < a\}, \qquad -\infty < a \le \infty$$

are called *open intervals* containing ∞ and $-\infty$, respectively. If $\pm\infty$ is the tag associated with an unbounded interval, then the term in the corresponding Riemann sum would be $0 \cdot \infty$; if $0 \cdot \infty$ was zero, the difficulty of infinite Riemann sums would be avoided. No problems are encountered by assuming this, and so, we define $0 \cdot (\pm\infty) = 0$. Recall also that for the extended real numbers, $\overline{\mathbb{R}} = \mathbb{R} \cup \{\pm\infty\}$, the following algebraic and order properties hold. (Where applicable, the upper and lower signs, respectively, in each equation are to be taken together.)

    Addition: $x \pm \infty = \pm\infty$ for all $x \in \mathbb{R}$, the upper signs go together as do the lower signs.

    Multiplication: $x \cdot (\pm\infty) = \pm\infty$ for all $x > 0$; the signs on the right are reversed if $x < 0$, $0 \cdot (\pm\infty) = 0$.

    Order: $-\infty < x < \infty$ for all $x \in \mathbb{R}$. (*Note:* $\infty = +\infty$.)

The algebraic and order properties assigned to $\pm\infty$ were generally motivated by the desire to extend the usual properties for finite limits to infinite limits. However, the definition $0 \cdot (\pm\infty) = 0$ is strictly motivated by the problems encountered above in associating the integral with an appropriate area.

If f is any real-valued function defined on an interval I in $\mathbb{R}$, we assume that f is extended so that it has the value zero at both $\pm\infty$, that is, $f(\pm\infty) = 0$. All functions defined on subsets of $\overline{\mathbb{R}}$ will be assumed to have this property.

With the conventions described above, the definitions of division and tagged division can be extended to intervals in $\overline{\mathbb{R}}$.

Definition 1. Let I be a closed interval in $\overline{\mathbb{R}}$. A *division* of I is a finite collection of closed subintervals $\{I_1, \ldots, I_n\}$ of I such that $\cup I_i = I$ and $I_i^0 \cap I_j^0 = \varnothing$, $(i \neq j)$. A *tagged division* of I is a finite collection of pairs $\mathcal{D} = \{(x_i, I_i): i = 1, \ldots, n\}$ such that $\{I_1, \ldots, I_n\}$ is a division of I and $x_i \in I_i$ for $i = 1, \ldots, n$. The elements I_i are called *subintervals* of $\mathcal{D}$ and x_i is called the *tag* of I_i.

Since open intervals containing $\pm \infty$ have been defined, the definition of a gauge is easily extended to subsets of $\overline{\mathbb{R}}$.

Definition 2. Let $E \subset \overline{\mathbb{R}}$. A *gauge* on E is a function γ defined on E that associates with each $x \in E$ an open interval in $\overline{\mathbb{R}}$ containing x.

Definition 3. If γ is a gauge on a closed interval I in $\overline{\mathbb{R}}$, then a tagged division $\mathcal{D} = \{(x_i, I_i): i = 1, \ldots, n\}$ is γ-*fine* if $\gamma(x_i) \supset I_i$ for each i.

The following is a generalization of Theorem 13.6 to unbounded intervals in $\overline{\mathbb{R}}$.

Theorem 4. Let I be a closed interval in $\overline{\mathbb{R}}$ and γ a gauge on I. Then there exists a γ-fine tagged division of I.

Proof. Consider the case $I = [a, \infty]$; the other cases are similar. Let $\gamma(\infty) = (b, \infty]$. If $b < a$, set $\mathcal{D} = \{(\infty, I)\}$. If $b \geq a$, there is a γ-fine tagged division $\mathcal{D}_0$ of $[a, b + 1]$. Then $\mathcal{D} = \mathcal{D}_0 \cup \{(\infty, [b + 1, \infty])\}$ is a γ-fine tagged division of I. $\square$

If I is an unbounded interval in $\mathbb{R}$ or $\overline{\mathbb{R}}$, we say that I has *infinite length* and write $\ell(I) = \infty$. The tools and nomenclature needed to define the gauge integral over unbounded intervals are now in place. We proceed with the definition.

Definition 5. Let I be a closed interval in $\overline{\mathbb{R}}$ and $f: I \to \mathbb{R}$. Then f is *integrable* over I if there is $A \in \mathbb{R}$ with the property that for each $\varepsilon > 0$ there is a gauge γ on I such that if

$$\mathcal{D} = \{(x_i, I_i): i = 1, \ldots, n\}$$

is a γ-fine tagged division of I, then

$$\left| \sum_{i=1}^n f(x_i) \ell(I_i) - A \right| < \varepsilon.$$

As before, the number A is unique when it exists and is called the *integral of f over I*; it is denoted by $\int_I f$. The sum $\sum_{i=1}^n f(x_i)\ell(I_i)$ is called a *Riemann sum* of f with respect to $\mathcal{D}$ and is denoted by $S(f, \mathcal{D})$. If one of the tags x_i is $\pm\infty$, then the term $f(x_i)\ell(I_i)$ in the Riemann sum is zero by our convention that $0 \cdot \infty = 0$. Note also that if a gauge γ has the property that $\gamma(x)$ is a bounded open interval for every $x \in \mathbb{R}$ and if $\mathcal{D}$ is a γ-fine tagged division of an unbounded interval I, then the subinterval in $\mathcal{D}$ containing $\pm\infty$ must have $\pm\infty$ as its tag.

The basic properties of the integral that were previously established extend, where applicable, to integrals over unbounded intervals. We do not repeat the properties, but invite the reader to review them and check a few of the proofs so that he or she may be convinced that they may be repeated with very little alteration.

In the development of the Riemann theory of integration in the elementary calculus, it was necessary to make a special definition for the integral over unbounded regions. This led to the "improper" or Cauchy–Riemann integrals. Just as with "improprieties" on finite intervals, no special treatment is needed for the gauge integral in this last case.

Theorem 6. Suppose that $f: I = [a, \infty] \to \mathbb{R}$ is integrable over $[a, b]$ for all $a < b < \infty$. Then f is integrable over I if, and only if,

$$\lim_{b \to \infty} \int_a^b f = A$$

exists; in this case, $A = \int_I f$.

Proof. To show the condition necessary, let $\varepsilon > 0$. There is a gauge γ on I such that $|S(f, \mathcal{D}) - \int_I f| < \varepsilon/2$ for each γ-fine tagged division $\mathcal{D}$ of I. For $a < c < \infty$, there is a gauge γ_c of $[a, c]$ such that $|S(f, \mathcal{E}) - \int_a^c f| < \varepsilon/2$ for every γ_c-fine tagged division $\mathcal{E}$ of $[a, c]$. We may assume that $\gamma_c(z) \subset \gamma(z)$ for all $z \in [a, c]$. Let $\gamma(\infty) = (T, \infty]$. For $c > T$, let $\mathcal{E}$ be a γ_c-fine tagged division of $[a, c]$. Set $\mathcal{D} = \mathcal{E} \cup \{(\infty, [c, \infty])\}$. Then $\mathcal{D}$ is a γ-fine tagged division of I and

$$\left| \int_I f - \int_a^c f \right| \leq \left| \int_I f - S(f, \mathcal{D}) \right| + \left| S(f, \mathcal{E}) - \int_a^c f \right|$$

$$+ |f(\infty)| \ell([c, \infty]) < \varepsilon/2 + \varepsilon/2 = \varepsilon.$$

Hence,

$$\lim_{c \to \infty} \int_a^c f = \int_I f.$$

To show that the condition is sufficient, choose a sequence $\{c_n\}$ such that $a = c_0 < c_1 < \cdots < c_n$ and $c_n \to \infty$. Choose a gauge γ_0 on $[c_0, c_1]$ such that

$$\left| S(f, \mathscr{D}) - \int_{c_0}^{c_1} f \right| < \varepsilon/2^2$$

for every γ_0-fine tagged division $\mathscr{D}$ of $[c_0, c_1]$. For $n \geq 1$, choose a gauge γ_n on $[c_{n-1}, c_{n+1}]$ such that

$$\left| S(f, \mathscr{D}) - \int_{c_{n-1}}^{c_{n+1}} f \right| \leq \varepsilon/2^{n+2}$$

for every γ_n-fine tagged division $\mathscr{D}$ of $[c_{n-1}, c_{n+1}]$.

Choose N such that $b \geq c_N$ implies $|\int_a^b f - A| < \varepsilon/2$. Now define a gauge γ on I by

$$\gamma(z) = \begin{cases} (c_N, \infty], & z = \infty, \\ \gamma_k(z) \cap (c_{k-1}, c_{k+1}), & c_k \leq z < c_{k+1}, \quad k = 1, 2, \ldots . \\ \gamma_0(z) \cap (c_0, c_1), & c_0 \leq z < c_1. \end{cases}$$

Suppose that $\mathscr{D} = \{(x_i, I_i): i = 1, \ldots, n\}$ is a γ-fine tagged division of I. Some subinterval of $\mathscr{D}$ must be unbounded; denote it by I_n. Then the tag associated with I_n must be $\infty = x_n$. Let $I_n = [t_n, \infty]$; then $t_n > c_N$. For $m \geq 0$, let $\mathscr{D}_m$ be those pairs (x_i, I_i) in $\mathscr{D}$ whose tags are in $[c_m, c_{m+1})$. Only a finite number of the $\mathscr{D}_m$ are nonvoid and the $\{\mathscr{D}_m\}$ do not have common elements. By the definition of γ, each $\mathscr{D}_m$ is a γ_m-fine partial tagged division of the union, J_m, of the subintervals in $\mathscr{D}_m$ and $J_0 \subseteq [c_0, c_1)$, $J_m \subseteq [c_{m-1}, c_{m+1})$ for $m \geq 1$. By Henstock's lemma,

$$\left| \int_{J_m} f - S(f, \mathscr{D}_m) \right| \leq \varepsilon/2^{m+2}.$$

Thus,

$$|A - S(f, \mathscr{D})| \leq \left| A - \int_a^{t_n} f \right| + \left| \int_a^{t_n} f - S(f, \mathscr{D}) \right|$$

$$\leq \varepsilon/2 + \left| \sum_{m=0}^{\infty} \int_{J_m} f - \sum_{m=0}^{\infty} S(f, \mathscr{D}_m) + f(\infty)\ell(I_n) \right|$$

$$\leq \varepsilon/2 + \sum_{m=0}^{\infty} \varepsilon/2^{m+2} = \varepsilon/2 + \varepsilon/2 = \varepsilon. \qquad \square$$

An analogous result holds for intervals of the form $[-\infty, b]$ and $[-\infty, \infty]$.

Thus, from Theorem **13**.42 and Theorem 6, we see that there are no "improper" integrals for the gauge integral.

The analog of Corollary **13**.43 for integrals on unbounded intervals is

Corollary 7. Let $f: [a, \infty] = I \to \mathbb{R}$ be integrable over $[a, b]$ for $a < b < \infty$.

   (i) If f is nonnegative, then f is integrable over I if, and only if, $\{ \int_a^b f: a < b < \infty \}$ is bounded.
  (ii) If $g: I \to \mathbb{R}$ is a nonnegative integrable function over I and $|f(t)| \le g(t)$ for $t \in I$, then f is (absolutely) integrable over I.

We conclude this chapter with some examples and applications. If a function f is defined on an interval $[a, \infty)$, then its integral over such an interval is naturally defined to be the integral of the extension of f over the interval $[a, \infty]$ and is denoted by $\int_a^\infty f$. Similar agreements are made concerning intervals of the form $(-\infty, b]$ and $(-\infty, \infty)$.

EXAMPLE 8. Let $f(t) = 1/t^p$, $t \neq 0$, $p \in \mathbb{R}$. Then

$$\int_1^b f = \frac{b^{-p+1} - 1}{1 - p}$$

for $p \neq 1$; if $p = 1$, $\int_1^b f = \log b$. Thus, f is integrable over $[1, \infty)$ iff $p > 1$. In this case, $\int_1^\infty 1/t^p = 1/(p - 1)$.   □

The reader is encouraged to compare this with the result concerning convergence of p-series (**4**.9). In fact, we have the following result relating convergence of series to integrals.

Proposition 9 (Integral Test). Let $f: [1, \infty] \to \mathbb{R}$ be positive, decreasing, and integrable on $[1, b]$ for $1 < b < \infty$. Then the integral $\int_1^\infty f = A$ exists if, and only if, the series $\sum_{k=1}^\infty f(k) = S$ converges. In this case, $A \le S \le A + f(1)$.

Proof. For $i \le x \le i + 1$, $f(i + 1) \le f(x) \le f(i)$. (See Figure 14.1.) So,

$$f(i + 1) \le \int_i^{i+1} f \le f(i)$$

and

$$\sum_{i=1}^{n-1} f(i + 1) \le \int_1^n f \le \sum_{i=1}^{n-1} f(i).$$

Letting $n \to \infty$ gives the first result and, in the case of convergence, yields

$$S - f(1) \le A \le S.$$ □

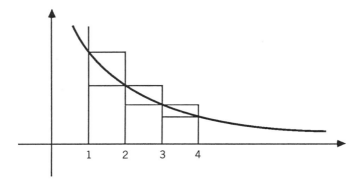

Figure 14.1

EXAMPLE 10. The integral

$$\int_1^\infty \frac{\sin x}{x}\, dx$$

exists. In fact, integration by parts gives

$$\int_1^b \frac{\sin x}{x}\, dx = -\left.\frac{\cos x}{x}\right|_1^b - \int_1^b \frac{\cos x}{x^2}\, dx,$$

where we use the customary notation $F(x)|_a^b = F(b) - F(a)$. The observation

$$\lim_{b\to\infty} \frac{\cos b}{b} = 0$$

and Exercise 6 give the desired result.

Note that $(\sin x)/x$ is not absolutely integrable over $[1, \infty)$. We have

$$\int_\pi^{n\pi} \left|\frac{\sin x}{x}\right|\, dx = \sum_{k=1}^{n-1} \int_{k\pi}^{(k+1)\pi} \left|\frac{\sin x}{x}\right|\, dx$$

$$\geq \sum_{k=1}^{n-1} \frac{1}{(k+1)\pi} \int_{k\pi}^{(k+1)\pi} |\sin x|\, dx = \sum_{k=1}^{n-1} \frac{2}{(k+1)\pi}$$

for all n. This function is another example of a function that is conditionally integrable. (Here, of course, we are extending the previous terminology to unbounded intervals.) □

As our last example, we define the gamma function. Some of its important properties will be developed later. (See also Exercise 14.9.)

EXAMPLE 11 (The Gamma Function). The *gamma function* is defined by

$$\Gamma(x) = \int_0^\infty t^{x-1} e^{-t} \, dt.$$

It will be shown that the integral exists for $x > 0$, and so, the domain of the gamma function is $\{x: x > 0\}$. For $0 < t \le 1$, $t^{x-1} e^{-t} \le t^{x-1}$. So, the comparison test (**13.43**) and Example **13.44** imply that

$$\int_0^1 t^{x-1} e^{-t} \, dt$$

exists for $x > 0$. For $t > 1$, $t^{x-1} e^{-t} = [t^{x+1} e^{-t}] t^{-2}$. For $x > 0$,

$$\lim_{t \to \infty} [t^{x+1} e^{-t}] = 0.$$

Consequently, the comparison result 7(ii) and Example 8 imply that

$$\int_1^\infty t^{x-1} e^{-t} \, dt = \int_1^\infty t^{-2} [t^{x+1} e^{-t}] \, dt$$

exists.   □

EXERCISES 14

1. Check the proofs of **13**.13, 14, 18, 20, 21, and 25 and verify that they hold for the situation where unbounded intervals are involved.

2. Examine the series $\sum_{n=1}^\infty [(-1)^n \ln n]/n$ for convergence and absolute convergence.

3. Let $\{a_n\}_{n=1}^\infty \subset \mathbb{R}$. Define $f: [1, \infty) \to \mathbb{R}$ by $f(t) = a_k$ if $k \le t < k + 1$. Show that f is integrable over $[1, \infty)$ if, and only if, the series Σa_n converges. When is f absolutely integrable?

4. Let $f: [1, \infty) \to \mathbb{R}$ be nonnegative, decreasing, and $\lim_{t \to \infty} f(t) = 0$. If $\int_1^\infty f$ exists, then $\lim_{t \to \infty} tf(t) = 0$.
   Hint: Apply Abel's Theorem **4.11** to the series $\sum_{i=1}^\infty \int_i^{i+1} f$.

5. Let $f: [a, \infty) \to \mathbb{R}$ be differentiable. Give necessary and sufficient conditions so that f' is integrable over $[a, \infty)$ and calculate $\int_a^\infty f'$.

6. Is the function $f(t) = (\cos t)/t^2$ integrable over $[1, \infty)$?

7. Show that $\int_0^\infty \sin(x^2)\,dx$ exists. (This is the Fresnel integral.)
   Hint: Try $t = x^2$. Is the integrand absolutely integrable?

8. Does the analogue for Theorem **13**.23 hold for integrals over unbounded intervals?

9. Show that
   (a) $\Gamma(1) = 1$.
   (b) $\Gamma(x + 1) = x\Gamma(x)$.
   Hint: Integrate by parts.
   (c) $\Gamma(n + 1) = n!$ for $n \in \mathbb{N}$.

10. Let $f: [1, \infty] \to \mathbb{R}$ be nonnegative and decreasing. Show that $u_n = f(n + 1) - f(n) - \int_1^n f$ converges to a limit between 0 and $f(1)$.
    Hint: Show that $0 \le f(n) \le u_n \le f(1)$ and $u_n \downarrow$.

11. Show that $1 + \frac{1}{2} + \frac{1}{3} + \cdots + 1/n - \ln n = \gamma_n$ converges. The limit of this sequence is called Euler's Constant.

12. (Abel) Let $f: [a, \infty) \to \mathbb{R}$ be continuous and such that the indefinite integral $F(x) = \int_a^x f$ is bounded. Let $k: [a, \infty) \to \mathbb{R}$ be nonnegative, differentiable, and decreasing. Show that $\int_a^\infty kf$ exists if either (a) $\lim_{x \to \infty} k(x) = 0$, or (b) $\int_a^\infty f$ exists.
    Hint: Integrate by parts.

13. Use Abel's test to show that $\int_1^\infty \sin x/x^p\,dx$ and $\int_1^\infty \cos x/x^p\,dx$ converge for $p > 0$.

14. If $0 < p \le 1$, show that $\int_1^\infty \sin x/x^p\,dx$ converges conditionally.
    Hint: Example **14**.10.

15. Give an example of a nonnegative continuous function $f: [1, \infty) \to \mathbb{R}$ such that the series $\sum_{k=1}^\infty f(k)$ converges (diverges) while the integral $\int_1^\infty f$ fails to exist (exists). (Compare with Proposition 9.)

16. Let $f: [a, \infty) \to \mathbb{R}$ be continuous with $F(x) = \int_a^x f$ bounded on $[a, \infty)$. Let $g: [a, \infty) \to \mathbb{R}$ be such that g' is nonpositive and continuous on $[a, \infty)$ and $\lim_{t \to \infty} g(t) = 0$. Show that $\int_a^\infty fg$ exists.

17. Use Exercise 16 to show that

$$\int_3^\infty \frac{\sin t}{\log t}\,dt$$

   exists.

18. Let

$$f(x) = \begin{cases} 0, & \text{if } |x| < 1, \\ (\sin x)/x^2, & \text{if } |x| \ge 1. \end{cases}$$

   Show that f is integrable over $\mathbb{R}$.

19. Show that

$$\int_0^\infty \frac{\sin x}{x}\,dx = \int_0^\infty \left[\frac{\sin x}{x}\right]^2 dx.$$

Hint: Integrate by parts.

20. Determine which of the following integrals exist

(a) $\displaystyle\int_1^\infty \sin\left(1/x^2\right) dx$ (b) $\displaystyle\int_1^\infty e^{-x}\ln x\,dx$

(c) $\displaystyle\int_3^\infty \frac{x^2+1}{x^2-4}\,dx$ (d) $\displaystyle\int_1^\infty \frac{\sin\left(1/x\right)}{x}\,dx$

21. If $f: [0, \infty) \to \mathbb{R}$ is continuous and absolutely integrable over $[0, \infty)$, does it follow that $\lim_{x\to\infty} f(x) = 0$? Is f necessarily bounded?

22. Show that $f: [a, \infty) \to \mathbb{R}$ is integrable over $[a, \infty)$ if, and only if, there is $A \in \mathbb{R}$ such that for every $\varepsilon > 0$, there exists $b > a$ and a gauge γ on $[a, b]$ such that $|A - S(f, \mathscr{D})| < \varepsilon$ for every γ-fine tagged division $\mathscr{D}$ of $[a, b]$.

CHAPTER 15

Convergence Theorems

One of the important advantages of the gauge integral is the way it allows us to deal with the convergence properties of the integral. In Example **12**.3, it was shown that there is a uniformly bounded sequence of Riemann integrable functions that converges pointwise to a function that is not Riemann integrable. This does not happen with the gauge integral. In fact, the gauge integral allows easy interchange between the operations of limit and integration. It is the easy interchange of these operations that makes the gauge integral (and the Lebesgue integral) so much easier to use in the applications. This is the reason that the Lebesgue integral has historically been preferred to the Riemann integral. We begin with the Monotone Convergence Theorem (MCT), which characterizes the integrability for limits of monotone sequences of integrable functions.

Theorem 1 (Monotone Convergence Theorem). Let I be a closed interval in $\overline{\mathbb{R}}$, and $f_k, f\colon I \to \mathbb{R}$ be such that

   (i) Each f_k is integrable over I,

   (ii) $f_1(x) \leq f_2(x) \leq \cdots \leq f_k(x) \leq \cdots$, for all $x \in I$,

   (iii) $\lim_k f_k(x) = f(x)$.

Then f is integrable over I if, and only if, $\lim_k \int_I f_k < \infty$; in this case,

$$\int_I f = \lim_k \int_I f_k.$$

Proof. Since $\int_I f_k \leq \int_I f_{k+1}$, it follows that $\lim \int_I f_k$ exists (it may be ∞). By replacing f_k by $f_k - f_1$, we may assume that $f_k \geq 0$.

The necessity of the condition follows immediately since $\int_I f_k \leq \int_I f < \infty$ implies $\lim \int_I f_k \leq \int_I f$.

To establish sufficiency, we assume first that $I = [a, b]$ is a bounded interval. Let $\varepsilon > 0$ and $A = \lim \int_I f_k$. For each k, there is a gauge γ_k on I such that

$$\left| \int_I f_k - S(f_k, \mathcal{D}) \right| < \varepsilon/2^{k+1}$$

for every γ_k-fine tagged division $\mathcal{D}$ of I. There is n_0 such that $k \geq n_0$ implies that $0 \leq A - \int_I f_k < \varepsilon/4$. For each $x \in I$, there is $n(x) \geq n_0$ such that $k \geq n(x)$ implies that $0 \leq f(x) - f_k(x) < \varepsilon/4(b - a)$. Define a gauge γ on I by $\gamma(x) = \gamma_{n(x)}(x)$. Suppose $\mathcal{D} = \{(x_i, I_i): i = 1, \ldots, m\}$ is a γ-fine tagged division of I. Then

$$|S(f, \mathcal{D}) - A| \leq \left| \sum_{j=1}^m f(x_j)\ell(I_j) - \sum_{j=1}^m f_{n(x_j)}(x_j)\ell(I_j) \right|$$

$$+ \left| \sum_{j=1}^m \left\{ f_{n(x_j)}(x_j)\ell(I_j) - \int_{I_j} f_{n(x_j)} \right\} \right|$$

$$+ \left| \sum_{j=1}^m \int_{I_j} f_{n(x_j)} - A \right|. \tag{1}$$

The first term on the right-hand side of (1) satisfies

$$\left| \sum_{j=1}^m f(x_j)\ell(I_j) - \sum_{j=1}^m f_{n(x_j)}(x_j)\ell(I_j) \right|$$

$$\leq \sum_{j=1}^m \left| f(x_j) - f_{n(x_j)}(x_j) \right| \ell(I_j)$$

$$< (\varepsilon/4(b - a)) \sum_{j=1}^m \ell(I_j) = \varepsilon/4.$$

For the second term on the right-hand side of (1), let $s = \max\{n(x_1), \ldots, n(x_m)\}$. Then, this term is

$$\left| \sum_{k=1}^{s} \left\{ \sum_{\substack{j \\ n(x_j)=k}} \left[f_{n(x_j)}(x_j)\ell(I_j) - \int_{I_j} f_{n(x_j)} \right] \right\} \right|$$

$$\leq \sum_{k=1}^{s} \left| \sum_{\substack{j \\ n(x_j)=k}} \left[f_{n(x_j)}(x_j)\ell(I_j) - \int_{I_j} f_{n(x_j)} \right] \right|$$

$$\leq \sum_{k=1}^{s} \varepsilon/2^{k+1} < \varepsilon/2,$$

by Henstock's lemma. For the third term on the right-hand side of (1), let $r = \min\{n(x_1), \ldots, n(x_m)\}$, so that $s \geq r \geq n_0$. Then

$$\int_a^b f_r = \sum_{j=1}^{m} \int_{I_j} f_r \leq \sum_{j=1}^{m} \int_{I_j} f_{n(x_j)} \leq \sum_{j=1}^{m} \int_{I_j} f_s = \int_a^b f_s \leq A.$$

and

$$\left| \sum_{j=1}^{m} \int_{I_j} f_{n(x_j)} - A \right| = A - \sum_{j=1}^{m} \int_{I_j} f_{n(x_j)} \leq A - \int_a^b f_r < \varepsilon/4.$$

It follows from (1) that $|S(f, \mathcal{D}) - A| < \varepsilon$ and the result is established for the case of a bounded interval.

Now assume that I is an unbounded interval and to be definite that $I = [a, \infty]$; the other cases are similar. Pick a sequence $\{c_k\}$ such that $c_{k+1} > c_k$, $c_1 > a$ and $c_k \to \infty$. For $a < c$, $\lim_k \int_a^c f_k \leq A$. It follows from the above discussion that f is integrable on each interval $[a, c]$. By Theorem **14.6** it suffices to show that

$$\lim_k \int_a^{c_k} f = A.$$

Note that if

$$a_{nk} = \int_a^{c_n} f_k,$$

then

$$\lim_n \lim_k a_{nk} = \lim_n \int_a^{c_n} f,$$

$$\lim_k \lim_n a_{nk} = \lim_k \int_a^\infty f_k = A,$$

$\{a_{nk}\}_{k=1}^\infty$ is increasing for each n, and $\{a_{nk}\}_{n=1}^\infty$ is increasing for each k. The result will follow if we can show that $\lim_n \lim_k a_{nk} = \lim_k \lim_n a_{nk}$; this is the content of the following lemma on infinite matrices.   □

Lemma 2. Let $a_{nk} \in \mathbb{R}$ for $n, k \in \mathbb{N}$. Assume that $\{a_{nk}: n, k \in \mathbb{N}\}$ is bounded with $A = \sup\{a_{nk}: n, k \in \mathbb{N}\}$. Then

(i) $\sup_n \sup_k a_{nk} = \sup_k \sup_n a_{nk} = A$.
(ii) If $\{a_{nk}\}_{n=1}^\infty$ is increasing for all k and if $\{a_{nk}\}_{k=1}^\infty$ is increasing for all n, then $A = \lim_n \lim_k a_{nk} = \lim_k \lim_n a_{nk}$.

Proof.

(i) Clearly $\sup_n \sup_k a_{nk} \le A$. But

$$a_{nk} \le \sup_k a_{nk} \le \sup_n \sup_k a_{nk}$$

for all n, k implies that $A \le \sup_n \sup_k a_{nk}$. The other part follows by symmetry.
(ii) For each n, $a_{nk} \uparrow \sup_k a_{nk}$ as $k \to \infty$ and $\sup_k a_{nk} \uparrow A$ as $n \to \infty$, by (i). Symmetry again yields the other part.   □

Note that the monotone convergence theorem gives a sufficient condition that justifies the interchange of the operations of limit and integral, that is,

$$\lim_k \int_I f_k = \int_I \left(\lim_k f_k \right) = \int_I f. \tag{2}$$

In many applications, the condition that $\{f_k\}$ be monotone increasing is not satisfied. The next result gives conditions that are sufficient for (2) to hold and are generally easier to use. It should be noted that (2) cannot be expected to hold without some hypothesis; for example, consider $f_n = nC_{[0,1/n]}$, where $\int_0^1 f_n = 1 \ \forall n$ and $\int_0^1 (\lim f_n) = 0$.

Theorem 3 (Lebesgue Dominated Convergence Theorem; DCT). Let I be a closed interval in $\mathbb{R}$ and $f_k: I \to \mathbb{R}$ be integrable over I for each $k \in \mathbb{N}$. Let $f, g: I \to \mathbb{R}$ be such that $\{f_k\}$ converges pointwise to f on I and g is integrable over I. If $|f_k(t)| \le g(t)$ for $k \in \mathbb{N}$ and $t \in I$, then f is integrable over I and (2) holds.

Proof. Since $|f_k(t)| \le g(t)$, each f_k is absolutely integrable. Set $U_1 = \sup\{f_k: k \in \mathbb{N}\}$. If $u_k = f_1 \vee \cdots \vee f_k$, then each u_k is integrable (**13.40**), $u_k \uparrow U_1$, and $\int_I u_k \le \int_I g$. It follows from the monotone convergence theorem that U_1 is integrable over I. Similarly, $U_k = \sup\{f_j: j \ge k\}$ is integrable over I. It is clear that $U_k \downarrow f$ pointwise (Exercise 3.18) and $\int_I U_k \ge -\int_I g$. Exercise 2 implies that f is integrable with $\lim_k \int_I U_k = \int_I f$.

Similarly, if $L_1 = \inf\{f_k: k \in \mathbb{N}\}$, then L_1 is integrable. (If $l_k = f_1 \wedge \cdots \wedge f_k$, then each l_k is integrable, $l_k \downarrow L_1$, and $\int_I l_k \ge -\int_I g$.) If $L_k = \inf\{f_j: j \ge k\}$, then, as above, L_k is integrable. Again $L_k \uparrow f$ pointwise and $\int_I L_k \le \int_I g$; the monotone convergence theorem implies that f is integrable with $\lim_k \int_I L_k = \int_I f$.

Finally, $L_k \le f_k \le U_k$, and since

$$\lim_k \int_I L_k = \lim_k \int_I U_k = \int_I f,$$

we have

$$\lim_k \int_I f_k = \int_I f. \qquad \square$$

We next establish another result that gives conditions under which we can "take the limit under the integral sign" as in (2). This result was *the* convergence theorem for the Riemann integral in the latter part of the nineteenth century. It was, and remains a very useful result; however, its utility is limited when working with unbounded intervals.

Proposition 4. Let $I = [a, b]$ be a bounded interval and $f_k, f: I \to \mathbb{R}$. If each f_k is integrable over I and $f_k \to f$ uniformly on I, then f is integrable over I and $\lim \int_I f_k = \int_I f$.

Proof. Let $\varepsilon > 0$. There is N such that $k \ge N$ implies that

$$|f_k(t) - f(t)| < \varepsilon/[2(b-a)] \qquad \text{for all } t \in I.$$

There is a gauge $\gamma (= \gamma_N)$ such that

$$\left| S(f_N, \mathcal{D}) - \int_I f_N \right| < \varepsilon/2$$

for every γ-fine tagged division $\mathcal{D}$ of I. Consequently the inequalities

$$-\varepsilon/[2(b-a)] + f_N \le f \le f_N + \varepsilon/[2(b-a)]$$

imply that if $\mathscr{D}$ is a γ-fine tagged division of I, then

$$-\varepsilon + \int_I f_N < -\varepsilon/2 + S(f_N, \mathscr{D}) \le S(f, \mathscr{D}) \le S(f_N, \mathscr{D}) + \varepsilon/2 < \int_I f_N + \varepsilon.$$

If $\mathscr{D}$ and $\mathscr{E}$ are γ-fine tagged divisions of I, it follows that

$$|S(f, \mathscr{D}) - S(f, \mathscr{E})| < 2\varepsilon,$$

and f is integrable by the Cauchy criterion.

Finally, since

$$-\varepsilon/[2(b - a)] + f_k < f < f_k + \varepsilon/[2(b - a)] \qquad \text{for } k \ge N,$$

we have

$$-\varepsilon/2 + \int_I f_k \le \int_I f \le \int_I f_k + \varepsilon/2, \qquad \text{or} \qquad \left| \int_I f_k - \int_I f \right| \le \varepsilon/2. \qquad \square$$

We conclude this section with an application of Theorem **11**.17 to the evaluation of the alternating harmonic series $\sum_{k=1}^{\infty}(-1)^k/k$. Recall that if $f(x) = \sum_{k=0}^{\infty} a_k x^k$ is a power series that converges in the interval $(-a, a)$, then the power series converges uniformly and absolutely in each closed subinterval of $(-a, a)$. It follows from Proposition 4 that the power series can be integrated termwise, that is,

$$\int_0^x f(t)\, dt = \sum_{k=0}^{\infty} a_k \frac{x^{k+1}}{k + 1},$$

for $-a < x < a$. Since

$$\sum_{k=1}^{\infty} t^{k-1} = \frac{1}{1 - t}$$

for $-1 < t < 1$,

$$\int_0^x \frac{1}{1 - t}\, dt = -\ln(1 - x) = \sum_{k=1}^{\infty} x^k/k \qquad \text{for } -1 < x < 1.$$

Since $\sum_{k=1}^{\infty} x^k/k$ has radius of convergence 1 and converges for $x = -1$, Abel's theorem implies that

$$\lim_{x \to -1^+} [-\ln(1 - x)] = -\ln 2 = \sum_{k=1}^{\infty} (-1)^k/k.$$

Integrals Containing Parameters

Let $E \subset \mathbb{R}^n$ and I be a closed subinterval in $\overline{\mathbb{R}}$. If $f \colon E \times I \to \mathbb{R}$ is such that the function $f(x, \cdot)$ is integrable over I for $x \in E$, then the integral of $f(x, \cdot)$ over I, denoted by $F(x) = \int_I f(x, t)\, dt$, defines a function $F \colon E \to \mathbb{R}$. Functions of this kind often occur in the applications of mathematics and it is usually important to know whether or not they are continuous and differentiable. Since these operations, along with the integral defining F, are infinite processes, there is the potential for problems in taking the various limits involved. Particularly interesting is the interchanging of infinite processes such as differentiation and integration. We begin with a result on the continuity of F.

Theorem 5. If $f(\cdot, t)$ is continuous at $x_0 \in E$, for each $t \in I$, each $f(x, \cdot)$ is integrable over I, and there is a function $g \colon I \to \mathbb{R}$ which is integrable over I such that $|f(x, t)| \le g(t)$ for all $x \in E$, $t \in I$, then the function F is continuous at x_0.

Proof. Let $\{x_k\}$ be a sequence in E that converges to x_0. Then $f(x_k, t) \to f(x_0, t)$ by the continuity of $f(\cdot, t)$. Since $|f(x_k, t)| \le g(t)$, the dominated convergence theorem implies that

$$\int_I f(x_k, t)\, dt \to \int_I f(x_0, t)\, dt.$$

This, of course, is equivalent to the statement that F is continuous at x_0. $\square$

EXAMPLE 6. The gamma function (**14**.11)

$$\Gamma(x) = \int_0^\infty t^{x-1} e^{-t}\, dt, \qquad x > 0,$$

is continuous. To show this, let $x_0 > 0$. If $0 \le t \le 1$ and $x_0/2 < x$, then $t^x \le t^{x_0/2}$ and $e^{-t} \le 1$. Therefore,

$$t^{x-1} e^{-t} \le t^{(x_0/2)-1}.$$

So, by Theorem 5,

$$\int_0^1 t^{x-1} e^{-t}\, dt$$

is continuous at $x = x_0$. For $t > 1$ and $x_0/2 \le x \le 2x_0$, there is B such that

$t^{x-1}e^{-t} \le Bt^{-2}$. It follows again that

$$\int_1^\infty t^{x-1}e^{-t}\,dt$$

is continuous at x_0. $\square$

The next result gives conditions under which the function F is differentiable and for which the derivative of the integral is the integral of the derivative.

Theorem 7 (Leibniz' Rule). Let $E = [a, b]$, $a, b \in \mathbb{R}$. Suppose

  (i) $\partial f/\partial x = D_1 f$ exists for all $x \in E$, $t \in I$,
  (ii) Each $f(x, \cdot)$ is integrable and there is an integrable function $g: I \to \mathbb{R}$ such that $|(\partial f/\partial x)(x, t)| = |D_1 f(x, t)| \le g(t)$ for all $x \in E$, $t \in I$.

Then, F is differentiable on E with

$$F'(x) = \int_I D_1 f(x, t)\,dt = \int_I \frac{\partial f}{\partial x}(x, t)\,dt.$$

Proof. Fix $x_0 \in E$ and let $\{x_k\} \subset E$ be a sequence converging to x_0, $x_k \ne x_0$. For $t \in I$,

$$\lim_k \frac{f(x_k, t) - f(x_0, t)}{x_k - x_0} = D_1 f(x_0, t) = \frac{\partial f}{\partial x}(x_0, t)$$

and the function

$$h_k(t) = \frac{f(x_k, t) - f(x_0, t)}{x_k - x_0}$$

is integrable over I. By the mean value theorem, for each pair k, t, there is $z_{k,t}$ between x_k and x_0 such that

$$\frac{f(x_k, t) - f(x_0, t)}{x_k - x_0} = D_1 f(z_{k,t}, t),$$

and so,

$$\left| \frac{f(x_k, t) - f(x_0, t)}{x_k - x_0} \right| \le g(t)$$

for all k, t. The dominated convergence theorem implies that

$$\lim_{k} \frac{F(x_k) - F(x_0)}{x_k - x_0} = \int_I D_1 f(x_0, t) \, dt = F'(x_0). \qquad \Box$$

Note that if I is a bounded interval and $D_1 f$ is continuous on $E \times I$, then the conclusion of Theorem 7 holds, for in this case the function g in Theorem 7 can be taken to be constant.

We give several applications of Leibniz' rule.

EXAMPLE 8. $\int_0^\infty e^{-x^2} \, dx = \sqrt{\pi}/2$. Since $e^{-x^2} \le e^{-x}$ for $x \ge 1$, the above integral exists. Define

$$f(t) = \left[\int_0^t e^{-x^2} \, dx \right]^2$$

and

$$g(t) = \int_0^1 e^{-t^2(x^2+1)} / (x^2 + 1) \, dx, \qquad t \ge 0.$$

We need to calculate $\lim_{t \to \infty} f(t)$. To do this, we show that $f + g = \text{constant} = c$, evaluate c, and then show that $\lim_{t \to \infty} g(t) = 0$. For $t \ge 0$,

$$f'(t) = 2e^{-t^2} \int_0^t e^{-x^2} \, dx.$$

Since

$$\left| \frac{\partial}{\partial t} \left(e^{-t^2(x^2+1)} / (x^2+1) \right) \right| = \left| -2(te^{-t^2}) e^{-t^2 x^2} \right| \le B$$

for $t > 0, 0 \le x \le 1$, Theorem 7 implies that g is differentiable with

$$g'(t) = -2te^{-t^2} \int_0^1 e^{-t^2 x^2} \, dx = -2e^{-t^2} \int_0^t e^{-u^2} \, du.$$

Hence, $f'(t) + g'(t) = 0$ for $t > 0$ and $f + g = \text{constant} = c$ on $[0, \infty)$, with $f(0) + g(0) = c = \int_0^1 (1 + x^2)^{-1} \, dx = \pi/4$. Since

$$\left| e^{-t^2(x^2+1)} / (x^2 + 1) \right| \le 1/(x^2 + 1),$$

the dominated convergence theorem implies that

$$\lim_{t \to \infty} (f(t) + g(t)) = \pi/4 = \left[\int_0^\infty e^{-x^2} \, dx \right]^2 + 0.$$

This integral is encountered in dealing with the normal distribution in probability and statistics.   □

The general Leibniz differentiation rule involves integrals in which either the upper or lower limit of integration can be variable. This form of the rule is easily obtained from Theorem 7 and the chain rule.

Corollary 9. Let I be a bounded interval, f and $D_1 f$ continuous on $I \times \mathbb{R}$, and $\alpha, \beta: I \to \mathbb{R}$ differentiable on $\mathbb{R}$. If

$$F(x) = \int_{\alpha(x)}^{\beta(x)} f(x, t) \, dt \qquad \text{for } x \in I,$$

then F is differentiable and

$$F'(x) = \int_{\alpha(x)}^{\beta(x)} D_1 f(x, t) \, dt + f(x, \beta(x))\beta'(x) - f(x, \alpha(x))\alpha'(x).$$

Proof. Define G on $I \times \mathbb{R} \times \mathbb{R}$ by $G(x, y, z) = \int_y^z f(x, t) \, dt$. By Theorem 7, $D_1 G(x, y, z) = \int_y^z D_1 f(x, t) \, dt$. Since $D_1 f$ is bounded on bounded sets, $D_1 G$ is continuous. By Theorem 13.28, $D_2 G(x, y, z) = -f(x, y)$ and $D_3 G(x, y, z) = f(x, z)$. Thus, G has continuous first partial derivatives and is, therefore, differentiable. Applying the chain rule to $F(x) = G(x, \alpha(x), \beta(x))$ yields the result.   □

A generalization of the formula in Example 8 is given in Example 10.

EXAMPLE 10. $\int_0^\infty e^{-x^2} \cos(2xt) \, dx = \sqrt{\pi} \, e^{-t^2}/2$ for $t \in \mathbb{R}$. Set

$$F(t) = \int_0^\infty e^{-x^2} \cos(2xt) \, dx.$$

The integral exists, since

$$\left| e^{-x^2} \cos(2xt) \right| \le e^{-x^2}.$$

Since

$$\left| \frac{\partial}{\partial t} \left[e^{-x^2} \cos(2xt) \right] \right| \le 2xe^{-x^2} = g(x)$$

and g is integrable over $[0, \infty)$, it follows that F is differentiable and

$$F'(t) = \int_0^\infty - 2xe^{-x^2} \sin(2xt) \, dx.$$

Integration by parts yields

$$F'(t) = e^{-x^2} \sin 2xt|_{x=0}^{x=\infty} - \int_0^\infty e^{-x^2} 2t \cos(2xt)\, dx$$

$$= -2t \int_0^\infty e^{-x^2} \cos(2xt)\, dx = -2tF(t).$$

Thus, F satisfies the differential equation $F'(t) = -2tF(t)$. The unique solution of this equation is

$$F(t) = F(0)e^{-t^2}.$$

But $F(0) = \sqrt{\pi}/2$ by Example 8. It follows that

$$F(t) = \frac{\sqrt{\pi}}{2}e^{-t^2}. \qquad \square$$

EXAMPLE 11. $F(x) = \int_0^\infty e^{-xt}(\sin t/t)\, dt = \pi/2 - \arctan x$ for $x \ge 0$. First, consider the case $x > 0$: Since

$$\left| e^{-xt}\frac{\sin t}{t} \right| \le e^{-xt},$$

the integral exists. We claim that $\lim_{x \to \infty} F(x) = 0$. Let $x_k \to \infty$ with $x_k > 1$. Then

$$\left| e^{-x_k t}\frac{\sin t}{t} \right| \le e^{-t}$$

for $t \ge 0$, and the dominated convergence theorem implies that $\lim_{x \to \infty} F(x) = 0$. We next claim that $F'(x) = -1/(1 + x^2)$. Since

$$\left| \frac{\partial}{\partial x}\left(e^{-xt}\frac{\sin t}{t} \right) \right| = |e^{-xt} \sin t| \le e^{-xt} \le e^{-at} \qquad \text{for } x \ge a > 0,$$

Theorem 7 applied to the interval $[a, \infty)$ implies that

$$F'(x) = e^{-xt}(x \sin t + \cos t)/(1 + x^2)|_0^\infty = \frac{-1}{(1 + x^2)},$$

for $x \ge a$; thus the formula is valid for $x > 0$. Hence,

$$F(x) - F(c) = \int_c^x F' = \arctan c - \arctan x.$$

Letting $c \to \infty$ yields $F(x) = \pi/2 - \arctan x$.

Next, consider the case $x = 0$: Recall that

$$F(0) = \int_0^\infty \frac{\sin t}{t} \, dt$$

exists (Example **14**.10). Define

$$F_k(x) = \int_0^k e^{-xt} \frac{\sin t}{t} \, dt.$$

Since $\lim_{k \to \infty} F_k(0) = F(0)$, we want to calculate $\lim_{k \to \infty} F_k(0)$; we do this by evaluating $\lim_k \{ F_k(k) - F_k(0) \}$. Note that

$$|F_k(k)| \le \int_0^k e^{-kt} \, dt = (1 - e^{-k^2})/k \le 1/k \to 0.$$

Since

$$\left| \frac{\partial}{\partial x} \left(e^{-xt} \frac{\sin t}{t} \right) \right| = |e^{-xt} \sin t| \le e^{-xt},$$

F_k is differentiable on $[0, \infty)$ with

$$F_k'(x) = -\int_0^k e^{-xt} \sin t \, dt = \frac{e^{-kx}(x \sin k + \cos k) - 1}{1 + x^2}.$$

If $g_k = C_{[0, k]} F_k'$, then

$$|g_k(x)| \le |F_k'(x)| \le \frac{e^{-x}(1 + x) + 1}{1 + x^2}$$

and $\lim_k g_k(x) = -1/(1 + x^2)$. The dominated convergence theorem implies that

$$\lim_k \int_0^\infty g_k = \lim_k \int_0^k F_k' = \lim_k \{ F_k(k) - F_k(0) \} = \lim_k (-F_k(0))$$

$$= -F(0) = \int_0^\infty -1/(1 + x^2) \, dx = -\pi/2.$$

Recall that the integrand defining $F(0)$ is not absolutely integrable (Example **14**.10).   □

Integrals over General Sets

Let $f: \mathbb{R} \to \mathbb{R}$ and $E \subseteq \mathbb{R}$. We say that F is *integrable over E* if $C_E f$ is integrable over $\mathbb{R}$. In this case, we define the *integral of f over E* by

$\int_E f = \int_{\mathbb{R}} C_E f$. If f is initially defined only on $F \supseteq E$, then F can be extended to $\mathbb{R}$ by defining $f(x) = 0$ for $x \in \mathbb{R} \setminus F$.

Proposition 12. Let $f: \mathbb{R} \rightarrow \mathbb{R}$.

(i) If $E \subseteq F$ and f is integrable over both E and F, then f is integrable over $F \setminus E$ with

$$\int_E f + \int_{F \setminus E} f = \int_F f.$$

(ii) If $E \cap F = \varnothing$ and f is integrable over both E and F, then f is integrable over $E \cup F$ with

$$\int_{E \cup F} f = \int_E f + \int_F f.$$

Proof. (i) and (ii) follow from the identities $C_F = C_E + C_{F \setminus E}$ and $C_{E \cup F} = C_E + C_F$ and the linearity of the integral.   □

It is not generally true that if a function is integrable over each of two sets, then it is also integrable over their union or their intersection.

EXAMPLE 13. Define a sequence $\{a_k\}$ in $\mathbb{R}$ by

$$\{a_k\} = \left\{ 1, -1, 1, \tfrac{1}{2}, -\tfrac{1}{2}, \tfrac{1}{2}, \tfrac{1}{3}, -\tfrac{1}{3}, \tfrac{1}{3}, \dots \right\}.$$

Define $f: \mathbb{R} \rightarrow \mathbb{R}$ by

$$f(x) = \begin{cases} 0, & \text{if } x < 1, \\ a_k, & \text{if } k \le x < k + 1,\ k = 1, 2, \dots . \end{cases}$$

By Exercise 14.3, f is not integrable over $[1, \infty)$ since the series Σa_k doesn't converge. If $E_k = [k, k + 1)$, then f is integrable over the set F_3, which is the union of the sets $E_1, E_2, E_4, E_5, \dots$, (i.e., we delete every third member of the sequence $\{E_i\}$), by Exercise 14.3; f is also integrable over the set F_1, which is the union of the sequence of sets $E_2, E_3, E_5, E_6, \dots$; that is, beginning with E_1, we delete every third member of the sequence $\{E_i\}$. However, f is not integrable over either of the sets $F_1 \cup F_3 = [1, \infty)$ or $F_1 \cap F_3$ (again Exercise 14.3).   □

Again, note that the function in Example 13 is not absolutely integrable. For absolutely integrable functions, the situation is much better. (As before, we say that a function f is *absolutely integrable over* E if both f and $|f|$ are integrable over E.)

Proposition 14. Suppose that $f: \mathbb{R} \to \mathbb{R}$ is absolutely integrable over E and F. Then f is absolutely integrable over both $E \cup F$ and $E \cap F$.

Proof. By considering $f = f^+ - f^-$, we may assume that $f \geq 0$. Then $C_E f \wedge C_F f = C_{E \cap F} f$ and $C_E f \vee C_F f = C_{E \cup F} f$. Proposition **13**.40 yields the desired result. $\square$

We also have the following general comparison result from Proposition **13**.33.

Proposition 15. Let f be integrable and g be nonnegative and integrable over the set E. If $|f(t)| \leq g(t)$ for all $t \in E$, then f is absolutely integrable over E.

The monotone convergence theorem yields the following two results relative to integration over sequences of sets.

Proposition 16. Let $f: \mathbb{R} \to \mathbb{R}$ and let $\{E_i\}$ be an increasing sequence of sets with

$$E = \bigcup_{i=1}^{\infty} E_i.$$

Suppose that f is integrable over each E_i and $f \geq 0$ on E. Then f is integrable over E if, and only if,

$$\lim \int_{E_i} f$$

is finite; in this case,

$$\int_E f = \lim \int_{E_i} f.$$

Proof. Since $C_{E_i} f \uparrow C_E f$, the monotone convergence theorem gives the result. $\square$

Proposition 17. Let $\{E_i\}$ be a sequence of pairwise disjoint subsets of $\mathbb{R}$ with

$$E = \bigcup_{i=1}^{\infty} E_i,$$

and suppose that $f: \mathbb{R} \to \mathbb{R}$ is absolutely integrable over each E_i. Then f is absolutely integrable over E if, and only if,

$$\sum_{i=1}^{\infty} \int_{E_i} |f| < \infty;$$

in this case,

$$\int_E f = \sum_{i=1}^{\infty} \int_{E_i} f.$$

Proof. Suppose that f is absolutely integrable over E. Then, for each $n \in \mathbb{N}$,

$$C_{\bigcup_{i=1}^{n} E_i} |f| \le C_E |f|,$$

and so,

$$\int_{\mathbb{R}} C_{\bigcup_{i=1}^{n} E_i} |f| = \sum_{i=1}^{n} \int_{E_i} |f| \le \int_{\mathbb{R}} C_E |f| = \int_E |f| < \infty.$$

It follows that

$$\sum_{i=1}^{\infty} \int_{E_i} |f| \le \int_E |f|.$$

For the converse, note that

$$\sum_{i=1}^{\infty} C_{E_i} |f| = C_E |f|$$

pointwise and that for every n,

$$\sum_{i=1}^{n} \int_{\mathbb{R}} C_{E_i} |f| \le \sum_{i=1}^{\infty} \int_{E_i} |f|.$$

The monotone convergence theorem implies that $|f|$ is integrable over E. Finally,

$$\sum_{i=1}^{n} C_{E_i} f \to C_E f$$

pointwise and

$$\left| \sum_{i=1}^{n} C_{E_i} f \right| \le C_E |f|.$$

So, the dominated convergence theorem implies that f is integrable over E with

$$\int_E f = \sum_{i=1}^{\infty} \int_{E_i} f. \qquad \square$$

EXERCISES 15

1. Show that the analogue of Theorem 1 is false for the Riemann integral.

2. State and prove a version of the monotone convergence theorem for decreasing sequences $\{f_k\}$.

3. Let $f_k \colon I \to \mathbb{R}$, $k \in \mathbb{N}$, be nonnegative and integrable over I and $f = \sum_{k=1}^{\infty} f_k$. Give the necessary and sufficient conditions for f to be integrable over I.

4. Give a power series expansion for arctan x.
   [*Hint:* Use the power series for $1/(1 + t^2)$.] Use this to show that

$$\sum_{k=0}^{\infty} (-1)^k \frac{1}{2k + 1} = \frac{\pi}{4}.$$

5. Show that the analogue of Theorem 3 is false for the Riemann integral.

6. (Bounded Convergence Theorem) Let I be a bounded, closed interval. Let $f_k, f \colon I \to \mathbb{R}$, f_k integrable over I for each $k \in \mathbb{N}$, and $f_k \to f$ pointwise on I. If there is $M > 0$ such that $|f_k(t)| \le M$ for all $k \in \mathbb{N}$ and all $t \in I$, show that f is absolutely integrable over I and that (2) holds.

7. Show that the analogue of Exercise 6 is false for the Riemann integral.

8. Show that

$$F(x) = \int_1^{\infty} \frac{\sin t}{x^2 + t^2} \, dt$$

   is continuous for all $x \in \mathbb{R}$.

9. (Laplace Transform) Let $f \colon [0, \infty) \to \mathbb{R}$ be integrable over every bounded interval. The *Laplace transform* of f is defined by

$$\mathscr{L}\{f\}(x) = \int_0^{\infty} e^{-xt} f(t) \, dt.$$

   A function f is said to be of *exponential order* if there are $M > 0$ and $a \in \mathbb{R}$ such that $|f(t)| \le M e^{at}$ for all $t > 0$. Show that the Laplace transform of such a function exists for $x > a$ and defines a continuous function.

10. Let f be of exponential order and differentiable on $[0, \infty)$. Show that $\mathscr{L}\{f'\}$ exists for $x > a$ and $\mathscr{L}\{f'\}(s) = s\mathscr{L}\{f\}(s) - f(0)$.

11. Show that $F(x) = \int_0^{\infty} e^{-xt} \sin t \, dt$ defines a continuous function for $x > 0$.

12. Show that $\Gamma(\frac{1}{2}) = \sqrt{\pi}$.

13. Calculate $F'(x)$ where

$$F(x) = \int_{-x}^{\sin x} \frac{dt}{x + t + 1}$$

    (a) Directly.

    (b) Using Leibniz' rule.

14. Show that the differential equation $F'(t) = -2tF(t)$ has a unique solution for $t \in \mathbb{R}$.

15. Show that

$$\int_0^\infty x^{2n} e^{-x^2}\, dx = \frac{(2n)!}{2^{2n} n!} \frac{\sqrt{\pi}}{2} \qquad \text{for } n = 0, 1, 2, \ldots .$$

    Hint: $n = 0$ is Example 8; use induction.

16. Show that

$$\int_0^\infty e^{-tx^2}\, dx = \tfrac{1}{2}\sqrt{\pi/t} \qquad \text{for } t > 0.$$

17. Show that the Laplace transform of a function of exponential order is differentiable and express the derivative in terms of the Laplace transform.

18. Show that the function $F(x) = \int_0^\infty e^{-xt}/(1 + t)\, dt$ is differentiable for $x > 0$.

19. Show that the function $F(x) = \int_0^\infty e^{-t}/(1 + xt)\, dt$ is differentiable for $0 \leq x \leq 1$.

20. If E is a closed interval, show that the definition of integrability over E, given in this chapter, agrees with the earlier definition.

21. Does a result analogous to Proposition 16 hold for the Riemann integral?

22. Does a result analogous to Proposition 17 hold for the Riemann integral?

23. A subset $E \subset \mathbb{R}$ is *null* if C_E is integrable over $\mathbb{R}$ and $\int_{\mathbb{R}} C_E = 0$. If $\mathcal{N}$ is the class of null sets, show that $\mathcal{N}$ is closed under countable unions, countable intersections, differences, and if $E \in \mathcal{N}$, $F \subset E$, then $F \in \mathcal{N}$.

24. Let $f: [0, 1] \to \mathbb{R}$ be continuous. Define g_n by $g_n(t) = f(t^n)$. Show that $\int_0^1 g_n \to f(0)$.

25. Let $f: [0, \infty) \to \mathbb{R}$ be continuous and $\lim_{x \to \infty} f(x) = L$. What can you say about $\lim_{n \to \infty} \int_0^2 f(nx)\, dx$?

26. Let $f, g: I \to \mathbb{R}$. Suppose f is absolutely integrable over I and g is such that there exists a sequence of step functions $\{s_n\}$ such that $s_n \to g$

pointwise and $|s_n(t)| \le M$ for all $t \in I$, $n \in \mathbb{N}$. Show that fg is absolutely integrable over I. (For conditions that ensure the existence of such a sequence, see Lemma **16**.9 and the remarks following it.)

27. Show that the product of integrable functions need not be integrable.

28. Evaluate $\varphi(x) = \int_0^1 (t^x - 1)/\ln t \, dt$ for $x > 0$.
    Hint: Find $\varphi'(x)$ and note that $\varphi(x) \to 0$ as $x \to 0^+$. Be sure to check the validity of using Leibniz' rule.

29. Let $f: [0, \infty) \to \mathbb{R}$ be continuous. Let $u(x) = \int_0^x (x - s)f(s) \, ds$. Find u''.

30. Let $f: [a, b] \times \mathbb{R} \times \mathbb{R} \to \mathbb{R}$ be continuous and have continuous partial derivatives $D_2 f$ and $D_3 f$. Let $\alpha, \beta: [a, b] \to \mathbb{R}$ have continuous first derivatives. Define $F: \mathbb{R} \to \mathbb{R}$ by $F(s) = \int_a^b f(t, \alpha(t) + s\beta(t), \alpha'(t) + s\beta'(t)) \, dt$. Calculate $F'(s)$.
    Hint: The chain rule is useful.

31. (Improved DCT) Show that if the hypotheses of Theorem 3 hold, then $\int_I |f_n - f| \to 0$.
    Hint: $|f_k - f| \le 2g$.

32. Let $f_n(t) = n/(t^2 + n^2)$. Show that $0 \le f_n \le 1$, $\lim f_n(t) = 0$, and $\int_\mathbb{R} f_n = \pi$. Why doesn't this contradict the DCT?

33. Let f_k, f satisfy the hypotheses of the DCT on $\mathbb{R}$. If $F_k(x) = \int_0^x f_k$, $F(x) = \int_0^x f$, show that $F_k \to F$ uniformly on $\mathbb{R}$.

34. Give an example in which $f_k \to 0$ pointwise and $\int f_k \to 0$, but $\{f_k\}$ is not dominated by any integrable function.

35. Let $f_k(t) = (1/k^2\sqrt{t}) \cos(k/t)$. Show that $\sum_{k=1}^\infty \int_0^1 f_k = \int_0^1 \sum_{k=1}^\infty f_k$.

36. Let $f_k(t) = e^{-kt} - 2e^{-2kt}$. Show that $\sum_{k=1}^\infty \int_0^\infty f_k \ne \int_0^\infty \sum_{k=1}^\infty f_k$.

37. Let $f: \mathbb{R} \to \mathbb{R}$ and set

$$f_k(t) = \begin{cases} 0 & \text{if } |f(t)| > k, \\ f(t) & \text{if } |f(t)| \le k. \end{cases}$$

Suppose that each f_k is absolutely integrable. Prove that if f is absolutely integrable, then $\int_\mathbb{R} f = \lim \int_\mathbb{R} f_k$. If $\lim \int_\mathbb{R} f_k$ exists, is f necessarily integrable?

38. Let $f: [-1, 1] \to \mathbb{R}$ be continuous. Show that

$$\lim_{h \to 0^+} \int_{-1}^1 \frac{h}{h^2 + x^2} f(x) \, dx = \frac{\pi}{2} f(0).$$

39. If $f: [0, \infty) \to \mathbb{R}$ is bounded and continuous, show that

$$\lim_k \int_0^\infty \frac{f(t)}{1 + kt^p} \, dt = 0, \qquad \text{for } p > 1.$$

40. (Fatou's Lemma) Let $f_k\colon I \to \mathbb{R}$ be nonnegative, integrable, and such that $\{f_k\}$ converges pointwise to the function $f\colon I \to \mathbb{R}$. Show that if $\varliminf \int_I f_k < \infty$, then f is integrable and

$$\int_I f \leq \varliminf \int_I f_k.$$

41. Let $f\colon [a, b] \times [c, d] \to \mathbb{R}$ be continuous. Show that $F(x) = \int_c^d f(x, y)\, dy$ is continuous on $[a, b]$.

42. Show that the kth derivative of the gamma function exists for all $k \in \mathbb{N}$. Give a formula for $\Gamma^{(k)}(x)$, $x > 0$.

43. (Riemann–Lebesgue Lemma) Let $f\colon [0, 1] \to \mathbb{R}$ be continuous. Show that $\int_0^1 f(t) \sin k\pi t\, dt \to 0$.
Hint: First prove the result for f a step function.

44. Show that

$$\lim_{x \to 0^+} \int_0^1 \frac{e^{-xt}}{1 + t}\, dt = \ln 2.$$

45. Evaluate

$$\lim_{x \to \infty} \int_0^\infty \frac{e^{-xt}}{1 + t}\, dt.$$

46. Let $f\colon \mathbb{R} \to \mathbb{R}$ be absolutely integrable. Show that if f is uniformly continuous on $\mathbb{R}$, then $\lim_{|x| \to \infty} f(x) = 0$. Can "uniform continuity" be replaced by "continuity"?

47. A function $f\colon \mathbb{R} \to \mathbb{R}$ is of *bounded variation over* $\mathbb{R}$ if f has bounded variation over every compact interval and if

$$\mathrm{Var}\,(f\colon \mathbb{R}) = \lim_{x \to \infty} \mathrm{Var}\,(f\colon [-x, x]) < \infty.$$

State and prove the analogue of Theorem **13**.31 for functions $f\colon \mathbb{R} \to \mathbb{R}$.

Lebesgue Measure in ℝ

For intervals in $\mathbb{R}$, there is a notion of length that arises through the introduction of the geometrical model, called the real line, for $\mathbb{R}$. [If the interval I has endpoints a and b, $a \le b$, then the *length* of I is $\ell(I) = b - a$.] It is natural to examine the possibility of extending this concept of length to subsets of $\mathbb{R}$ that are more general than intervals and to do it so that the extension will enjoy the geometric properties of the length function. At the very least, such an extension λ should satisfy the following properties:

   (i) $\lambda(\varnothing) = 0$ and $\lambda(E) \ge 0$.

   (ii) $\lambda(E + a) = \lambda(E)$ for $a \in \mathbb{R}$ (translation invariance).

 (iii) $\lambda(E \cup F) = \lambda(E) + \lambda(F)$ if $E \cap F = \varnothing$ (finite additivity).

It is often advantageous to require a stronger version of (iii):

 (iv) $\lambda(\bigcup_{i=1}^{\infty} E_i) = \sum_{i=1}^{\infty} \lambda(E_i)$ if $\{E_i\}$ are pairwise disjoint (countable additivity).

In his construction of the integral, Lebesgue gave a geometric description of an extension satisfying (i) to (iv). Using the properties of the gauge integral, we too can easily construct such an extension.

A subset $E \subseteq \mathbb{R}$ is said to be *Lebesgue integrable* or *integrable* if the characteristic function C_E is integrable over $\mathbb{R}$; in this case, the *Lebesgue measure* or *measure* of the set E is defined by $\lambda(E) = \int_{\mathbb{R}} C_E$. The class of Lebesgue integrable sets will be denoted by $\mathscr{L}$. It follows from the properties of the gauge integral that $\mathscr{L}$ contains the class of bounded intervals; Lebesgue measure λ then extends the length function to the class of sets $\mathscr{L}$. The basic properties of $\mathscr{L}$ and λ are summarized below.

Proposition 1.

    (i) All bounded intervals belong to $\mathscr{L}$.
    (ii) $E, F \in \mathscr{L}$, implies $E \setminus F \in \mathscr{L}$, $E \cup F \in \mathscr{L}$.
    (iii) $E_i \in \mathscr{L}$, $i \in \mathbb{N}$, implies $E = \bigcap_{i=1}^{\infty} E_i \in \mathscr{L}$.

Proof. (i) has been observed above and (ii) follows from Propositions **15.**12 and **15.**14. For (iii), let $F_i = \bigcap_{k=1}^{i} E_k$. Then $E = \bigcap_{i=1}^{\infty} F_i$ and each F_i is in $\mathscr{L}$ by Proposition **15.**14. Since $C_{F_i} \downarrow C_E$, $E \in \mathscr{L}$ by the monotone convergence theorem.   □

$\mathscr{L}$ is not closed under countable unions (see Exercise 1), that is, there exist sequences of sets in $\mathscr{L}$ whose union is not in $\mathscr{L}$. However, we do have the following proposition.

Proposition 2. Let $\{E_i\}$ be a sequence of pairwise disjoint sets in $\mathscr{L}$. Then $E = \bigcup_{i=1}^{\infty} E_i \in \mathscr{L}$ if, and only if, $\sum_{i=1}^{\infty} \lambda(E_i) < \infty$. In this case, $\lambda(E) = \sum_{i=1}^{\infty} \lambda(E_i)$. This property is called *countable additivity*.

Proof. Since $C_E = \sum_{i=1}^{\infty} C_{E_i}$, the conclusion follows from the monotone convergence theorem.   □

From Proposition **15.**17 and the the remarks above we obtain Proposition 3.

Proposition 3. The Lebesgue measure $\lambda \colon \mathscr{L} \to \mathbb{R}$ satisfies

    (i) $\lambda(\varnothing) = 0$.
    (ii) If I is a bounded interval, then $\lambda(I) = \ell(I)$.
    (iii) If $E_i \in \mathscr{L}$, $i \in \mathbb{N}$, are pairwise disjoint and $E = \bigcup_{i=1}^{\infty} E_i \in \mathscr{L}$. Then
        $\lambda(E) = \sum_{i=1}^{\infty} \lambda(E_i)$.

A subset $E \subseteq \mathbb{R}$ is *Lebesgue measurable* or, simply *measurable* if $E \cap I$ is integrable for every compact interval I. The family of all measurable sets is denoted by $\mathscr{M}$; its properties are summarized in Proposition 4.

Proposition 4.

(i) $\varnothing \in \mathscr{M}$.

(ii) $E \in \mathscr{M}$ implies $E^c \in \mathscr{M}$.

(iii) $E_i \in \mathscr{M}$, $i \in \mathbb{N}$, implies $E = \bigcup_{i=1}^{\infty} E_i \in \mathscr{M}$.

(iv) $\mathscr{L} \subset \mathscr{M}$.

(v) All intervals belong to $\mathscr{M}$.

Proof. (i) is clear. For (ii), if I is a compact interval, then $I = (I \cap E) \cup (I \cup E^c)$ and $I \in \mathscr{L}$, $I \cap E \in \mathscr{L}$ imply $I \cap E^c \in \mathscr{L}$ by Proposition 1(ii).

For (iii), first note that if I is a compact interval, then $I \cap (E \cup F) = (I \cap E) \cup (I \cap F)$ and $I \cap (E \cap F) = (I \cap E) \cap (I \cap F)$. Consequently, $E, F \in \mathscr{M}$ implies that $E \cup F$ and $E \cap F$ belong to $\mathscr{M}$. If $\{E_i\} \subset \mathscr{M}$, we construct a sequence of pairwise disjoint sets $\{F_i\} \subset \mathscr{M}$ such that $E = \bigcup_{i=1}^{\infty} F_i$. Define $F_1 = E_1$ and $F_{k+1} = E_{k+1} \setminus \bigcup_{i=1}^{k} E_i$. Then the F_i are pairwise disjoint, $F_i \in \mathscr{M}$ by (ii) and the observation above, and $E = \bigcup_{i=1}^{\infty} E_i = \bigcup_{i=1}^{\infty} F_i$. If I is a compact interval, $F_i \cap I \in \mathscr{L}$ and $\sum_{i=1}^{\infty} \lambda(F_i \cap I) \leq \lambda(I) < \infty$. So Proposition 2 implies that $\bigcup_{i=1}^{\infty} F_i \cap I = E \cap I \in \mathscr{L}$. Hence $E \in \mathscr{M}$.

(iv) and (v) follow from Proposition 3. □

We extend λ from $\mathscr{L}$ to $\mathscr{M}$ by setting $\lambda(E) = \infty$ if $E \in \mathscr{M} \setminus \mathscr{L}$ and continue to refer to λ as Lebesgue measure. The basic properties of this extension are given in the following proposition.

Proposition 5.

(i) $\lambda(\varnothing) = 0$.

(ii) $\lambda(E) \geq 0$ for all $E \in \mathscr{M}$.

(iii) If I is an interval (possibly unbounded), $\lambda(I) = \ell(I)$.

(iv) If $\{E_i\}$ is a sequence of pairwise disjoint sets in M, then $\sum_{i=1}^{\infty} \lambda(E_i) = \lambda(\bigcup_{i=1}^{\infty} E_i)$ (countable additivity).

(v) $E \in \mathscr{M}$ if, and only if, $E + a \in \mathscr{M}$ for every $a \in \mathbb{R}$ and $\lambda(E) = \lambda(E + a)$ (translation invariance).

Proof. (i) through (iv) follow from Proposition 4. The translation invariance for elements $E \in \mathscr{L}$ follows from Exercise 5. If E is measurable, but not integrable, then $\lambda(E + a) = \infty$, since otherwise $E = (E + a) - a$ would be integrable. □

The size of the class of measurable sets is not clear at this point. From Proposition 4 it follows that open intervals are measurable and that a countable union of pairwise disjoint open intervals is also measurable. It turns out, as is shown in the next example, that every open set in $\mathbb{R}$ is just such a union and is therefore measurable.

EXAMPLE 6. Let $G \subseteq \mathbb{R}$ be open and $x \in G$. We associate with x an open interval I_x as follows: let $\{ J_\alpha = (a_\alpha, b_\alpha): \alpha \in A \}$ be the set of all open intervals J_α, indexed by the set A, such that $x \in J_\alpha \subseteq G$. Set $I_x = \bigcup_{\alpha \in A} J_\alpha$. If $b = \sup \{ b_\alpha: \alpha \in A \}$ and $a = \inf \{ a_\alpha: \alpha \in A \}$, then we claim that $I_x = (a, b)$ so that I_x is indeed an open interval. For if $y \notin (a, b)$, then either $y \geq b$ or $y \leq a$, and so, either $y \geq b_\alpha$ for all $\alpha \in A$, or $y \leq a_\alpha$ for all $\alpha \in A$. It follows that $y \notin J_\alpha$ for all $\alpha \in A$ and that $y \notin I_x$. Thus, $(a, b) \supseteq I_x$. If $y \in (a, b)$, then either $y = x$, or $a < y < x$, or $x < y < b$. If $y = x$, clearly $y \in I_x$. If $a < y < x$, then there is $\alpha \in A$ such that $a < a_\alpha < y < x < b_\alpha$; this implies that $y \in J_\alpha \subseteq I_x$. Similarly, if $x < y < b$, $y \in I_x$. Hence, $(a, b) \subseteq I_x$, and so $(a, b) = I_x$.

Next we claim that $\{ I_x: x \in G \}$ is pairwise disjoint. Suppose $z \in I_x \cap I_y$. Then $I_x \cup I_y$ is a open interval and $x \in I_x \cup I_y$ implies $I_x \cup I_y \subseteq I_x$ by construction. Thus, $I_x = I_x \cup I_y$ and, similarly, $I_x \cup I_y = I_y$. It follows that the family $\{ I_x: x \in G \}$ is pairwise disjoint.

Finally, $\mathscr{I} = \{ I_x: x \in G \}$ is at most countable since we may associate with each I_x a rational number $r_x \in I_x$. Then $x \to r_x$ is a one-to-one map from $\mathscr{I}$ into $\mathbb{Q}$. If follows that $\mathscr{I}$ is at most countable. Therefore, $G = \bigcup_{x \in G} I_x$ gives G as a countable union of pairwise disjoint open intervals. Thus, all open sets are measurable and it follows from Proposition 4 that all closed sets are measurable. □

A family Σ of subsets of a set S that satisfies properties (i), (ii), and (iii) of Proposition 4 is called a *σ-algebra*. The smallest σ-algebra of subsets of $\mathbb{R}$ that contains the open sets is called the family of *Borel sets* [after the French mathematical analyst Emile Borel (1871–1938)] in $\mathbb{R}$ and is denoted by $\mathscr{B} = \mathscr{B}(\mathbb{R})$. It follows from Proposition 4 that $\mathscr{B} \subset \mathscr{M}$. This containment is proper; that is, there exist measurable sets that are not Borel sets. Assuming the axiom of choice, a fundamental axiom of set theory, it can be shown that there exist subsets of $\mathbb{R}$ that are not measurable. Consequently, $\mathscr{B} \subset \mathscr{M} \subset \mathscr{P}$, (the set inclusions here are proper), where $\mathscr{P}$ is the collection of all subsets of $\mathbb{R}$, M. E. Munroe, Measure and Integration, 1953.

A subset $E \subset \mathbb{R}$ is said to be *null* or a *null set* if $E \in \mathscr{M}$ and $\lambda(E) = 0$. It follows from Example **13**.8 and Proposition 5(iv) that any countable set is null. It might appear that the converse holds, but the Cantor set, described below in Example 7, demonstrates that there is an uncountable set that is null.

EXAMPLE 7 (Cantor Set). Let $I = [0, 1]$. The Cantor set K is constructed by removing open intervals from I. The first step, which for notational conve-

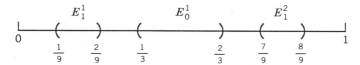

E_1^1 E_0^1 E_1^2

0    $\frac{1}{9}$ $\frac{2}{9}$ $\frac{1}{3}$ $\frac{2}{3}$ $\frac{7}{9}$ $\frac{8}{9}$ 1

Figure 16.1

nience we call the *zero*th step, is the removal of the "middle third", $E_0^1 = (\frac{1}{3}, \frac{2}{3})$, of I. At the next step (the first step) we remove the "middle thirds" of the remaining two intervals. We continue in this way, at each step removing the "middle thirds" of the remaining intervals. At the kth step there will be 2^k intervals remaining and we remove 2^k "middle thirds"; our notation for the deleted open intervals is then $E_k^1, E_k^2, \ldots, E_k^{2^k}$. See Figure 16.1.

Note that $\ell(E_k^j) = \frac{1}{3^{k+1}}$. The set remaining after removal of the middle thirds is called the *Cantor set*, K. Explicitly,

$$K = I \setminus \bigcup_{n=0}^{\infty} \bigcup_{k=1}^{2^n} E_n^k.$$

It may appear that the Cantor set consists only of endpoints of the deleted intervals, but we will show that K is actually uncountable so that it consists of much more than these endpoints.

First, it is clear that K is closed since its complement is open. Therefore K is measurable. In fact $\lambda(K) = 0$. This follows from

$$\lambda\left[\bigcup_{n=0}^{\infty} \bigcup_{k=1}^{2^n} E_n^k \right] = \sum_{n=0}^{\infty} \sum_{k=1}^{2^n} \frac{1}{3^{n+1}} = \left(\tfrac{1}{3}\right) \sum_{n=0}^{\infty} \left(\tfrac{2}{3}\right)^n = 1$$

and $\lambda(I) = 1$.

Finally, K is uncountable. Note that a point $x \in I$ is in K iff in one of its ternary expansions, $x = .a_1 a_2 \ldots$ (base 3), the digit 1 does not occur. That is, x has a ternary expansion that consists only of 0s and 2s. Thus, K is uncountable (See Proposition 1.8 and Exercise 3.46).   □

The utility of the concept of measurable sets is demonstrated in Proposition 8.

Proposition 8. Let $f: \mathbb{R} \to \mathbb{R}$ be absolutely integrable over the interval I. Then f is absolutely integrable over every measurable subset E of I.

Proof. By writing $f = f^+ - f^-$, we may assume that f is nonnegative. Now $f_n = f \wedge (nC_{[-n, n] \cap E})$ is integrable for every n and $f_n \uparrow fC_E$ with $f_n \leq f$. Thus, fC_E is integrable by the dominated convergence theorem.   □

The notion of a null set or set of measure zero can be used to give a necessary and sufficient condition for a function to be Riemann integrable. A statement concerning the points of a subset $E \subseteq \mathbb{R}$ holds *almost everywhere* (often abbreviated to a.e.) in E if the statement is true for all the points of E except perhaps for a null subset of E. Lebesgue's characterization of Riemann integrable functions is then the following: A bounded function $f: [a, b] \to \mathbb{R}$ is Riemann integrable iff f is continuous almost everywhere on $[a, b]$. An excellent discussion of this is given in Section 7.26 of Apostol, (1974).

EXERCISES 16

1. Show that $\mathscr{L}$ is not closed under countable unions.

2. If $E, F \in \mathscr{L}$, $E \subseteq F$, show that $\lambda(F \setminus E) = \lambda(F) - \lambda(E)$.

3. Show that $\mathscr{M}$ is closed under countable intersections and that $\mathscr{L}$ is a proper subset of $\mathscr{M}$.

4. If $E, F \in \mathscr{M}$, $E \subseteq F$, show that $\lambda(E) \leq \lambda(F)$.

5. Prove Proposition 5(v).
   Hint: Exercise **13**.6.

6. If $\{E_i\} \subseteq \mathscr{M}$, show that $\lambda(\bigcup_{i=1}^{\infty} E_i) \leq \sum_{i=1}^{\infty} \lambda(E_i)$.

7. Give an example of a measurable set that is neither open nor closed.

8. If $\mathscr{F}$ is any family of subsets of a set S, show that there is a smallest σ-algebra of subsets of S containing $\mathscr{F}$.

9. Show that a σ-algebra is closed under finite unions and countable intersections.

10. Can absolute integrability be replaced by integrability in Proposition 8?

11. Let $f: \mathbb{R} \to \mathbb{R}$ be absolutely integrable over every bounded interval. Let $E = \{t: f(t) \neq 0\}$. Show that E is measurable.
    Hint: If J is a compact interval, $f_n = (n|f|) \wedge C_J \uparrow C_{E \cap J}$.

12. A subset $E \subseteq [a, b]$ is called *R-integrable* if C_E is Riemann integrable. Do Propositions 1 and 2 hold for the family of *R*-integrable sets? If the *content* of an *R*-integrable set E is defined to be $c(E) = R\int_a^b C_E$, does the analogue of Proposition 3 hold for c?

13. Assuming the existence of a nonmeasurable subset of $\mathbb{R}$, give an example of a function f that is not integrable and such that $|f|$ is integrable.

14. A function $f: \mathbb{R} \to \mathbb{R}$ is said to be (Lebesgue) *measurable* if the set $\{t: f(t) < a\}$ is measurable for each $a \in \mathbb{R}$. Show that the following are equivalent:
    (a) f is measurable.

(b) $\{t\colon f(t) \geq a\}$ is measurable for each $a \in \mathbb{R}$.

(c) $\{t\colon f(t) > a\}$ is measurable for each $a \in \mathbb{R}$.

(d) $\{t\colon f(t) \leq a\}$ is measurable for each $a \in \mathbb{R}$.

(e) $f^{-1}(G)$ is measurable for each open set $G \subseteq \mathbb{R}$.

(f) Same as (b)–(d) with "for each $a \in \mathbb{R}$" replaced by "for each $a \in \mathbb{Q}$."

15. Let $f, g\colon \mathbb{R} \to \mathbb{R}$ be measurable. Show that $f + g$, fg, $1/f$, $|f|$, $f \wedge g$, and $f \vee g$ are measurable.

16. If f is measurable and $f = g$ a.e. in $\mathbb{R}$, show that g is measurable.

17. Let f_k be measurable for each $k \in \mathbb{N}$. If $\sup_k f_k(t) = f(t)$ $[\inf_k f_k(t) = g(t)]$ is finite for each $t \in \mathbb{R}$, show that $f(g)$ is measurable.

18. Let f_k be measurable for each k and $f\colon \mathbb{R} \to \mathbb{R}$. Show that if $f_k \to f$ a.e., then f is measurable.

19. Let $E \subseteq \mathbb{R}$. Show that E is measurable iff C_E is measurable.

20. Let $f\colon \mathbb{R} \to \mathbb{R}$ be absolutely integrable over $\mathbb{R}$. Show that the set function $\mu\colon \mathcal{M} \to \mathbb{R}$, defined by $\mu(E) = \int_E f$, is countably additive.

21. Let $f\colon \mathbb{R} \to \mathbb{R}$ be continuous and absolutely integrable. If $E_n = \{t\colon |f(t)| \geq n\}$, show that $\lim n\lambda(E_n) = 0$.

CHAPTER 17

Multiple Integrals

By appropriate extension of the definitions of interval, division, tagged division, and gauge, the integral is easily defined for functions f on intervals I in $\mathbb{R}^p$. The resulting integral is commonly referred to as a "multiple integral" of f over I. We begin our discussion of multiple integrals with these extensions.

Let $\overline{\mathbb{R}}^p = \overline{\mathbb{R}} \times \cdots \times \overline{\mathbb{R}}$, where there are p factors in the cartesian product, and $\overline{\mathbb{R}}$ denotes the extended real numbers. An interval in $\overline{\mathbb{R}}^p$ is a cartesian product $I = I_1 \times \cdots \times I_p$, where each I_j is an interval in $\overline{\mathbb{R}}$. I is an *open* (*closed*, *bounded*) interval if each I_j is an open (closed, bounded) interval. The *interior* of the interval I, denoted by I^0, is the product of the interiors of the I_j.

Definition 1. If I is an interval in $\overline{\mathbb{R}}^p$, a *division* of I is a finite collection of closed subintervals in $\{I_j: j = 1, \ldots, n\}$ such that $\bigcup_{j=1}^n I_j = I$, $I_i^0 \cap I_j^0 = \varnothing$ for $i \neq j$. A *tagged division* $\mathscr{D}$ of I is a collection of ordered pairs $\{(x_i, I_i): i = 1, \ldots, n\}$ such that $\{I_i: i = 1, \ldots, n\}$ is a division of I and $x_i \in I$. The elements I_i are called *subintervals* of the tagged division $\mathscr{D}$ and each x_i is called a *tag* of I_i.

Definition 2. A *gauge* γ on a subset $E \subseteq \overline{\mathbb{R}}^p$ is a function on E that associates with each point $x \in E$ an open interval in $\overline{\mathbb{R}}^p$ containing x.

Definition 3. If γ is a gauge on an interval I in $\overline{\mathbb{R}}^p$, then a tagged division $\mathcal{D}$ of I is said to be γ-*fine* if $\gamma(x) \supseteq J$ for every $(x, J) \in \mathcal{D}$.

As before, in order to make a meaningful definition of the gauge integral, it's necessary to show that any gauge has a γ-fine tagged division. This turns out to be more difficult in higher dimensions. The proof in $\mathbb{R}$ relied on the completeness of the order property in $\mathbb{R}$; however, this is not available in $\overline{\mathbb{R}}^p$.

Existence of γ-Fine Tagged Divisions

For clarity of exposition, the following proof of the existence of γ-fine tagged divisions is written in $\mathbb{R}^2$. By taking the intervals G and H in $\mathbb{R}^n$ and $\mathbb{R}^m$, the general construction in $\mathbb{R}^{n+m}$ should be clear.

Let G, H be intervals in $\overline{\mathbb{R}}$ and $I = G \times H$. Let γ be a gauge on I. If P_1 and P_2 are the projections on the first and second coordinates, respectively, then $\gamma_1(x, y) = P_1\gamma(x, y)$ $(\gamma_2(x, y) = P_2\gamma(x, y))$ has the property that $\gamma_1(\cdot, y)(\gamma_2(x, \cdot))$ defines a gauge on $G(H)$ for each $y \in H$ $(x \in G)$. (Figure 17.1).

Lemma 4. For $y \in H$, let $\mathcal{G}_y$ be a $\gamma_1(\cdot, y)$-fine tagged division of G.

   (i) If $\gamma_2'(y) = \bigcap\{\gamma_2(x, y): (x, J) \in \mathcal{G}_y\}$, then γ_2' is a gauge on H.
   (ii) If $\mathcal{H} = \{(y_j, K_j): j = 1, \ldots, m\}$ is a γ_2'-fine tagged division of H and if $\mathcal{G}_y = \{(x_j^y, J_j^y): j = 1, \ldots, m_y\}$, then

$$\mathcal{I} = \left\{ \left(\left(x_i^{y_j}, y_j \right), J_i^{y_j} \times K_j \right) \right): j = 1, \ldots, m; i = 1, \ldots, m_{y_j} \right\}$$

is a γ-fine tagged division of $I = G \times H$. (See Figure 17.2; $\mathcal{I}$ is called a *compound division* of the $\{\mathcal{G}_y\}$ and $\mathcal{H}$.)

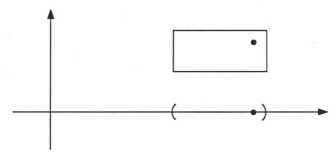

Figure 17.1

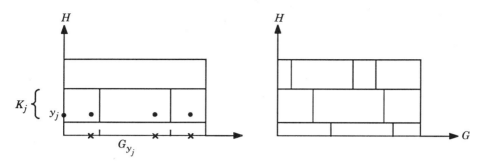

Figure 17.2

Proof. Since the intersection defining γ_2' is a finite intersection of open intervals containing y, (i) follows immediately.

For (ii), if

$$\left(x_i^{y_j}, y_j\right) \in J_i^{y_j} \times K_j,$$

then

$$y_j \in K_j \subseteq \gamma_2'(y_j) \subseteq \gamma_2\left(x_i^{y_j}, y_j\right) \qquad \text{and} \qquad x_i^{y_j} \in J_i^{y_j} \subseteq \gamma_1\left(x_i^{y_j}, y_j\right).$$

Hence,

$$J_i^{y_j} \times K_j \subseteq \gamma_1\left(x_i^{y_j}, y_j\right) \times \gamma_2\left(x_i^{y_j}, y_j\right) = \gamma\left(x_i^{y_j}, y_j\right). \qquad \square$$

Theorem **14.4** and Lemma 4 imply the following corollary.

Corollary 5. Every gauge γ on I has a γ-fine tagged division.

Definition 6. If $I = I_1 \times \cdots \times I_p$ is an interval in $\overline{\mathbb{R}}^p$, its *volume* is defined to be $v(I) = \ell(I_1) \cdots \ell(I_p)$. (In $\mathbb{R}^1$ and $\mathbb{R}^2$, the volume is usually called the *length* and the *area*, respectively.)

If $f: I \to \mathbb{R}$ and $\mathscr{I}$ is a tagged division of I, then the *Riemann sum* of f with respect to $\mathscr{I}$ is defined to be

$$S(f, \mathscr{I}) = \sum_{(x, I) \in \mathscr{I}} f(x) v(I),$$

where we continue to use the convention $0 \cdot \infty = 0$.

We now have the tools and vocabulary necessary for the definition of the integral in $\mathbb{R}^p$.

Definition 7. Let I be an interval in $\overline{\mathbb{R}}^p$ and $f: I \to \mathbb{R}$. Then f is *integrable over I* if there is $A \in \mathbb{R}$ with the property that for each $\varepsilon > 0$, there is a gauge γ on I such that $|S(f, \mathscr{I}) - A| < \varepsilon$ for every γ-fine tagged division $\mathscr{I}$ of I.

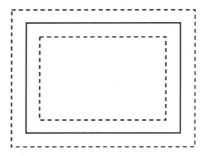

Figure 17.3

As before, the number A is unique. It is called the *integral of f over I* and is denoted by $\int_I f = A$.

The basic properties of the integral in $\overline{\mathbb{R}}$ carry over to the integral in $\overline{\mathbb{R}}^p$. Often, the proofs can be repeated with very little change; however, occasionally, as should be expected, the notation and the proofs become more complicated. We will not attempt to rewrite the basic properties for the $\overline{\mathbb{R}}^p$-setting, but will tacitly assume them as we proceed. It is expected that the reader will have little difficulty with this procedure.

EXAMPLE 8. Let I be a bounded interval in $\overline{\mathbb{R}}^p$. Then C_I is integrable over $\overline{\mathbb{R}}^p$ and $\int_{\overline{\mathbb{R}}^p} C_I = v(I)$.

We sketch the proof for $p = 2$. Let $\varepsilon > 0$. Build an "open strip" around ∂I (the boundary of I) as in Figure 17.3, with the volume of the strip S, indicated by the dashed lines, less than ε. Let $\bar{I}$ be the smallest closed interval containing I.

Define a gauge γ on $\mathbb{R}^2$ by choosing $\gamma(x)$ such that $\gamma(x) \cap \bar{I} = \varnothing$ if $x \notin \bar{I}$; $\gamma(x) \subseteq I^0$ if $x \in I^0$ and $\gamma(x) \subseteq S$ if $x \in \partial(I)$. Let $\mathscr{I}$ be a γ-fine tagged division of $\mathbb{R}^2$. Then

$$|v(I) - S(C_I, \mathscr{I})| = \left| v(I) - \sum_{\substack{(x, J) \in \mathscr{I} \\ x \in I^0}} v(J) - \sum_{\substack{(x, J) \in \mathscr{I} \\ x \in \partial I}} C_I(x) v(J) \right|$$

$$\leq v(S) < \varepsilon,$$

where we are assuming that the volume function v is additive over rectangles. $\square$

It follows from Example 8 and the linearity of the integral that if f is a *step function*, that is, a linear combination of characteristic functions of intervals, then f is integrable over every bounded interval in $\overline{\mathbb{R}}^p$ and its integral is given in the natural way.

Integrability of Continuous Functions

As one would expect, continuous functions on compact sets in $\mathbb{R}^p$ are integrable. To establish this, a preliminary result is needed.

Lemma 9. Let $K \subset \overline{\mathbb{R}}^p$ be compact and I a compact interval containing K. If $f: K \to \mathbb{R}$ is continuous and nonnegative on K and if f is extended to I by setting $f(x) = 0$ for $x \in I \setminus K$, then there is a sequence of nonnegative step functions $\{s_n\}$ vanishing outside I and such that $s_n(x) \downarrow f(x) \; \forall \, x \in I$, where the convergence is uniform for $x \in K$.

Proof. For each $n \in \mathbb{N}$, divide I into a finite collection $\mathscr{F}_n$ of pairwise disjoint intervals such that the diameter of each subinterval in $\mathscr{F}_n$ is smaller than $1/n$. Make the choice such that $\mathscr{F}_{n+1}$ is a refinement of $\mathscr{F}_n$, that is, each subinterval in $\mathscr{F}_{n+1}$ is contained in a subinterval in $\mathscr{F}_n$.

Construct s_n from $\mathscr{F}_n$ as follows: for $x \in I$, there is a unique $J \in \mathscr{F}_n$ such that $x \in J$; set $s_n(x) = \sup \{ f(x): \; x \in J \}$ and $s_n(x) = 0$ for $x \notin I$. Note that $s_n(x) \downarrow$, since $\mathscr{F}_{n+1}$ is a refinement of $\mathscr{F}_n$.

We claim that $s_n(x) \downarrow f(x)$: First, suppose that $x \in I \setminus K$. Then, there is an open interval I_x containing x and such that $I_x \cap K = \varnothing$. Any interval of sufficiently small diameter containing x will be contained in I_x. Since f is zero outside K, this implies that $s_n(x) = 0$ for large n. Thus, in this case, $s_n(x) \downarrow f(x)$. For $x \in K$, let $\varepsilon > 0$. There is an open interval I_x containing x such that $|f(x) - f(y)| < \varepsilon$ for all $y \in K \cap I_x$. Thus, $0 \le f(x)$ and $f(y) \le f(x) + \varepsilon$ for all $y \in I_x$, since f is zero outside K. For large n, the interval of $\mathscr{F}_n$ that contains x will be a subset of I_x; so, $s_n(x) \le f(x) + \varepsilon$ for large n. Since $f(x) \le s_n(x)$ for all x, it follows that $s_n(x) \downarrow f(x)$.

The uniform convergence on K follows from the uniform continuity of f on K, that is, the estimate $s_n(x) \le f(x) + \varepsilon$ holds uniformly for $x \in K$ and large n. $\qquad\qquad\qquad\qquad\qquad\qquad\qquad\qquad\qquad\qquad\qquad\square$

It should be noted that when the function f is extended to I, it will not generally be continuous on I, nor for that matter, on all of K.

If the function in Lemma 9 is not nonnegative, then consideration of $f = f^+ - f^-$ shows that there exists a sequence of step functions $\{s_n\}$ vanishing outside I such that $s_n \to f$ pointwise on I; the convergence is uniform on K.

For later use in the Fubini theorems, we also note that if f is a continuous function defined on an unbounded closed interval I, then there is a sequence of step functions $\{s_n\}$ that converges pointwise to f on I. For let $\{K_n\}$ be a sequence of compact intervals such that $K_n \subseteq K_{n+1} \subseteq I$ and $\bigcup_{n=1}^{\infty} K_n = I$. By the observation above, for each n there is a step function s_n that vanishes outside K_n and is such that $|s_n(x) - f(x)| < 1/n$ for $x \in K_n$. The sequence $\{s_n\}$ then converges to f pointwise.

Theorem 10. Let $K \subset \overline{\mathbb{R}}^p$ be compact and $f \colon K \to \mathbb{R}$ be continuous on K. If f is extended to $\mathbb{R}^p$ by setting $f(x) = 0$ for $x \notin K$, then f is absolutely integrable over $\mathbb{R}^p$.

Proof. By considering $f = f^+ - f^-$, we may assume that f is nonnegative. Choose a compact interval I containing K and let $\{s_n\}$ be the sequence of step functions given by Lemma 9. Since these functions vanish outside of I, each s_n is integrable. Now $s_n \downarrow f$ and $\int_I s_n \geq 0$; thus, the monotone convergence theorem implies that f is integrable over $\mathbb{R}^p$. $\square$

Theorem 10 gives a reasonably broad class of functions that are integrable; these functions suffice in a large number of applications. We now need some means of actually evaluating these multiple integrals in $\mathbb{R}^p$. As in the case of the elementary calculus, these multiple integrals are usually evaluated by computing iterated integrals. Consequently, we direct our attention to conditions under which the multiple and iterated integrals are equal.

Iterated Integrals

Let G and H be closed intervals in $\overline{\mathbb{R}}$, $I = G \times H$, and $f \colon I \to \mathbb{R}$. Following the method first encountered in the elementary calculus, we attempt to evaluate the integral of f over I by considering the iterated integral $\int_H \int_G f(x, y)\, dx\, dy$. As usual, the iterated integral means that we first evaluate $\int_G f(x, y)\, dx = g(y)$ to obtain a function g, which we assume to exist for all $y \in H$. If this function g exists for all $y \in H$ and is integrable over H, we then evaluate $\int_H g(y)\, dy$ to obtained the value of the iterated integral $\int_H \int_G f(x, y)\, dx\, dy$. A theorem that gives conditions for the equality of a multiple integral and its corresponding iterated integral is usually called a Fubini theorem in honor of the Italian mathematician G. Fubini (1879–1943), the first to establish a general result of this nature. Our first Fubini theorem is the following.

Theorem 11. Let G and H be bounded intervals, $I = G \times H$, and $f \colon I \to \mathbb{R}$ be bounded. Suppose there is a sequence of step functions $\{s_n\}$ such that $s_n \to f$ pointwise on I. Then

  (i) For each $y \in H$, the function $f(\cdot, y)$ is integrable over G.
  (ii) The function $g(y) = \int_G f(\cdot, y)$ is integrable over H.
  (iii) The function f is integrable over I and

$$\int_I f = \int_H \int_G f(x, y)\, dx\, dy.$$

Proof. First, suppose f is a characteristic function of an interval, $f = C_{A \times B}$, A and B intervals in $\mathbb{R}$. For (i) and (ii) we have

$$\int_G f(\cdot, y) = \int_G C_A C_B(y) = C_B(y) \ell(G \cap A)$$

and

$$\int_H \int_G f(\cdot, y)\, dy = \int_H C_B(y) \ell(G \cap A)\, dy = \ell(B \cap H) \ell(G \cap A).$$

Since $\int_I f = v(I \cap A \times B) = \ell(B \cap H)\ell(G \cap A)$, (iii) is immediate.

The case where f is a step function, that is, a linear combination of characteristic functions of intervals in $\mathbb{R}^2$ follows immediately from the linearity of the integral. Note also that if f is a step function, then the function $g(y) = \int_G f(\cdot, y)$ is also a step function (on $\mathbb{R}$).

Now we consider the general case. Let $M > 0$ be such that $|f(x)| \leq M$ for $x \in I$. For $x \in I$, set

$$\bar{s}_n(x) = \begin{cases} M, & s_n(x) > M, \\ s_n(x), & -M \leq s_n(x) \leq M, \\ -M, & s_n(x) < -M. \end{cases}$$

Then $\{\bar{s}_n\}$ is a sequence of step functions that converges pointwise to f on I and is uniformly bounded by M. By the first part of the proof, $\bar{s}_n(\cdot, y)$ is integrable over G for each $y \in H$ and $\{\bar{s}_n(\cdot, y)\}$ converges pointwise to $f(\cdot, y)$ on G. By the bounded convergence theorem (Exercise **15.6**), $f(\cdot, y)$ is integrable over G with

$$\lim \int_G \bar{s}_n(\cdot, y) = \int_G f(\cdot, y).$$

Also by the first part, each function $\int_G \bar{s}_n(\cdot, y)$ is integrable over H, and we have just shown that this sequence of functions converges pointwise on H to the function $\int_G f(\cdot, y)$. Since $|\int_G \bar{s}_n(\cdot, y)| \leq M\ell(G)$, the bounded convergence theorem implies that the function $\int_G f(\cdot, y)$ is integrable over H with

$$\lim \int_H \int_G \bar{s}_n(\cdot, y)\, dy = \int_H \int_G f(\cdot, y)\, dy.$$

Since $\{\bar{s}_n\}$ converges pointwise to f on I, the bounded convergence theorem yields $\lim \int_I \bar{s}_n = \int_I f$. Again, by the first part,

$$\int_I \bar{s}_n = \int_H \int_G \bar{s}_n(x, y)\, dx\, dy.$$

Consequently, $\int_I f = \int_H \int_G f(x, y)\, dx\, dy$, and the result is established. □

Theorem 11 is also applicable to the other iterated integral $\int_G \int_H f(x, y)\, dy\, dx$; moreover, Theorem 11 implies that these two iterated integrals are equal. This is, in general, not true (see Example 14).

Note that Lemma 9 gives a sufficient condition for a function to satisfy the hypothesis of Theorem 11. This result is general enough to cover a wide variety of situations encountered in applications; however, we do require a result that covers the situation when unbounded intervals are encountered.

Theorem 12. Let I be an interval in $\mathbb{R}^2$ and $f, g: I \to \mathbb{R}$. Suppose that there is a sequence of step functions $\{s_n\}$ such that $s_n \to f$ pointwise on I, that $|f(z)| \le g(z)$ for all $z \in I$, and that the iterated integral, $\int_H \int_G g(x, y)\, dx\, dy$, exists. Then

(i) For each $y \in H$, the function $f(\cdot, y)$ is integrable over G.
(ii) The function $F(y) = \int_G f(\cdot, y)$ is integrable over H.
(iii) The function f is integrable over I and

$$\int_I f = \int_H \int_G f(x, y)\, dx\, dy.$$

Proof. Note that if f satisfies the hypotheses of the theorem, then so do the functions f^+ and f^-. By considering $f = f^+ - f^-$, we may assume that f is nonnegative on I.

For each $n \in \mathbb{N}$, set $I_n = \{(x, y): |x| \le n, |y| \le n\} \cap I$ and define f_n by $f_n = f \wedge (nC_{I_n})$, that is, f is altered by setting $f_n = 0$ outside I_n and truncating f at n. Each function f_n satisfies the hypotheses of Theorem 11 over the interval I_n. Consequently, if $G_n = [-n, n] \cap G$ and $H_n = [-n, n] \cap H$, then

$$\int_{I_n \cap I} f_n = \int_{H_n} \int_{G_n} f_n(x, y)\, dx\, dy.$$

Since f_n vanishes outside I_n, we have

$$\int_I f_n = \int_H \int_G f_n(x, y)\, dx\, dy.$$

The sequence $\{f_n\}$ increases pointwise on I to the function f, so both of the sequences $\{\int_I f_n\}$ and $\{\int_G f_n(x, y)\, dx\}$, $y \in H$, are increasing. Using the inequalities

$$\int_I f_n = \int_H \int_G f_n(x, y) \, dx \, dy \le \int_H \int_G g(x, y) \, dx \, dy,$$

and

$$\int_G f_n(x, y) \, dx \le \int_G g(x, y) \, dx, \qquad y \in H,$$

the monotone convergence theorem yields that f is integrable over I with $\lim \int_I f_n = \int_I f$ and the function $f(\cdot, y)$ is integrable over G with

$$\lim \int_G f_n(x, y) \, dx = \int_G f(x, y) \, dx.$$

The sequence of functions $F_n(y) = \int_G f_n(x, y) \, dx$ is increasing and

$$\int_H F_n \le \int_H \int_G g(x, y) \, dx \, dy.$$

Again, the monotone convergence theorem is applicable and yields

$$\lim \int_H F_n = \lim \int_H \int_G f_n(x, y) \, dx \, dy = \lim \int_I f_n = \int_H \int_G f(x, y) \, dx \, dy = \int_I f,$$

and the result is established. $\qquad\qquad\qquad\qquad\qquad\qquad\qquad\qquad\square$

In applying Theorem 12 to a function f, a good candidate for the function g is $|f|$. In particular, if $\int_H \int_G |f(x, y)| \, dx \, dy$ exists, then both f and $|f|$ are integrable and the conclusion of Theorem 12 holds for both f and $|f|$. This gives a useful criterion for the integrability of a function f.

We have the following corollary to Theorem 12.

Corollary 13. Let $f: I \to \mathbb{R}$ be absolutely integrable on I. Suppose that there is a sequence of step functions $\{s_n\}$ converging pointwise to f on I. Then the iterated integral $\int_H \int_G f(x, y) \, dx \, dy$ exists and is equal to $\int_I f$.

Proof. Since f is absolutely integrable, we may assume that it is nonnegative. Using the same notation as in the proof of Theorem 12, we see from Theorem 11 that $\int_H \int_G f_n(x, y) \, dx \, dy = \int_I f_n$ for each n, and that $\{\int_G f_n(\cdot, y)\}$ is an increasing sequence for each $y \in H$. Since $\int_I f_n \uparrow \int_I f$, the monotone convergence theorem implies that $\int_H \int_G f(x, y) \, dx \, dy = \lim \int_H \int_G f_n(x, y) \, dx \, dy = \lim \int_I f_n = \int_I f$. $\qquad\square$

The following example shows that it is not in general true that the iterated integrals are equal, even though both may exist.

EXAMPLE 14. The iterated integral

$$\int_0^1 \int_1^\infty (e^{-xy} - 2e^{-2xy})\, dx\, dy$$

is positive since the "inner integral" is positive for each y. On the other hand, the iterated integral

$$\int_1^\infty \int_0^1 (e^{-xy} - 2e^{-2xy})\, dy\, dx$$

is negative since the "inner integral" is negative for each x. ☐

This example also shows that it is important that the function g in Theorem 12 be nonnegative.

The following example shows that it's possible for both iterated integrals to exist and be equal and yet have the function fail to be integrable.

EXAMPLE 15. Let

$$f(x, y) = \begin{cases} xy/(x^2 + y^2)^2, & (x, y) \neq (0,0), \\ 0, & (x, y) = (0,0). \end{cases}$$

Then $\int_{-1}^1 f(x, y)\, dx = 0$, and so, $\int_{-1}^1 \int_{-1}^1 f(x, y)\, dx\, dy = 0$. By symmetry, $\int_{-1}^1 \int_{-1}^1 f(x, y)\, dy\, dx = 0$. However, f is not integrable on $E = [-1,1] \times [-1,1]$, for, if f were integrable over E, then it would be integrable over $F = [0,1] \times [0,1]$. But

$$\int_{1/n}^1 \int_0^1 f(x, y)\, dx\, dy = -\frac{1}{4}\ln 2 + \frac{1}{4}\ln\left(1 + \frac{1}{n^2}\right) - \frac{1}{2}\ln\left(\frac{1}{n}\right) = \int_{E_n} f,$$

where $E_n = \{(x, y): 0 \le x \le 1,\ 1/n \le y \le 1\}$. Since $C_{E_n} f \le fC_F$, we have $\int_{E_n} f \le \int_F f$, and so, $\{\int_{E_n} f_n\}$ should be bounded. This contradiction shows that f is not integrable. ☐

The general limit theorems such as the monotone and dominated convergence theorems are extremely useful in many of the applications involving the Lebesgue integral. It has been shown that the gauge integral also possesses these general limit interchange properties. Another useful property of the Lebesgue integral is that very general Fubini-type theorems of the nature of Theorems 11 and 12 hold. Theorems 11 and 12 have been established for a much more restricted class of functions than one usually sees in the analogous situation for the Lebesgue integral. These more general Fubini theorems can also be shown to hold for the gauge integral. Development of the tools necessary to establish these more general results would be well outside the goals of this book; therefore, we leave these topics to more advanced discussions of real analysis.

There is a general theory for making a change of variables in multiple integrals that parallels Theorem 13.12. Because of the length of time required for this development, we omit it. The interested reader will find a thorough discussion of this topic in Section 8.45 of Bartle (1976).

Finally, for later use in the presentation of the Riesz–Fischer theorem, we give slightly more general forms of the monotone and dominated convergence theorems.

As before, a subset $E \subset \mathbb{R}^p$, is *null* or a *null set* if C_E is integrable and $\int_{\mathbb{R}^p} C_E = 0$. A statement or property concerning the points of a subset $E \subseteq \mathbb{R}^p$ is said to hold *almost everywhere in* E; we write "a.e. in E," or *for almost all points in* E if it holds for all of the points in E except possibly those points that belong to a null set.

Lemma 16. Let $f: \mathbb{R}^p \to \mathbb{R}$ and $E \subset \mathbb{R}^p$ be a null set.

   (i) f is absolutely integrable over E with $\int_E |f| = 0$.
   (ii) If $g: \mathbb{R}^p \to \mathbb{R}$ is integrable and $f = g$ a.e. in $\mathbb{R}^p$, then f is integrable and $\int_{\mathbb{R}^p} g = \int_{\mathbb{R}^p} f$.

Proof. For (i), let $F_k = \{x: |f(x)| \le k\}$. Then

$$C_{E \cap F_k}|f| \le kC_{E \cap F_k}$$

implies that $|f|$ is integrable over $E \cap F_k$ and

$$\int_{E \cap F_k} |f| \le \int_{E \cap F_k} k = 0.$$

Since

$$C_{E \cap F_k}|f| \uparrow C_E|f|,$$

the monotone convergence theorem implies that $C_E|f|$ is integrable with $\int_E |f| = 0$.
(ii) follows from (i). $\qquad\qquad\qquad\qquad\qquad\qquad\qquad\qquad\qquad\square$

Theorem 17. Let $f_k: \mathbb{R}^p \to \mathbb{R}$ be nonnegative and $\{f_k(x)\}$ increasing for each $x \in \mathbb{R}^p$. Suppose each f_k is integrable over $\mathbb{R}^p$ and $\{\int_{\mathbb{R}^p} f_k\}$ is bounded.

Then $\lim f_k$ exists and is finite for almost all $x \in \mathbb{R}^p$. If

$$f(x) = \begin{cases} \lim f_k(x) & \text{when the limit is finite,} \\ 0 & \text{otherwise,} \end{cases}$$

then f is integrable and $\int_{\mathbb{R}^p} f = \lim_k \int_{\mathbb{R}^p} f_k$.

Proof. Let $E = \{x: \lim f_k(x) = \infty\}$ and $\int_{\mathbb{R}^p} f_k \leq M$ for all k. Let $f_{ik} = 1 \wedge (1/i) f_k$. For fixed i, $f_{ik} \uparrow h_i$ as $k \to \infty$, where

$$h_i(x) = \begin{cases} 1 \wedge (1/i) f(x) & \text{if } x \notin E, \\ 1 & \text{if } x \in E. \end{cases}$$

Since $\int_{\mathbb{R}^p} f_{ik} \leq (1/i) \int_{\mathbb{R}^p} f_k \leq M/i$, h_i is integrable by the monotone convergence theorem. Since $h_i \downarrow C_E$, the monotone convergence theorem implies that

$$\int_{\mathbb{R}^p} C_E = \lim \int_{\mathbb{R}^p} h_i \leq \lim M/i = 0.$$

So, E is null.
   Set

$$g_k = f_k C_{\mathbb{R}^p \setminus E}.$$

Then each g_k is integrable and Lemma 16 implies that

$$\int_{\mathbb{R}^p} g_k = \int_{\mathbb{R}^p} f_k.$$

Moreover, $g_k \uparrow f$ pointwise on $\mathbb{R}^p$. It follows from the monotone convergence theorem that f is integrable and

$$\int_{\mathbb{R}^p} f = \lim \int_{\mathbb{R}^p} f_k. \qquad \qquad \square$$

   Similarly, using null sets, we have the following more general form of the dominated convergence theorem.

Theorem 18. Let $f_k: \mathbb{R}^p \to \mathbb{R}$ be integrable for each $k \in \mathbb{N}$ and let $f, g: \mathbb{R}^p \to \mathbb{R}$. Suppose that $f_k \to f$ pointwise a.e. in $\mathbb{R}^p$ and that g is integrable. If $|f_k| \leq g$ a.e. in $\mathbb{R}^p$ for each k, then f is integrable in $\mathbb{R}^p$ and

$$\int_{\mathbb{R}^p} f = \lim \int_{\mathbb{R}^p} f_k.$$

The proof is left to Exercise 16.

EXERCISES 17

1. Let G, H be closed intervals in $\overline{\mathbb{R}}$ and let $\mathscr{I}, \mathscr{J}$ be tagged divisions of G, H, respectively. Show that

$$\{((x, y), I \times J): (x, I) \in \mathscr{I}, (y, J) \in \mathscr{J}\}$$

   is a tagged division of $G \times H$. Give an example of a tagged division of $G \times H$ that is not of this form.

2. Let $E, F \subseteq \overline{\mathbb{R}}$ and γ_1, γ_2 be gauges on E and F, respectively. Show that $\gamma(x, y) = \gamma_1(x) \times \gamma_2(y)$ defines a gauge on $E \times F$.

3. State and prove Corollary 5 in $\overline{\mathbb{R}}^p$ for $p \geq 2$.

4. State and prove the analogues of **13**.13, 14, 20, 25 in $\overline{\mathbb{R}}^p$.

5. Show that v is additive over rectangles in $\mathbb{R}^2$, that is, if I is a rectangle and $I = \bigcup_{j=1}^n I_j$, where each I_j is a rectangle and $I_i^0 \cap I_j^0 = \varnothing$ for $i \neq j$, then $v(I) = \sum_{j=1}^n v(I_j)$.

6. Using only the definition of the integral, show that $C_{[0,1]}$ is integrable over $\mathbb{R}^2$.

7. Let $f: [0, 1] \times [0, 1] \to \mathbb{R}$ be defined by

$$f(x, y) = \begin{cases} 1, & \text{if } x, y \in \mathbb{Q}, \\ 0, & \text{otherwise.} \end{cases}$$

   Using only the definition of the integral, show that f is integrable over $[0, 1] \times [0, 1]$.

8. Suppose f satisfies the hypotheses of Lemma 9. Show that there is a sequence of nonnegative step functions $\{s_n\}$ such that $s_n(x) \uparrow f(x)$, where the convergence is uniform on K.

9. Let $\alpha, \beta: [a, b] \to \mathbb{R}$ be continuous with $\alpha(x) \leq \beta(x)$ for all $x \in [a, b]$. Let $E = \{(x, y): a \leq x \leq b, \alpha(x) \leq y \leq \beta(x)\}$. Suppose $f: \mathbb{R}^2 \to \mathbb{R}$ is continuous. Show that f is integrable over E and

$$\int_E f = \int_a^b \int_{\alpha(x)}^{\beta(x)} f(x, y)\, dy\, dx.$$

10. Find the integral of the function $f(x, y) = x^3 y^2$ over the region in the first quadrant that is inside the ellipse $9x^2 + 25y^2 = 25 \cdot 9$ and outside the circle $x^2 + y^2 = 1$.

11. Let $f: [0, 1] \times [0, 1] \to \mathbb{R}$ be defined by

$$f(x, y) = \begin{cases} 1, & \text{if } x, y \in \mathbb{Q}, \\ 0, & \text{otherwise.} \end{cases}$$

Verify Theorem 12 for f. Do any of the integrals in Theorem 12 exist as Riemann integrals? What do you conclude?

12. Show that the function $f(x, y) = e^{-x} \sin y$ is integrable over $[0, \infty) \times [0, 2\pi)$ and calculate the integral.

13. Is the function $f(x, y) = 1/(x + y)$ integrable over $[0, 1] \times [0, 1]$? What about $f(x, y) = 1/(x^2 + y^2)$?

14. Describe what is meant by $f = g$ a.e. in E.

15. Show that countable unions of null sets and subsets of null sets are null sets.

16. Prove Theorem 18.
    Hint: Use Exercise 15 and Lemma 16.

17. Give an example of a function $f: G \times H \to \mathbb{R}$ such that $\int_H \int_G f(x, y)\, dx\, dy$ exists and $\int_G \int_H f(x, y)\, dy\, dx$ does not exist.

18. Let $f: G \to \mathbb{R}$, $g: H \to \mathbb{R}$ be integrable and such that both f and g are the pointwise limits of step functions. Show that the function $h(x, y) = f(x)g(y)$ is integrable over $I = G \times H$ with $\int_I h = \int_G f \int_H g$.

19. Evaluate

$$\int_0^1 \int_{3y}^3 e^{x^2}\, dx\, dy.$$

Hint: Change the order of integration.

20. Let $f: [a, b] \times [c, d] \to \mathbb{R}$ be continuous. For $(x_1, x_2) \in (a, b) \times (c, d)$, define

$$F(x_1, x_2) = \int_a^{x_1} \int_c^{x_2} f(x, y)\, dy\, dx.$$

Calculate $D_{12} F(x_1, x_2)$ and $D_{21} F(x_1, x_2)$.

21. Prove the analog of Example **13**.8 over a rectangle $[a, b] \times [c, d]$ in $\mathbb{R}^2$.

22. Define Lebesgue integrability (measurability) for subsets of $\mathbb{R}^n$. Define Lebesgue measure in $\mathbb{R}^n$. State and prove the analogues of Propositions **16**.1 to 5 for $\mathbb{R}^n$.

23. If A and B are measurable subsets of $\mathbb{R}$, show that $A \times B$ is a measurable subset of $\mathbb{R}^2$. What is its Lebesgue measure?

24. Evaluate $I = \int_0^\infty e^{-x^2}\,dx$ by writing I^2 as a double integral and by using polar coordinates. (Assume the usual calculus change of variable formula for polar coordinates.)

25. Let $f(x, y) = e^{-xy} \sin x \sin y$. Show that both iterated integrals $\int_0^\infty \int_0^\infty f(x, y)\,dx\,dy$ and $\int_0^\infty \int_0^\infty f(x, y)\,dy\,dx$ exist and are equal, but that f is not integrable over $[0, \infty) \times [0, \infty)$.

26. Let $E_n = [0, n] \times [0, n]$. Show that

$$\int_0^\infty \frac{\sin x}{x}\,dx = \frac{\pi}{2}$$

by evaluating

$$\lim_n \int_{E_n} e^{-xy} \sin x$$

in two different ways.

27. Let $f(x, y) = e^{-xy}/(1 + x^2)$. Show that f is integrable over $[0, \infty) \times [0, 1]$.

28. Let $I = [0, \infty) \times [1, 2]$ and $f(x, y) = e^{-xy} \sin xy$. Show that f is integrable over I.

29. Let $f(x, y) = ye^{-(1+x^2)y^2}$. Show that

$$\int_0^\infty \int_0^\infty f(x, y)\,dx\,dy = \int_0^\infty \int_0^\infty f(x, y)\,dy\,dx.$$

Use this equality to show that

$$\int_0^\infty e^{-x^2}\,dx = \frac{\sqrt{\pi}}{2}.$$

CHAPTER 18

Convolution and Approximation

As an application of some of the results of the previous chapter on multiple integration, we consider an important "product" of functions called the convolution. After defining and developing the concept of convolution, we shall examine some of its applications to approximation theory and integral transforms.

Let $f, g: \mathbb{R} \to \mathbb{R}$. The *convolution product*, $f * g$, of f and g is defined by

$$f * g(x) = \int_{-\infty}^{\infty} f(x - y)g(y)\, dy,$$

provided the integral exists. It can be shown that the convolution product exists for a large class of functions; however, we shall restrict our discussion to the class of continuous functions so that we can use Theorem **17.12** and Corollary **17.13**.

Suppose that f and g are continuous and absolutely integrable over $\mathbb{R}$. Then $f * g$ exists for every $x \in \mathbb{R}$. For consider the interated integral

$$\int_{-\infty}^{\infty}\int_{-\infty}^{\infty}|f(x-y)||g(y)|\,dx\,dy = \int_{-\infty}^{\infty}|g(y)|\int_{-\infty}^{\infty}|f(x)|\,dx\,dy$$

$$= \int_{-\infty}^{\infty}|f|\int_{-\infty}^{\infty}|g|.$$

By Theorem **17.12**, the function $h(x, y) = f(x-y)g(y)$ is absolutely integrable over $\mathbb{R}^2$ and, by Corollary **16.13**, the iterated integral

$$\int_{-\infty}^{\infty}\int_{-\infty}^{\infty} f(x-y)g(y)\,dx\,dy$$

exists. Thus, the convolution product $f * g$ exists for every $x \in \mathbb{R}$. Since

$$|f * g(x)| \le \int_{-\infty}^{\infty}|f(x-y)g(y)|\,dy,$$

the convolution product is absolutely integrable over $\mathbb{R}$ with

$$\int_{-\infty}^{\infty}|f * g| \le \int_{-\infty}^{\infty}|f|\int_{-\infty}^{\infty}|g|.$$

Definition 1. A sequence $\varphi_n : \mathbb{R} \to \mathbb{R}$ is a *delta sequence (Dirac sequence,* or *approximate identity)* if

(i) $\varphi_n(t) \ge 0$ for all $t \in \mathbb{R}$,
(ii) φ_n is continuous and integrable over $\mathbb{R}$ with $\int_{-\infty}^{\infty}\varphi_n = 1$,
(iii) Given any $\delta > 0$, $\lim_n \int_{-\infty}^{-\delta} + \int_{\delta}^{\infty}\varphi_n(t)\,dt = 0$.

EXAMPLE 2. Let

$$\varphi_n(t) = \frac{1}{\pi}\frac{1/n}{t^2 + 1/n^2}.$$

Observing that

$$\int_a^b \varphi_n = \frac{1}{\pi}[\arctan(nb) - \arctan(na)]$$

shows that φ_n is a delta sequence. $\qquad\square$

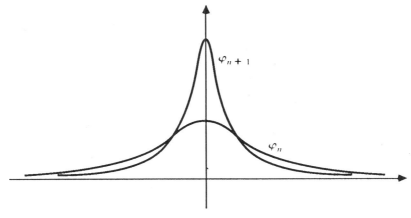

Figure 18.1

EXAMPLE 3. $\varphi_n(t) = \sqrt{n/\pi}\, e^{-nt^2}$ defines a delta sequence since

$$\int_a^b \varphi_n = 1/\sqrt{\pi} \int_{a\sqrt{n}}^{b\sqrt{n}} e^{-y^2}\, dy$$

(Example **15**.8). See Figure 18.1.   □

EXAMPLE 4. Let φ be continuous, nonnegative, $\varphi(t) = 0$ for $|t| \geq 1$ and $\int_{-1}^1 \varphi = 1$. Set $\varphi_n(t) = n\varphi(nt)$. Then $\{\varphi_n\}$ is a delta sequence.   □

EXAMPLE 5. Taking

$$\varphi(t) = \begin{cases} ce^{-1/(1-t^2)}, & |t| < 1, \\ 0, & |t| \geq 1, \end{cases}$$

where c is chosen so that $\int_{-1}^1 \varphi = 1$, in Example 4, gives a delta sequence with each term of the sequence being infinitely differentiable and zero outside the interval $[-1, 1]$.   □

EXAMPLE 6. Let

$$\varphi_n(t) = \begin{cases} (1 - t^2)^n/c_n, & |t| < 1, \\ 0, & |t| \geq 1, \end{cases}$$

where c_n is chosen so that $\int_{-1}^1 \varphi_n = 1$, that is, $c_n = \int_{-1}^1 (1 - t^2)^n\, dt$. Then $\{\varphi_n\}$

clearly satisfies (i) and (ii). For (iii), note that

$$c_n/2 = \int_0^1 (1 - t^2)^n \, dt = \int_0^1 (1 + t)^n (1 - t)^n \, dt$$

$$\geq \int_0^1 (1 - t)^n \, dt = \frac{1}{n + 1}.$$

Thus, if $0 < \delta < 1$ is fixed,

$$\int_\delta^1 \varphi_n = \int_\delta^1 \frac{(1 - t^2)^n}{c_n} \, dt \leq \int_\delta^1 \frac{n + 1}{2} (1 - \delta^2)^n \, dt$$

$$\leq \frac{n + 1}{2} (1 - \delta^2)^n (1 - \delta) \to 0 \text{ as } n \to \infty.$$

By symmetry, $\int_{-1}^{-\delta} = \int_\delta^1$, and (iii) follows.   □

The φ_n are called *Landau kernels* and will be used in the proof of the Weierstrass approximation theorem (see Theorem 8).

Theorem 7. Let $\{\varphi_n\}$ be a delta sequence, and $f: \mathbb{R} \to \mathbb{R}$ be bounded and continuous. Then $f * \varphi_n \to f$ uniformly on compact subsets of $\mathbb{R}$.

Proof. The existence of the convolution product $f * \varphi_n$ is established in Exercise 3. Let $K \subset \mathbb{R}$ be compact and let $\varepsilon > 0$. Then there is $\delta > 0$ such that $|f(x - t) - f(x)| < \varepsilon/2$ for $x \in K$ and $|t| \leq \delta$. Suppose $|f(t)| \leq M$ for all t. There is N such that $n \geq N$ implies $\int_{-\infty}^{-\delta} + \int_\delta^\infty \varphi_n < \varepsilon/4M$. Then, for $x \in K$ and $n \geq N$,

$$|f * \varphi_n(x) - f(x)| = \left| \int_{-\infty}^\infty \{f(x - t) - f(x)\} \varphi_n(t) \, dt \right|$$

$$\leq \int_{-\infty}^{-\delta} + \int_{-\delta}^\delta + \int_\delta^\infty |f(x - t) - f(x)| \varphi_n(t) \, dt$$

$$\leq 2M \int_{-\infty}^{-\delta} \varphi_n + \frac{\varepsilon}{2} \int_{-\delta}^\delta \varphi_n + 2M \int_\delta^\infty \varphi_n$$

$$\leq 2M \frac{\varepsilon}{4M} + \frac{\varepsilon}{2} \int_{-\infty}^\infty \varphi_n = \varepsilon.$$ □

By choosing an appropriate delta sequence, we obtain the classical Weierstrass approximation theorem. This result shows that every continuous function can be uniformly approximated by a polynomial.

Theorem 8 (Weierstrass Approximation Theorem). Let $f: [a, b] \to \mathbb{R}$ be continuous. Then for each $\varepsilon > 0$, there is a polynomial p such that $|f(x) - p(x)| < \varepsilon$ for all $x \in [a, b]$.

Proof. We first simplify the problem by making some changes of variables. Let $u = (x - a)/(b - a)$. Then $x = (b - a)u + a$ and $0 \le u \le 1$. Set $g(u) = f((b - a)u + a) = f(x)$. If we can find a polynominal p such that $|p(u) - g(u)| < \varepsilon$ for $0 \le u \le 1$, then

$$\left| p\left(\frac{x - a}{b - a} \right) - f(x) \right| < \varepsilon$$

for $a \le x \le b$, and the function defined by $p((x - a)/(b - a))$ is a polynomial. So, we may assume that $[a, b] = [0, 1]$.

Next, let $h(x) = f(x) - f(0) - x[f(1) - f(0)]$. If we can approximate h uniformly by a polynomial on $[0\ 1]$, then f can be uniformly approximated by a polynomial on $[0, 1]$. Since $h(0) = h(1) = 0$, we may assume that $f(0) = f(1) = 0$.

Assuming this is the case, extend f continuously to $\mathbb{R}$ by setting $f(x) = 0$ for $x \notin [0, 1]$. Then the extension, which we still denote by f, is bounded and continuous on $\mathbb{R}$.

Let $\{\varphi_n\}$ be the Landau kernels of Example 6. By Theorem 7, $f * \varphi_n \to f$ uniformly on $[0, 1]$. But, $\varphi_n(x - t) = [1 - (x - t)^2]^n/c_n$ is a polynomial in the variables x and t; so, by Exercise 1,

$$\int_{-\infty}^{\infty} \varphi_n(x - t)f(t)\, dt = \varphi_n * f(x) = f * \varphi_n(x)$$

is a polynomial in the variable x. The result now follows from Theorem 7. □

For a constructive proof of the Weierstrass approximation theorem using Bernstein polynomials, see Bartle (1976, p. 171).

For applications of the convolution product to integral transform, see Exercise 7.

EXERCISES 18

1. If $f * g$ exists for all $x \in \mathbb{R}$, show that $g * f$, exists and $f * g = g * f$.

2. If f, g, h are continuous and absolutely integrable over $\mathbb{R}$, show that $(f * g) * h = f * (g * h)$.

3. If f is continuous and bounded and g is continuous and absolutely integrable, show that $f * g$ exists, is absolutely integrable, and $\sup\{|f * g(x)|: x \in \mathbb{R}\} \le \sup \{|f(x)|:x \in \mathbb{R}\}\int_{-\infty}^{\infty}|g|$.

4. If f and g are continuous and vanish on $(-\infty, 0)$, show that $f * g$ exists and that $f * g(x) = \int_0^x f(x-y)g(y)\,dy$. Show also that $f * g$ is continuous.

5. Let f be continuous on $[a, b]$ and suppose that $\int_a^b f(t)t^n\,dt = 0$ for $n = 0, 1, 2 \ldots$. Show that $f \equiv 0$.
   Hint: Consider $\int_a^b f^2$.

6. A function $f: \mathbb{R} \to \mathbb{R}$ is of exponential order if f vanishes on $(-\infty, 0)$ and there are $M > 0$, $a \in \mathbb{R}$ such that $|f(t)| \le Me^{-at}$ for all $t \ge 0$. If f and g are continuous and of exponential order, show that $f * g$ is also of exponential order.
   Hint: See Exercise 4.

7. If f and g are continuous and of exponential order, show that $\mathcal{L}\{f * g\} = \mathcal{L}\{f\}\mathcal{L}\{g\}$, that is, the Laplace transform transforms convolution products into ordinary pointwise products.

8. If $f_k \in BV[a, b]$ and $f_k \to f$ uniformly on $[a, b]$, is it necessarily true that $f \in BV[a, b]$?

9. Let $f(t) = e^{-a|t|}$, $g(t) = e^{-b|t|}$, where $a, b > 0$. Calculate $f * g$.

CHAPTER 19

Metric Spaces

The reader has, by now, an appreciation of the way in which the distance concept helps in the discussion of limit, convergence, and other mathematical properties in $\mathbb{R}^n$. For example, the definition of convergence of sequences $\{\mathbf{x}_k\}$ to $\mathbf{x}$ in $\mathbb{R}^n$ relied on our being able to use the Euclidean distance in $\mathbb{R}^n$ to quantify the "nearness" of $\mathbf{x}_k$ to $\mathbf{x}$. Extension of the notion of distance will help in the discussion of similar properties in more general settings. The distance defined in $\mathbb{R}^n$ is a nonnegative valued function on $\mathbb{R}^n \times \mathbb{R}^n$. The properties of the distance function in $\mathbb{R}^n$ that were useful in the discussion of limit, convergence, and so on, and the description of open and closed sets are introduced in a more general setting in the following definition of an abstract metric space.

Definition 1. Let S be a nonempty set. A *metric* on S is a function $d: S \times S \to \mathbb{R}$ with the properties

   (1) $d(x, y) \geq 0$ for all $x, y \in S$ and $d(x, y) = 0$ iff $x = y$.

(2) $d(x, y) = d(y, x)$ for all $x, y \in S$ (symmetry).

(3) $d(x, y) \leq d(x, z) + d(z, y)$ for all $x, y, z \in S$ (triangle inequality).

The pair (S, d) is called a *metric space*. If d satisfies (2), (3) and

(1′) $d(x, y) \geq 0$ for all $x, y \in S$ and $d(x, x) = 0$,

then d is called a *semimetric* and the pair (S, d) is a *semimetric space*. Most of our work will be with metric spaces; however, semimetric spaces arise naturally in the "space" of integrable functions (see Exercise **19**.16). Exercise **19**.18 shows a way to "convert" a semimetric space into a metric space. We carry our development far enough to establish the important Riesz–Fischer theorem (**12**.7) for the gauge integral.

The appellation "symmetry property" for (2) should be clear. The property (3) is the extension of the $\mathbb{R}^2$ property that the "sum of the lengths of any two sides of a triangle is greater than the length of the remaining side."

Of course, $\mathbb{R}^n$ with the usual Euclidean distance function is an example of a metric space. It will be seen later that it's possible to introduce other metrics on $\mathbb{R}^n$, many of which may, in some sense, be considered "natural."

EXAMPLE 2. Every nonempty set S has at least one metric, the "*distance-1*" (or *discrete*) metric, defined on it by setting

$$d(x, y) = 1 \quad \text{if } x \neq y,$$

$$d(x, x) = 0. \qquad \qquad \square$$

This metric is often useful in the construction of counterexamples.

Definition 3. Let (S, d) be a metric space. If $x \in S$ and $r > 0$, the *sphere* with center x and radius r is defined by

$$S(x, r) = \{ y \in S : d(x, y) < r \}.$$

A sphere in $\mathbb{R}$ is an open interval, in $\mathbb{R}^2$ an open disk, and in $\mathbb{R}^3$ a "ball" without its boundary, that is, a sphere. In more general spaces, spheres can be rather strange (see Exercise 1 and Example 5).

Many of the examples and applications of metrics on $\mathbb{R}^n$ involve the concept of a "norm," which turns out to be a particular kind of metric or distance function on $\mathbb{R}^n$.

Definition 4. Let X be a vector space over $\mathbb{R}$ or $\mathbb{C}$, the set of complex numbers. A *norm* on X is a function $\|\cdot\|$: $X \to \mathbb{R}$ such that

(4) $\|x\| \geq 0$ for all $x \in X$ and $\|x\| = 0$ iff $x = 0$.
(5) $\|tx\| = |t| \|x\|$ for all t in the scalar field and $x \in X$ (homogeneity).
(6) $\|x + y\| \leq \|x\| + \|y\|$ for all $x, y \in X$ (triangle inequality).

The pair $(X, \|\cdot\|)$ is called a *normed linear space* (*nls*) or *normed space*. If the function $\|\cdot\|$ satisfies (5), (6) and

(4′) $$\|x\| \geq 0 \qquad \text{for all } x \in X,$$

then $\|\cdot\|$ is called a *seminorm* on X and $(X, \|\cdot\|)$ is called a *seminormed linear space* or *seminormed space*.

If X is a seminormed space (nls), then $d(x, y) = \|x - y\|$ induces a semimetric (metric) on S (Exercise 4). The distance function d induced by $\|\cdot\|$ is translation invariant in the sense that

$$d(x + z, y + z) = d(x, y).$$

When we are dealing with a seminormed space, we always assume that it has the semimetric induced by the seminorm.

Of course, the Euclidean norm

$$\|\mathbf{x}\| = \left[\sum_{i=1}^{n} x_i^2 \right]^{1/2}$$

induces the Euclidean metric on $\mathbb{R}^n$, but we show that $\mathbb{R}^n$ has many natural norms defined on it.

EXAMPLE 5. For $\mathbf{x} = (x_1, \ldots, x_n) \in \mathbb{R}^n$, set

$$\|\mathbf{x}\|_2 = \left[\sum_{i=1}^{n} x_i^2 \right]^{1/2}, \tag{7}$$

$$\|\mathbf{x}\|_1 = \sum_{i=1}^{n} |x_i|, \tag{8}$$

$$\|\mathbf{x}\|_\infty = \max \{|x_i| : 1 \leq i \leq n\}. \tag{9}$$

Each of these formulas induces a norm on $\mathbb{R}^n$ (Exercise 5); $\|\cdot\|_2$ is just the

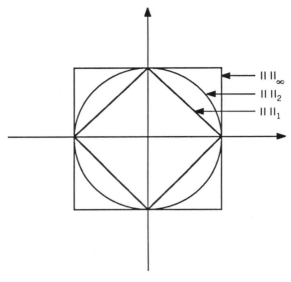

Figure 19.1

usual Euclidean norm. The "unit spheres" with respect to these norms are shown in Figure 19.1.   □

Formulas (7) and (8) are special cases of a family of norms on $\mathbb{R}^n$ defined by

$$\|\mathbf{x}\|_p = \left(\sum_{i=1}^{n} |x_i|^p \right)^{1/p}, \qquad 1 \le p < \infty. \qquad (10)$$

To show that $\| \cdot \|_p$ is a norm on $\mathbb{R}^n$ we observe first that properties (4) and (5) are immediate. To establish the triangle inequality (6) we require

Lemma 6 (Hölders Inequality). If $a_i, b_i \in \mathbb{R}, 1 < p < \infty$, and $1/p + 1/q = 1$, then

$$\sum_{i=1}^{n} |a_i b_i| \le \left(\sum_{i=1}^{n} |a_i|^p \right)^{1/p} \left(\sum_{i=1}^{n} |b_i|^q \right)^{1/q}. \qquad (11)$$

Note that for $p = 2$, this is just the Cauchy–Schwarz inequality.

Proof. Equation (11) is homogeneous, that is, if it holds for $\mathbf{a} = (a_1, \ldots, a_n)$, $\mathbf{b} = (b_1, \ldots, b_n)$, it also holds for $\lambda \mathbf{a}, \mu \mathbf{b}$ where $\lambda, \mu \in \mathbb{R}$. So we may assume that $\sum_{i=1}^{n}|a_i|^p = \sum_{i=1}^{n}|b_i|^q = 1$, and it suffices to show that $\sum_{i=1}^{n}|a_ib_i| \le 1$ in this case.

Consider the diagram in Figure 19.2:

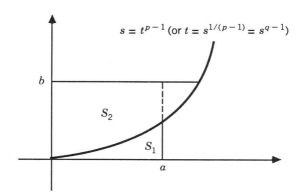

Figure 19.2

The area $S_1 = \int_0^a t^{p-1}\, dt = a^p/p$, and $S_2 = \int_0^b s^{q-1}\, ds = b^q/q$. Since $S_1 + S_2 \ge ab$ for any $a, b > 0$, $ab \le a^p/p + b^q/q$. Set $a = |a_i|$, $b = |b_i|$ and sum to obtain

$$\sum_{i=1}^{n}|a_ib_i| \le \frac{1}{p}\sum_{i=1}^{n}|a_i|^p + \frac{1}{q}\sum_{i=1}^{n}|b_i|^q = \frac{1}{p} + \frac{1}{q} = 1. \qquad \square$$

Now, to show that the triangle inequality holds, let $\mathbf{x}, \mathbf{y} \in \mathbb{R}^n$. Then, by Lemma 6,

$$\sum_{i=1}^{n}(|x_i + y_i|)^p \le \sum_{i=1}^{n}(|x_i + y_i|)^{p-1}|x_i| + \sum_{i=1}^{n}(|x_i + y_i|)^{p-1}|y_i|$$

$$\le \left\{\sum_{i=1}^{n}(|x_i + y_i|)^{(p-1)q}\right\}^{1/q}\left(\left[\sum_{i=1}^{n}|x_i|^p\right]^{1/p} + \left[\sum_{i=1}^{n}|y_i|^p\right]^{1/p}\right)$$

Since $(p - 1)q = p$ and $1 - 1/q = 1/p$, dividing both sides of this inequality by

$$\left\{\sum_{i=1}^{n}(|x_i + y_i|)^p\right\}^{1/q}$$

gives the desired result. The triangle inequality for $\|\cdot\|_p$ is called the *Minkowski inequality*. We have established the following.

Theorem 7. $\|\cdot\|_p$ is a norm on $\mathbb{R}^n$ for $1 \le p \le \infty$.

There is a rich variety of abstract metric spaces with the underlying set being other than $\mathbb{R}^n$. Important in many of the applications are metric spaces whose elements are functions. It was the consideration of such function spaces that led to the introduction of the concept of "metric space" by the French mathematician M. Frechet (1878–1973). These will be visited again and discussed in greater detail in this and later chapters.

We begin our collection of examples of metric spaces with some spaces in which the elements of the space are sequences. Let s be the set of all real-valued sequences. Then s is a vector space if we define addition of vectors to be coordinatewise addition, that is, $\{s_i\} + \{t_i\} = \{s_i + t_i\}$, and scalar multiplication to be likewise coordinatewise, $\lambda\{s_i\} = \{\lambda s_i\}$. We first consider vector subspaces of s that carry natural norms.

EXAMPLE 8. Let $1 \le p < \infty$ and

$$\ell^p = \left\{ \{s_i\}: \sum_{i=1}^{\infty} |s_i|^p < \infty \right\}.$$

We show below that the sum of any two elements in ℓ^p is also in ℓ^p so that it is a vector subspace of s. Define a norm $\|\cdot\|_p$ on ℓ^p by

$$\|\{s_i\}\|_p = \left(\sum_{i=1}^{\infty} |s_i|^p \right)^{1/p}.$$

This is a natural extension of the analogous norm on $\mathbb{R}^n$. To show that $\|\cdot\|_p$ is a norm on ℓ^p, it suffices to check the triangle inequality. If $\{s_i\}$ and $\{t_i\}$ belong to ℓ^p, then for each $n \in \mathbb{N}$,

$$\left(\sum_{i=1}^{n} |s_i + t_i|^p \right)^{1/p} \le \left(\sum_{i=1}^{n} |s_i|^p \right)^{1/p} + \left(\sum_{i=1}^{n} |t_i|^p \right)^{1/p}$$

$$\le \|\{s_i\}\|_p + \|\{t_i\}\|_p, \tag{12}$$

by Minkowski's inequality. Letting $n \to \infty$ in (12) shows both that ℓ^p is closed under vector addition and that $\|\cdot\|_p$ satisfies the triangle inequality. □

Relationships among the various ℓ^p spaces are given in the exercises.

For $p = \infty$, we define $\ell^{\infty} = \{ \{t_i\} \in s: \sup|t_i| < \infty \}$ and set $\|\{t_i\}\|_{\infty} = \sup_i |t_i|$. It is easily seen that ℓ^{∞} is a vector space, $\|\cdot\|_{\infty}$ is a norm (exercise 9), and each ℓ^p, $1 \le p < \infty$, is contained in ℓ^{∞} (Exercise 10).

On s, there is a natural metric, called the *Frechet metric*, defined by

$$d(\{s_i\}, \{t_i\}) = \sum_{i=1}^{\infty} |s_i - t_i| / \left[(1 + |s_i - t_i|)2^i \right], \tag{13}$$

for $\{s_i\}, \{t_i\} \in s$. To show that d is a metric on s, we again need only show that the triangle inequality is satisfied. It will follow immediately from Lemma 9.

Lemma 9. For $a, b \in \mathbb{R}$,

$$|a + b| / (1 + |a + b|) \le \frac{|a|}{1 + |a|} + \frac{|b|}{1 + |b|}.$$

Proof. Consider the function $h(t) = t/(1 + t)$. For $t > -1$, h is increasing (examine h'). Consequently,

$$|a + b| / (1 + |a + b|) \le (|a| + |b|) / (1 + |a| + |b|)$$

$$\le \frac{|a|}{1 + |a|} + \frac{|b|}{1 + |b|}. \qquad \square$$

We next turn our attention to examples of metric spaces in which the elements are functions. For any space of functions, we assume that the operations of addition and multiplication by scalars are defined pointwise. We give first the general analogue of ℓ^{∞}.

EXAMPLE 10. Let $S \ne \varnothing$. Then $\mathscr{B}(S)$ is the space of all bounded real-valued functions defined on S. The norm on $\mathscr{B}(S)$ is defined by

$$\|f\|_{\infty} = \sup \{ |f(t)| : t \in S \}. \tag{14}$$

It is left to Exercise 13 to show that $\| \cdot \|_{\infty}$, called the *sup-norm*, is actually a norm on $\mathscr{B}(S)$. For $S = \mathbb{N}$, we have $\mathscr{B}(\mathbb{N}) = \ell^{\infty}$. $\qquad \square$

Many interesting function spaces arise as subspaces, so we formalize the terminology.

Definition 11. If (S, d) is a metric space and $Q \subset S$, then Q with the distance function d restricted to Q is also a metric space called a *subspace* of S. If $(S, \| \cdot \|)$ is a normed linear space and Q is a vector subspace of S, then Q is called a *linear subspace* of S.

EXAMPLE 12. Let $E \subset \mathbb{R}^n$ and $\mathscr{BC}(E)$ be the subspace of $\mathscr{B}(E)$, which consists of the bounded continuous functions on E. If $\mathscr{BC}(E)$ is equipped with the sup-norm from $\mathscr{B}(E)$, then it is a linear subspace of $\mathscr{B}(E)$. $\qquad \square$

EXAMPLE 13. Let $K \subset \mathbb{R}^n$ be compact and $\mathscr{C}(K)$ be the space of continuous functions on K. In this case, we write $\mathscr{BC}(K) = \mathscr{C}(K)$. □

If K is an interval in $\mathbb{R}$, then $\mathscr{C}(K)$ has many interesting subspaces.

EXAMPLE 14. Let $K = [a, b]$ and $k \in \mathbb{N}$. Set $\mathscr{C}^k[a, b] = \{ f \in \mathscr{C}[a, b]: f$ has k continuous derivatives on $[a, b]\}$. Then $\mathscr{C}^k[a, b]$ is a linear subspace of $\mathscr{C}[a, b]$ with $\mathscr{C}^{k+1}[a, b] \subset \mathscr{C}^k[a, b]$.

Besides the sup-norm, $\mathscr{C}^k[a, b]$ inherits another natural norm from $\mathscr{C}[a, b]$, given by

$$\|f\|_{\infty, k} = \sum_{i=0}^{k} \|f^{(i)}\|_\infty, \qquad \text{where } f^{(0)} = f. \tag{15}$$

□

Another interesting subspace of $\mathscr{C}[a, b]$ is the space of (real) polynomials, $\mathscr{P}$. Since $\mathscr{P} \subset \mathscr{C}^k[a, b]$, for $k = 0, 1 \ldots$, $\mathscr{P}$ can be considered as a linear subspace of any $\mathscr{C}^k[a, b]$.

Similarly, ℓ^∞ has several interesting subspaces.

EXAMPLE 15. The space of convergent sequences, $c = \{ \{t_i\} \in s: \lim t_i \text{ exists}\}$, is a linear subspace of ℓ^∞.

The space of sequences that converge to 0, $c_0 = \{ \{t_i\} \in S: \lim t_i = 0\}$, is a linear subspace of $c(\ell^\infty)$.

The space of sequences that are eventually 0, $c_{00} = \{ \{t_i\} \in S: t_i = 0 \text{ for large } i \}$, is a linear subspace of c_0. □

Finally, there are some interesting function spaces associated with the integral.

EXAMPLE 16. Let I be an interval in $\mathbb{R}^n$ and $\mathscr{L}^1(I)$ be the vector space of all real-valued functions that are absolutely integrable over I. We define a seminorm on $\mathscr{L}^1(I)$ by

$$\|f\|_1 = \int_I |f|. \tag{16}$$

The use of the same notation in both (16) and (10) (with $p = 1$) should cause no difficulty since the underlying spaces are quite different. It is easily checked that (16) defines a seminorm that is not a norm (Exercise 16). □

There are at least two interesting subspaces $\mathscr{L}^1(I)$ when $I = [a, b]$ is a compact interval. The first is the subspace of Riemann integrable functions, $\mathscr{R}[a, b]$, and the second is the space of continuous functions, $\mathscr{C}[a, b]$. As we shall see later, these spaces are of greatest interest when they are equipped with the seminorm $\| \cdot \|_1$.

Note that $\| \cdot \|_1$ is actually a norm on $\mathscr{C}[a, b]$ (Exercise 17) so that $\mathscr{C}[a, b]$ carries both the sup-norm from (14) and the integral norm from (16). We shall see later that these norms are quite different.

Convergence

The concept of convergence can be generalized from $\mathbb{R}^n$ to semimetric, and consequently also to metric spaces, in the following way.

Definition 17. Let (S, d) be a semimetric space. The sequence $\{x_k\}$ in S *converges, with respect to d, to $x \in S$ if $\lim_k d(x_k, x) = 0$. If the metric d is understood, we say simply that $\{x_k\}$ *converges* to x and write $\lim x_k = x$ or $x_k \to x$; x is called the *limit* of the sequence $\{x_k\}$.

Proposition 18. Let (S, d) be a metric space. If $x_k \to x$ and $x_k \to y$, then $x = y$.

Proof. Since $d(x, y) \le d(x, x_k) + d(x_k, y)$ for all k, it follows that $d(x, y) = 0$ and that $x = y$. $\square$

Recall that a sequence in $\mathbb{R}^n$ converges with respect to the euclidean norm $\| \cdot \|_2$ iff it converges coordinatewise. It will be shown later that this holds for any norm on $\mathbb{R}^n$. However, this is not the case in the sequence spaces ℓ^p. If a sequence converges in ℓ^p, it converges coordinatewise (Exercise 23). The converse of this is not true; let $e_i = \{\delta_{ij}\}_{j=1}^\infty$, where δ_{ij} is the *Kronecker delta function* defined by

$$\delta_{ij} = \begin{cases} 1, & \text{if } i = j, \\ 0, & \text{if } i \ne j. \end{cases}$$

The sequence $\{e_i\}$ converges coordinatewise to 0; but it is not convergent with respect to $\| \cdot \|_p$ for any p, $1 \le p \le \infty$.

For the space s of real-valued sequences with the Frechet metric, however, the following proposition holds.

Proposition 19. A sequence $\{x_k\}$ in s converges to a point $x \in s$ if, and only if, the sequence $\{x_k\}$ converges coordinatewise to x.

Proof. Let $x_k = \{t_{kj}\}_{j=1}^\infty$ and $x = \{t_j\}$. For each j,

$$|t_{kj} - t_j| / 2^j (1 + |t_{kj} - t_j|) \le d(x_k, x).$$

So, if $x_k \to x$ with respect to d, then $\lim_k t_{kj} = t_j$ and $\{x_k\}$ converges to x coordinatewise.

Now suppose that $\{x_k\}$ converges coordinatewise to x. Let $\varepsilon > 0$. There is $M > 0$ such that $\sum_{k=M+1}^{\infty} 1/2^k < \varepsilon/2$. There is $N > M$ such that if $k \geq N$, then $|t_{kj} - t_j| < \varepsilon/2M$ for $1 \leq j \leq M$. Thus, if $k \geq N$, then $d(x_k, x) \leq \sum_{j=1}^{M} \varepsilon/2M + \varepsilon/2 = \varepsilon$.   □

From the definition of the sup-norm (14) and Proposition **11**.2, it follows that a sequence $\{f_j\}$ in $\mathcal{B}(S)$ converges to f with respect to the sup-norm iff $f_j \to f$ uniformly on S. Similarly, a sequence $\{f_j\}$ in $\mathscr{C}^k[a, b]$ converges to $f \in \mathscr{C}^k[a, b]$ with respect to the norm $\| \cdot \|_{\infty, k}$ iff the sequence $\{f_j\}$ converges to f uniformly on $[a, b]$ and, likewise, the sequence of derivatives $\{f_j^{(i)}\}$ converges to $f^{(i)}$ uniformly on $[a, b]$ for $1 \leq i \leq k$.

Convergence in a metric space depends on the particular metric defined on the space. For example, the sequence $f_k(t) = t^k$ in $\mathscr{C}[0, 1]$ converges to 0 with respect to the norm $\| \cdot \|_1$ of (16), but does not converge to 0 with respect to the sup-norm $\| \cdot \|_{\infty}$. (See, however, Exercise 24 for the converse.)

Again generalizing the situation in $\mathbb{R}^n$, we make the following definition of bounded sets in a metric space.

Definition 20. Let (S, d) be a metric space. A subset $E \subset S$ is *bounded* (with respect to d) if there is $R > 0$ and $x_0 \in S$ such that $E \subset S(x_0, R)$, the sphere of radius R about x_0.

Proposition 21. If $x_k \to x$, then $\{x_k : k \in \mathbb{N}\}$ is bounded.

Proof. There is N such that $d(x_k, x) < 1$ for all $k \geq N$. Let $R = \max\{1, d(x, x_1), \ldots, d(x, x_{N-1})\} + 1$. Then $x_k \in S(x, R)$ for all $k \in \mathbb{N}$.   □

Proposition 22. Let $(X, \| \cdot \|)$ be a seminormed space.

(i) If $x_k \to x$ and $y_k \to y$ in X, then $x_k + y_k \to x + y$.
(ii) If $t_k \to t$ in the scalar field and $x_k \to x$ in X, then $t_k x_k \to tx$.

Proof. (i) follows from $\|(x_k + y_k) - (x + y)\| \leq \|x_k - x\| + \|y_k - y\|$, and (ii) follows from $\|t_k x_k - tx\| \leq |t - t_k| \|x\| + |t_k| \|x_k - x\|$.   □

Since we have an operation of addition in seminormed spaces, the definition of convergence of infinite series in $\mathbb{R}^n$ can be immediately extended to the current situation.

Definition 23. Let $(X, \| \cdot \|)$ be a seminormed space and $\{x_k\} \subseteq X$. The (formal) series $\sum_{k=1}^{\infty} x_k$ *converges* to $x \in X$ if the sequence of partial sums $s_n = \sum_{k=1}^{n} x_k$ converges to x. In this case, we write $\sum_{k=1}^{\infty} x_k = x$, and call x the *sum* of the series.

As was the case in $\mathbb{R}^n$, a necessary condition for convergence of an infinite series is that the kth term approach zero.

Proposition 24. If $\sum_{k=1}^{\infty} x_k$ converges, then $x_k \to 0$.

Proof. The result is an immediate consequence of $s_{n+1} - s_n = x_{n+1}$. □

Several examples of infinite series in normed linear spaces are given in Exercise 30. See also Exercise 31.

EXERCISES 19

1. Describe the spheres in $\mathbb{R}, \mathbb{R}^2, \mathbb{R}^3$, for the distance-1 metric.

2. If (S, d) is a metric space, show that $|d(x, z) - d(z, y)| \leq d(x, y)$ and $|d(x, z) - d(y, u)| \leq d(x, y) + d(z, u)$.

3. Show that in a seminormed space, $\|0\| = 0$ and

$$| \|x\| - \|y\| | \leq \|x - y\|.$$

4. If X is a seminormed space (normed linear space), show that $d(x, y) = \|x - y\|$ defines a semimetric (metric) on X.

5. Show that (7), (8) and (9) define norms on $\mathbb{R}^n$.

6. Does the proof of Lemma 6 work for $0 < p \leq 1$?

7. Show that $\lim_{p \to \infty} \|x\|_p = \|x\|_\infty$ for $x \in \mathbb{R}^n$.

8. Show that ℓ^1 is a proper subset of ℓ^2 and that if $\{t_i\} \in \ell^1$, then $\|\{t_i\}\|_2 \leq \|\{t_i\}\|_1$. Can you give a more general statement for arbitrary $1 \leq p < \infty$?

9. Show that ℓ^∞ is a vector space and that $\|\cdot\|_\infty$ is a norm on ℓ^∞.

10. For $1 \leq p < \infty$, show that ℓ^p is a proper subset of ℓ^∞.

11. For $1 \leq p < \infty$, show that ℓ^p is infinite dimensional.
    Hint: Consider $e_i = \{\delta_{ij}\}_{j=1}^{\infty}$, where δ_{ij} is the Kronecker delta function.

12. Show that ℓ^∞ is a proper subset of s.

13. Show that $\|\cdot\|_\infty$ defines a norm on $\mathcal{B}(S)$.

14. Show that $\|f\|_{\infty, k} \geq \|f\|_\infty$ for $f \in \mathscr{C}^k[a, b]$ and that $\|f\|_{\infty, k+1} \geq \|f\|_{\infty, k}$ in $\mathscr{C}^{k+1}[a, b]$.

15. Show that $c_{00} \subset c_0 \subset c \subset \ell^\infty$, where the set inclusions are proper in each case.

16. Show that on $\mathscr{L}^1(I)$, $\|\cdot\|_1$ is a seminorm but not a norm.

17. On $\mathscr{C}[a, b]$, show that $\| \cdot \|_1$ is a norm and $\|f\|_1 \leq \|f\|_\infty (b - a)$.

18. Let (S, d) be a semimetric space. Show that "$x \sim y$ iff $d(x, y) = 0$" defines an equivalence relation on S. For $x \in S$, let x^* be the equivalence class determined by x and let $S^* = \{x^*: x \in S\}$. Show that $d^*(x^*, y^*) = d(x, y)$ defines a metric on S^*.

19. Does Proposition 18 hold for semimetric spaces?

20. Characterize the convergent sequences in the distance-1 metric.

21. If $x_k \to x$ with respect to d, show that every subsequence of $\{x_k\}$ also converges to x.

22. If $x_k \to x$ and $y_k \to x$, show that the "interlaced" sequence $x_1, y_1, x_2, y_2, \ldots$ also converges to x.

23. Show that if a sequence converges in ℓ^p, then it converges coordinatewise.

24. If $\{f_k\}$ converges to f with respect to $\| \cdot \|_\infty$ in $\mathscr{C}[a, b]$, show that $\{f_k\}$ converges to f with respect to $\| \cdot \|_1$.

25. Let $f_k(t) = (1/\sqrt{k}) \sin kt$ for $0 \leq t \leq 2\pi$. Show that $f_k \to 0$ in $\mathscr{C}[0, 2\pi]$ with respect to $\| \cdot \|_\infty$. Does $f_k \to 0$ in $\mathscr{C}^1[0, 2\pi]$ with respect to $\| \cdot \|_{\infty, 1}$?

26. In Definition 20, show that the point x_0 can be chosen arbitrarily in S.

27. For $E \subset S$, the *diameter* of E is defined by

$$\text{diam}(E) = \sup \{d(x, y): x, y \in E\}.$$

Show that E is bounded if, and only if, $\text{diam}(E) < \infty$.

28. Characterize the bounded subsets in the distance-1 metric.

29. Is the converse of Proposition 20 true?

30. Does the series $\sum_{i=1}^\infty i^{-1} e_i$ converge in ℓ^∞? In ℓ^p, for $1 \leq p < \infty$? What about the series $\sum_{i=1}^\infty e_i$?

31. Define convergence of infinite series in s. Does the series $\sum_{i=1}^\infty e_i$ converge in s?

32. If $x_n \to x$ and $y_n \to y$, show that $d(x_n, y_n) \to d(x, y)$.
    Hint: Exercise 2.

33. Show that a semimetric space is a metric space if, and only if, convergent sequences have unique limits.

34. If $x = \{t_k\} \in c_0$, show that there is k such that $\|x\|_\infty = |t_k|$.

35. Let $\{x_{jk}\}$ be a double sequence in a metric space and suppose that $\lim_k x_{jk} = x_j$ and $\lim_j x_j = x$ both exist. Show that there is a subsequence $\{n_k\}$ such that $\lim x_{kn_k} = x$.

36. Let $f: \mathbb{R} \to \mathbb{R}$ be bounded and continuous. Define $d: \mathbb{R} \times \mathbb{R} \to \mathbb{R}$ by $d(x, y) = \sup \{ |f(t - x) - f(t - y)| : t \in \mathbb{R} \}$. Show that d is a semimetric on $\mathbb{R}$. Show that d is a metric if, and only if, f is not periodic.

37. Let X be a nls and $f: [a, b] \to X$. How would you define the gauge integral of f over $[a, b]$?

38. How should one define absolute convergence for a series Σx_k in a nls? Do the series in Exercise 30 converge absolutely?

39. Fix $\mathbf{x} \in \mathbb{R}^n$. Show that the function $\mathbf{y} \to |\mathbf{x} \cdot \mathbf{y}|$ defines a seminorm on $\mathbb{R}^n$. When is this seminorm a norm?

40. If $f_k \in \mathscr{L}^1(I)$ and $f_k \to f$ pointwise on I and there is $g \in \mathscr{L}^1(I)$ such that $|f_k(t)| \le g(t) \; \forall \, t \in I$, show that $\|f_k - f\|_1 \to 0$ (Exercise **15**.31).

41. Show that $BV[a, b]$ is a seminormed linear space under $\|\varphi\| = \mathrm{Var}(\varphi: [a, b])$. Is $\| \cdot \|$ a norm? (Recall Exercises **13**.25 and **13**.30.)

CHAPTER 20

Topology in Metric Spaces

Analysis of convergence in $\mathbb{R}^n$ is dependent on the structure imposed on $\mathbb{R}^n$ by the family of neighborhoods of its points. This same structure can also be described in terms of the family of open subsets of $\mathbb{R}^n$ or in terms of the family of closed subsets of $\mathbb{R}^n$. It is this structure, which can be described in several different equivalent ways, that is referred to as a *topology* for $\mathbb{R}^n$. The ideas involved can be extended to yield a similar structure, a topology, for general metric spaces. In fact, the generalization can be carried much further, but the field that emerges is so large that it requires a separate course of study.

By using the distance function, we can easily extend the definitions of open and closed sets in $\mathbb{R}^n$ to metric spaces (S, d). If $x \in S$ and $\varepsilon > 0$, then the sphere $S(x, \varepsilon) = \{ y: \ d(x, y) < \varepsilon \}$ is called an *ε-neighborhood* of x (with respect to d). Occasionally, an ε-neighborhood of x is referred to simply as a *neighborhood* of x. A neighborhood is yet another device for describing the proximity of a point to the point x. If $E \subseteq S$, a point x is said to be an *interior point* of E if there is an ε-neighborhood $S(x, \varepsilon)$ such that $S(x, \varepsilon) \subseteq E$. The set of all interior points of E is called the *interior* of E and is denoted by

255

E^0. A set E is *open* if $E = E^0$. Equivalently, a set is open if it contains a neighborhood of each of its points. The empty set $\varnothing$ and the set S are open subsets of the metric space (S, d). This definition of open agrees with that given earlier for sets in $\mathbb{R}^n$, which are open with respect to the Euclidean metric.

EXAMPLE 1. Any sphere $S(x, r)$ is open. To see this, if $y \in S(x, r)$, set $\varepsilon = r - d(x, y)$ and note that $S(y, \varepsilon) \subseteq S(x, r)$.   □

Proposition 2. Let (S, d) be a metric space.

(i) If G_α is an open set in S for each $\alpha \in A$, then $G = \bigcup_{\alpha \in A} G_\alpha$ is open.
(ii) If $G_1, \ldots, G_k$ are open subsets of S, then $G = \bigcap_{i=1}^k G_i$ is open.

Proof. For (i), let $x \in G$. Then $x \in G_\alpha$ for some $\alpha \in A$. There is then an ε-neighborhood $S(x, \varepsilon)$ such that $S(x, \varepsilon) \subseteq G_\alpha \subseteq G$. Hence, G is open.

For (ii), let $x \in G$. For each $i = 1, \ldots, k$, there is an ε_i-neighborhood $S(x, \varepsilon_i)$ such that $S(x, \varepsilon_i) \subseteq G_i$. If $\varepsilon = \min\{\varepsilon_1, \ldots, \varepsilon_k\}$, then $S(x, \varepsilon) \subseteq G$ and so G is open.   □

A closed set will be defined by a device that is related to the concept of convergence. If $E \subseteq S$, then a point $x \in S$ is an *accumulation point* of E if every ε-neighborhood of x contains infinitely many points of E. A set $E \subseteq S$ is *closed* (with respect to d) if E contains all of its accumulation points. Again, both $\varnothing$ and S are closed subsets of (S, d). In the following propositions it will be shown that a set E being closed is equivalent to its complement $S \setminus E$ being open and also equivalent to every convergent sequence in E having its limit in E. In fact, any one of these can be chosen as a definition for closed and the equivalence of the others to it and to each other can then be established. Where one begins is largely a matter of taste and toilet training.

Proposition 3. A point x is an accumulation point of E if, and only if, every ε-neighborhood of x contains a point of E that is distinct from x.

Proof. Necessity is clear. Suppose x is not an accumulation point of E. Then there is an ε-neighborhood that contains only a finite number of points of E, say, $x_1, \ldots, x_k$, distinct from x. Set $r = \min\{d(x, x_i): 1 \le i \le k\}$. Then the sphere $S(x, r)$ is an r-neighborhood of x that contains no point of E distinct from x.   □

Closed sets and accumulation points in a metric space can also be characterized in terms of convergent sequences as follows.

Proposition 4.

(i) E is closed if, and only if, every convergent sequence in E has its limit in E.

(ii) x is an accumulation point of E if, and only if, there is a sequence $\{x_k\}$ of distinct points from E such that $x_k \to x$.

Proof. (i) First, suppose that E is closed, $\{x_k\} \subset E$, and $x_k \to x$. If $H = \{x_k: k \in \mathbb{N}\}$ is infinite, then every ε-neighborhood of x contains an infinite number of points of H. Thus, x is an accumulation point of E and, since E is closed, $x \in E$. If H is finite, then $x_k = x$ for large k so that $x = x_k \in E$.

For the converse, let x be an accumulation point of E. Pick $x_1 \in S(x, 1) \cap E$. Suppose that $x_1, \dots, x_k$ have been chosen to be distinct and such that $x_j \in S(x, 1/j) \cap E$, $1 \le j \le k$. Pick $x_{k+1} \in S(x, 1/(k+1)) \cap E$ such that $x_{k+1} \ne x_j$ for $1 \le j \le k$. Then $\{x_k\}$ is a sequence of distinct points such that $x_k \to x$; it follows that $x \in E$ and that E is closed.

(ii) The proof above establishes the necessity. If $\{x_k\}$ is a sequence of distinct elements of E such that $x_k \to x$, then every ε-neighborhood of x contains an infinite number of the $\{x_k: k \in \mathbb{N}\}$ so that x is an accumulation point of E. □

Analogous to Proposition 8.1 we have the following.

Proposition 5. $E \subseteq S$ is closed if, and only if, the complement of E, $S \setminus E$, is open.

Proof. Suppose E is closed and let $x \in S \setminus E$. Then x is not an accumulation point of E and so there is an ε-neighborhood of x such that $S(x, \varepsilon) \cap E = \varnothing$, that is, $S(x, \varepsilon) \subseteq S \setminus E$. Thus, x is an interior point of $S \setminus E$ and $S \setminus E$ is open.

Suppose next that $S \setminus E$ is open. If $x \notin E$, then there is an ε-neighborhood of x such that $S(x, \varepsilon) \subseteq S \setminus E$ so that no point in $S \setminus E$ can be an accumulation point of E. Hence, E contains all of its accumulation points and is closed. □

From Propositions 2 and 5 we obtain Proposition 6.

Proposition 6.

(i) If F_α is closed for all α in an index set A, then $F = \bigcap_{\alpha \in A} F_\alpha$ is closed.

(ii) If $F_1, \dots, F_k$ are closed, then $H = \bigcup_{i=1}^{k} F_i$ is closed.

Proof. The results are direct consequences of Propositions 2 and 5 after observing that DeMorgan's laws imply $S \setminus F = \bigcup_{\alpha \in A}(S \setminus F_\alpha)$ and

$$S \setminus H = \bigcap_{i=1}^{k}(S \setminus F_i).$$ □

If $E \subseteq S$, then the set of all accumulation points of E is called the *derived set* of E and is denoted by E'. The *closure* of E is defined to be $E \cup E'$ and is denoted by $\overline{E}$.

Proposition 7. $\overline{E}$ is closed and is the smallest closed set containing E.

Proof. Let x be an accumulation point of $\overline{E}$. Then every ε-neighborhood of x contains a point $y \in \overline{E}$ distinct from x, and so, must contain a point of E, distinct from x. Thus x is an accumulation point of E and $x \in \overline{E}$.

Clearly, if F is a closed set containing E, then $F \supseteq E'$ and so $F \supseteq \overline{E}$, that is, $\overline{E}$ is the smallest closed set containing E.   □

Finally, we say that a subset $E \subseteq S$ is *dense* in S if $\overline{E} = S$. For example, the rationals are dense in $\mathbb{R}$.

EXERCISES 20

1. Characterize the open sets in the distance-1 metric.
2. Let $g \in \mathscr{C}[a, b]$. Show that $E = \{ f \in \mathscr{C}[a, b] \colon f(t) < g(t)\ \forall\, t \in [a, b]\}$ is open in $\mathscr{C}[a, b]$ with respect to $\| \cdot \|_\infty$. Is the same statement true if $\mathscr{C}[a, b]$ is replaced by $\mathscr{B}[a, b]$?
3. Show that E^0 is open for every $E \subseteq S$.
4. Let $x \in S$ and $r > 0$. Show that $\{ y \colon d(x, y) > r \}$ is open.
5. Does (ii) of Proposition 2 hold for infinite families of open sets?
6. If $x \in S$ and $r > 0$, show that $E = \{ y \colon d(x, y) \leq r \}$ and $F = \{ y \colon d(x, y) \geq r \}$ are closed.
7. Let $g \in \mathscr{C}[a, b]$ and $r > 0$. Show that

$$E = \{ f \in \mathscr{C}[a, b] \colon f(t) \leq g(t)\ \forall\, t \in [a, b]\}$$

and

$$F = \{ f \in \mathscr{C}[a, b] \colon |f(t)| \leq r\ \forall\, t \in [a, b]\}$$

are closed in $\mathscr{C}[a, b]$ with respect to $\| \cdot \|_\infty$.

8. What are the closed sets in the distance-1 metric?

9. Is (ii) of Proposition 6 valid for infinite families of closed sets?

10. What is the closure of c_{00} in ℓ^∞ with respect to $\|\cdot\|_\infty$? What is the closure of c_{00} in ℓ^p with respect to $\|\cdot\|_p$ if $1 \le p < \infty$?

11. Show that E' is always closed.

12. Show that $\overline{E \cup F} = \overline{E} \cup \overline{F}$. What about $\overline{E \cap F} = \overline{E} \cap \overline{F}$?

13. If $E \subset S$, then x is a *boundary point* of E if every ε-neighborhood of x contains points of both E and $S \setminus E$. The set of all boundary points of E is denoted by ∂E. Show that ∂E is closed.

14. If S has the distance-1 metric and $E \subseteq S$, what are E' and ∂E?

15. In $\mathbb{R}^N$, find $\partial S(x, \varepsilon)$. Is your answer the same for the distance-1 metric?

16. Show that $\partial(\partial S) \subseteq \partial S$ for every S. Give an example in which ∂S is nonvoid, but $\partial(\partial S) = \varnothing$.

17. Show that $\mathscr{C}_0[a, b] = \{ f \in \mathscr{C}[a, b] : f(a) = f(b) = 0 \}$ is a closed linear subspace of $\mathscr{C}[a, b]$.

18. Show that $x \in \overline{E}$ if, and only if, there is a sequence $\{x_k\} \subseteq E$ such that $x_k \to x$.

19. Show that diam $E = $ diam $\overline{E}$. (Recall Exercise **19**.27.)

20. Give an example in which $\overline{\{x : d(x, x_0) < 1\}} \neq \{x : d(x, x_0) \le 1\}$. Show that this cannot happen in a normed space.

21. Let $x = \{t_j\} \in \ell^p$, $1 \le p \le \infty$, and set

$$E = \{ y = \{s_j\} \in \ell^p : |s_j| \le |t_j| \text{ for } j \in \mathbb{N} \}.$$

Show that E is closed. Is the analogous statement true in s?

22. Show that F is closed iff $F = \overline{F}$.

23. Show that $\overline{\overline{E}} = \overline{E}$.

24. Show that the polynomials are dense in $\mathscr{C}[a, b]$ with respect to $\|\cdot\|_\infty$.

25. Let Γ be the space of all $\mathbb{Q}$-valued sequences. Show that $\Gamma \cap \ell^p$ is dense in ℓ^p for $1 \le p \le \infty$. Is the analogous statement true for the sequence spaces c_{00}, c_0, c, and s?

26. Show that c_{00} is dense in s.

27. Show that $E^0 \cap F^0 = (E \cap F)^0$. What about $E^0 \cup F^0 = (E \cup F)^0$?

28. (This exercise requires some linear algebra. See Exercises **21**.40 and **22**.26 for continuations of this exercise.) Let X be a normed space and L a linear subspace. For $x \in X$, let $\hat{x} = x + L$ be the coset in the quotient space X/L induced by x. Show that $\|\hat{x}\| = \inf \{ \|x + \ell\| : \ell \in L \}$ defines a seminorm on X/L and that this semi-norm is a norm iff L is closed.

CHAPTER 21

Continuity

The distance function is a useful vehicle for generalizing the concept of continuity for mappings on $\mathbb{R}^n$ to mappings between metric spaces. Throughout the current chapter, unless otherwise specified, (S, d) and (S_i, d_i), $i = 1, 2$, are metric spaces and f is a function from S_1 into S_2. We begin with a definition of limit for a function from one metric space to another.

Definition 1. Let $E \subseteq S_1$, x an accumulation point of E, and f a function from E into S_2. The function f has a *limit* $y \in S_2$ at x if for each $\varepsilon > 0$ there is $\delta > 0$ such that $d_2(f(z), y) < \varepsilon$ whenever $0 < d_1(z, x) < \delta$ and $z \in E$; in this case we write $\lim_{z \to x} f(z) = y$.

It is shown below that the limit is unique whenever it exists. Clearly, the limit depends on the metrics defined on the sets S_1 and S_2 (see Exercise **21**.1).

Proposition 2. $\lim_{z \to x} f(z) = y$ if, and only if, $\lim_{k \to \infty} f(x_k) = y$ for every sequence $\{x_k\} \subseteq E$ such that $x_k \neq x$ and $x_k \to x$.

Proof. Suppose that $\lim_{z \to x} f(z) = y$ and $\{x_k\}$ is a sequence in E that converges to x and $x_k \neq x$. Let $\varepsilon > 0$ and $\delta > 0$ be as in Definition 1. Then there is N such that $0 < d_1(x_k, x) < \delta$ for $k \geq N$. So, for $k \geq N$, $d_2(f(x_k), y) < \varepsilon$ and $f(x_k) \to y$.

Suppose next that f does not have a limit y at x. Then there is $\varepsilon > 0$ such that for each k there is x_k with $0 < d_1(x_k, x) < 1/k$ and $d_2(f(x_k), y) \geq \varepsilon$. But then $x_k \to x$, $x_k \neq x$, while $\{f(x_k)\}$ does not converge to y. $\square$

Corollary 3. If f has a limit at x, it is unique.

Proof. Propositions 2 and **19**.18. $\square$

The restriction $0 < d(z, x)$ emphasizes the irrelevance of the value, if any, of f at x to the definition of limit. The concept of limit relates only to the behaviour of f at points near, in the sense of the metric d, to x. However, the case in which f has a value at x which coincides with the limit at x is often mathematically beneficial and f is then said to be continuous at x. We formalize this with the following definition.

Definition 4. Let $E \subseteq S_1$, $x \in E$ and $f \colon E \to S_2$. The function f is *continuous at* x if for each $\varepsilon > 0$, there is $\delta > 0$ such that $d(f(z), f(x)) < \varepsilon$ whenever $d(z, x) < \delta$ and $z \in E$; f is *continuous on* E if it is continuous at each point of E.

Again note that continuity depends on the metrics involved (see Exercise 2).

If $E \subseteq S$, a point $x \in E$ is called an *isolated point* of E if x is not an accumulation point of E. Thus, x is an isolated point of E if there is a δ-neighborhood of x such that $S(x, \delta) \cap E = \{x\}$. Any function defined on E is continuous at each isolated point of E. If x is an accumulation point of E, then continuity of f at x is equivalent to the existence of the limit at x, that is,

Proposition 5. If $E \subseteq S_1$ and $x \in E$ is an accumulation point of E, then f is continuous at x if, and only if, $\lim_{z \to x} f(z) = f(x)$.

Proposition 6. Let (S_i, d_i), $i = 1, 2, 3$ be metric spaces, $f \colon S_1 \to S_2$, $g \colon S_2 \to S_3$, and $x \in S_1$.

    (i) If f is continuous at x and g is continuous at $f(x)$, then $g \circ f$ is continuous at x.

    (ii) If f is continuous on S_1 and g is continuous on S_2, then $g \circ f$ is continuous on S_1.

Proof. For (i), let $\varepsilon > 0$. There is $\delta_1 > 0$ such that $d_2(w, f(x)) < \delta_1$ implies that $d_3(g(w), g(f(x))) < \varepsilon$. There is $\delta > 0$ such that $d_1(z, x) < \delta$ implies that $d_2(f(z), f(x)) < \delta_1$. Hence, $d_1(z, x) < \delta$ implies that $d_2(g(f(z)), g(f(x))) < \varepsilon$.

(ii) follows immediately from (i). $\qquad\square$

Global continuity, that is, continuity on the whole space S_1, can be characterized in terms of either the open or the closed sets in S_1.

Proposition 7. Let (S_i, d_i), $i = 1, 2$, be metric spaces and $f\colon S_1 \to S_2$. The following are equivalent:

    (i) f is continuous on S_1.
    (ii) $f^{-1}(U)$ is open in S_1 for each open set U in S_2.
    (iii) $f^{-1}(F)$ is closed in S_1 for each closed set F in S_2.

Proof. Suppose (i) holds and let U be an open subset of S_2. If $x \in f^{-1}(U)$, then there is an ε-neighborhood of $f(x)$ such that $S(f(x), \varepsilon) \subset U$. Let $\delta > 0$ be as in Definition 4. Then $f(S(x, \delta)) \subset S(f(x), \varepsilon)$ or $S(x, \delta) \subset f^{-1}(U)$. Hence, $f^{-1}(U)$ is open and (ii) holds.

We next show that (ii) and (iii) are equivalent. We write out the proof that (ii) implies (iii); the proof that (iii) implies (ii) is given parenthetically. Let $F \subset S_2$ be closed (open). Then $S_2 \setminus F$ is open (closed). Hence, $f^{-1}(S_2 \setminus F) = S_1 \setminus f^{-1}(F)$ is open (closed) in S_1. It follows that $f^{-1}(F)$ is closed (open).

Finally, (ii) implies (i). Let $x \in S_1$ and $\varepsilon > 0$. Then $S(f(x), \varepsilon)$ is open in S_2; so $V = f^{-1}(S(f(x), \varepsilon))$ is open in S_1 and $x \in V$. Thus, there is $\delta > 0$ such that $S(x, \delta) \subset V$, that is, $d_1(z, x) < \delta$ implies $d_2(f(z), f(x)) < \varepsilon$. $\qquad\square$

The criteria (ii) and (iii) concern the inverse images of open and closed sets; it is not true in general that the direct images of open or closed sets under continuous mappings are open or closed, respectively. For example, if $f\colon \mathbb{R} \to \mathbb{R}$ ($g\colon [1, \infty) \to \mathbb{R}$) is given by $f(x) = x^2$ ($g(x) = 1/x$), then

$$f((-1, 1)) = [0, 1) \qquad (g([1, \infty)) = (0, 1]).$$

Definition 8. Let $(X, \|\cdot\|)$ be a normed linear space. If $f, g\colon E \to X$ and t is a scalar, we define the *sum* and *scalar multiple* by $(f + g)(s) = f(s) + g(s)$ and $(tf)(s) = tf(s)$.

Proposition 9. If (S, d) is a metric space, $(X, \|\cdot\|)$ is a normed linear space, and $f, g\colon S \to X$ are continuous at $x \in S$, then $f + g$ and tf are continuous at x.

Proof. Proposition **19**.22 and Exercise 3. $\qquad\square$

Definition 10. The function $f\colon S_1 \to S_2$ is *uniformly continuous* on S_1 if for each $\varepsilon > 0$ there is $\delta > 0$ such that $d_2(f(x), f(y)) < \varepsilon$ whenever $d_1(x, y) < \delta$.

The many examples of continuous and uniformly continuous functions that were discussed in our study of functions on $\mathbb{R}^n$ are, of course, also examples of such functions in metric spaces. In Proposition 13 we give an example of a uniformly continuous function on a general metric space.

Definition 11. Let $E \subseteq S$ and $x \in S$. The *distance* from x to E is $d(x, E) = \inf\{d(x, y): y \in E\}$.

Proposition 12. $d(x, E) = 0$ if, and only if, $x \in \overline{E}$.

   Proof. Suppose $x \in \overline{E}$. Then there is a sequence $\{x_k\} \subseteq E$ such that $x_k \to x$ (Exercise **20**.18). Hence, $d(x, E) = 0$.
   If $d(x, E) = 0$, there is a sequence $\{x_k\} \subseteq E$ such that $d(x, x_k) < 1/k$. Hence, $x_k \to x$ and $x \in \overline{E}$ (Exercise **20**.18).   □

Proposition 13. The distance function $d(\cdot, E)$ is uniformly continuous on S.

   Proof. Let $x, y \in S$. For any $z \in E$, $d(x, E) \leq d(x, z) \leq d(x, y) + d(y, z)$, and so, $d(x, E) - d(y, E) \leq d(x, y)$. By symmetry, $d(y, E) - d(x, E) \leq d(x, y)$. It follows that $|d(x, E) - d(y, E)| \leq d(x, y)$. Thus, $d(\cdot, E)$ is uniformly continuous.   □

Linear Maps

If X and Y are vector spaces, recall that a map $T: X \to Y$ is *linear* if $T(sx + ty) = sT(x) + tT(y)$ for all $x, y \in X$ and all scalars s, t. For linear mappings it is customary to write $T(x) = Tx$. Some of the properties of linear maps between normed linear spaces are given in the following.

Proposition 14. Let X, Y be normed linear spaces and $T: X \to Y$ linear. The following are equivalent:

   (i) T is continuous on X.
   (ii) T is continuous at $x = 0$.
   (iii) $\{\|Tx\|: \|x\| \leq 1\}$ is bounded in $\mathbb{R}$.
   (iv) There is $M \geq 0$ such that $\|Tx\| \leq M\|x\|$ for all $x \in X$.
   (v) T is uniformly continuous on X.

(The norms on X and Y are in general different; however, the fact that we do not use different notation for them should cause no confusion.)

   Proof. We establish the equivalence of the above statements by showing that each statement implies the one that follows and that (v) implies (i).

That (i) implies (ii) is clear.

Suppose (ii) holds and that (iii) fails to hold. Then there is a sequence $\{x_k\} \subset X$ such that $\|x_k\| \leq 1$ and $\|Tx_k\| \geq k$. Set $y_k = x_k/k$. Then $y_k \to 0$. However, $\|Ty_k\| = (1/k)\|Tx_k\| \geq 1$, and so, $\{Ty_k\}$ doesn't converge to zero, that is, T is not continuous at 0.

Suppose (iii) holds and $M = \sup\{\|Tx\|: \|x\| \leq 1\}$. If $x = 0$, the conclusion in (iv) holds trivially. If $x \neq 0$, set $y = x/\|x\|$. Then $\|y\| = 1$ and $\|Ty\| \leq M$, that is, $(1/\|x\|)\|Tx\| \leq M$ and (iv) holds.

If (iv) holds, then for $x, y \in X$,

$$\|Tx - Ty\| = \|T(x - y)\| \leq M\|x - y\|.$$

It follows that T is uniformly continuous and that (v) holds.

That (v) implies (i) is immediately clear. $\square$

Let $L(X, Y)$ be the set of all continuous linear mappings from the vector space X into the vector space Y. Then $L(X, Y)$ is a vector space under the operations of addition and multiplication by scalars given in Definition 8. $L(X, Y)$ is also a normed linear space under a norm, called the *operator* or *uniform norm*, induced by the norms defined on X and Y. The elements of $L(X, Y)$ are called *linear operators* or, if there's little chance of confusion, just *operators*.

Definition 15. Let $T \in L(X, Y)$. Define $\|T\| = \sup\{\|Tx\|: \|x\| \leq 1\}$.

Note that $\|Tx\| \leq \|T\|\,\|x\|$ for all $x \in X$ [the proof of (iii) implies (iv) above]; also note that we do not use a special notation for the operator norm.

Proposition 16. $L(X, Y)$ with the operator norm is a normed linear space.

Proof. We need to establish only the triangle inequality. If $T_1, T_2 \in L(X, Y)$, then

$$\|(T_1 + T_2)x\| \leq \|T_1x\| + \|T_2x\| \leq (\|T_1\| + \|T_2\|)\|x\|.$$

This implies that $\|T_1 + T_2\| \leq \|T_1\| + \|T_2\|$. $\square$

If Y is the scalar field, then $L(X, Y)$ is called the *dual* (or *adjoint*, or *conjugate*) of X and is denoted by X^* (or X'). The elements of X^* are called *continuous linear functionals* and the operator or uniform norm on X^* is called the *dual norm*.

If Z is also a normed linear space and $U \in L(X, Y)$, $V \in L(Y, Z)$, then the composition $V \circ U$ is usually just denoted by VU. Of course, $VU \in L(X, Z)$, and it is easily checked that $\|VU\| \leq \|V\| \cdot \|U\|$ (Exercise **21**.11).

EXAMPLE 17. If $T: \mathbb{R}^n \to Y$ is linear, then T is continuous with respect to $\|\cdot\|_1$ on $\mathbb{R}^n$ since

$$\|T\mathbf{x}\| = \left\| T\left(\sum_{i=1}^n x_i \mathbf{e}_i \right) \right\| \leq \sum_{i=1}^n |x_i| \|T\mathbf{e}_i\| \leq \sup_{1 \leq i \leq n} \|T\mathbf{e}_i\| \cdot \|\mathbf{x}\|_1. \qquad \square$$

It will follow from Proposition 23 below that such a linear map on $\mathbb{R}^n$ is continuous with respect to any norm on $\mathbb{R}^n$.

EXAMPLE 18. Let $k: [a, b] \times [c, d] \to \mathbb{R}$ be continuous. Define $K(f)(s) = \int_c^d k(s, t) f(t) \, dt$ for $f \in \mathscr{C}[c, d]$, $a \leq s \leq b$. Then, Proposition **13**.13 implies that $K: \mathscr{C}[c, d] \to \mathscr{C}[a, b]$ is linear and

$$\|K(f)\|_\infty \leq \|f\|_\infty \sup_{a \leq s \leq b} \int_c^d |k(s, t)| \, dt$$

implies that K is continuous with respect to $\|\cdot\|_\infty$. Operators that have the form of the operator K are called *integral operators*; the function k is called the *kernel* of K. Such operators are encountered in the study of integral equations; see Examples **23**.2 and 5. $\qquad \square$

EXAMPLE 19. Define $T: \mathscr{C}[0, 1] \to \mathscr{C}[0, 1]$ by $T(f)(s) = \int_0^s f$. Then T is linear. Since $\|T(f)\|_\infty \leq \|f\|_\infty$, T is also continuous. It follows that $\|T\| \leq 1$ and, actually, that $\|T\| = 1$, since $T(1) = 1$. $\qquad \square$

The fact that continuity depends on the norms is again illustrated by the following example.

EXAMPLE 20. Let $X = \mathscr{C}^1[0, 1]$, the vector space of continuously differentiable functions, and define $D: X \to \mathscr{C}[0, 1]$ by $Df = f'$. Then D is clearly linear. If both X and $\mathscr{C}[0, 1]$ are equipped with $\|\cdot\|_\infty$, then D is not continuous. This follows from the example $f_k(t) = t^k$; then $\|f_k\|_\infty = 1$ and $\|Df_k\|_\infty = k$. If X is equipped with $\|\cdot\|_{\infty, 1}$ and $\mathscr{C}[0, 1]$ is equipped with $\|\cdot\|_\infty$, then D is continuous, since

$$\|Df\|_\infty = \|f'\|_\infty \leq \|f\|_{\infty, 1}. \qquad \square$$

EXAMPLE 21. Let X be the subspace of $\mathscr{L}^1(\mathbb{R})$, which consists of the continuous functions. Let $f \in X$ and define the *convolution operator* $T_f = T: X \to X$ by $Tg = f * g$. If X is equipped with $\|\cdot\|_1$, then $\|Tg\|_1 \leq \|f\|_1 \|g\|_1$, and so, $T \in L(X, X)$. $\qquad \square$

Equivalent Norms, Homeomorphisms, and Isometries

Definition 22. Two norms $\| \cdot \|_1$ and $\| \cdot \|_2$ on a vector space X are *equivalent* if there are $a, b > 0$ such that $a\|x\|_1 \leq \|x\|_2 \leq b\|x\|_1$ for all $x \in X$.

By Proposition 14, this is equivalent to the identity operator I: $(X, \| \cdot \|_1) \to (X, \| \cdot \|_2)$ being continuous and having a continuous inverse. Thus, if $\| \cdot \|_1$ and $\| \cdot \|_2$ are equivalent, they induce the same open and closed sets and the same convergent sequences on X. We give a generalization of this idea in Definition 25.

In $\mathbb{R}^n$, we have the following interesting result with respect to norms.

Proposition 23. Any two norms on $\mathbb{R}^n$ are equivalent.

Proof. From Exercise 18 it suffices to show that if $\| \cdot \|$ is a norm on $\mathbb{R}^n$, then $\| \cdot \|$ is equivalent to $\| \cdot \|_2$, the euclidean norm.

First, $\|x\| = \|\sum_{i=1}^n x_i e_i\| \leq \sum_{i=1}^n |x_i| \|e_i\| \leq \|x\|_2 (\sum_{i=1}^n \|e_i\|^2)^{1/2}$. Next, let $E = \{x : \|x\|_2 = 1\}$ and define $f: E \to \mathbb{R}$ by $f(x) = \|x\|$. Since $I: (\mathbb{R}^n, \| \cdot \|_2) \to (\mathbb{R}^n, \| \cdot \|)$ is continuous and the mapping $x \to \|x\|$ from $(\mathbb{R}^n, \| \cdot \|)$ into $\mathbb{R}$ is continuous (Exercise **21**.9), it follows that f is continuous with respect to $\| \cdot \|_2$. Since E is compact with respect to $\| \cdot \|_2$, f attains its minimum on E, say at $x_0 \in E$. Note that $f(x_0) = \|x_0\| = m > 0$. If $x \in \mathbb{R}^n$, $x \neq 0$, then $x/\|x\|_2 \in E$ so that $f(x/\|x\|_2) = \|x\|/\|x\|_2 \geq m$, that is, $\|x\| \geq m\|x\|_2$. □

Thus, it follows from Proposition 23 and Example 17 that a linear map $T: \mathbb{R}^n \to Y$ is continuous with respect to any norm on $\mathbb{R}^n$. In particular, it should also be observed that all of the norms, $\| \cdot \|_p$, $1 \leq p \leq \infty$, on $\mathbb{R}^n$ are equivalent.

EXAMPLE 24. The norms $\| \cdot \|_\infty$ and $\| \cdot \|_{\infty,1}$ on $C^1[0,1]$ are not equivalent (see Example 20). □

If $\| \cdot \|_1$ and $\| \cdot \|_2$ are equivalent norms on a vector space X, then the topological properties (i.e., open sets, closed sets, etc.) induced on X by these norms are equivalent. For general metric spaces, the corresponding notion is that of a homeomorphism.

Definition 25. A function $f: S_1 \to S_2$ is a *homeomorphism* from S_1 onto S_2 if f is 1-1, onto, continuous, and has a continuous inverse. Two spaces S_1 and S_2 are *homeomorphic* if there is a homeomorphism from S_1 onto S_2.

EXAMPLE 26. $f(t) = (2/\pi)\arctan(t)$ defines a homeomorphism from $\mathbb{R}$ onto $(-1, 1)$. □

For metric spaces there is a class of mappings, called isometries, which are stronger than homeomorphisms in the sense that every isometry is also a homeomorphism.

Definition 27. Let (S_i, d_i), $i = 1, 2$, be metric spaces. A function $f: S_1 \rightarrow S_2$ is an *isometry* if $d_1(x, y) = d_2(f(x), f(y))$ for all $x, y \in S_1$. S_1 and S_2 are *isometric* if there is an isometry from S_1 onto S_2.

If two metric spaces, S_1, and S_2, are isometric, then, as metric spaces they can be considered to be the same; in this case, we usually write $S_1 = S_2$.

Any isometry onto a space is obviously a homeomorphism; Example 26 gives an example of a homeomorphism that is not an isometry. Note that an isometry preserves distances and that this is not necessarily so for homeomorphisms. The following is an example of two spaces whose definitions are quite different; however, they are linearly isometric.

EXAMPLE 28. $(\ell^1)^*$ and ℓ^∞ are linearly isometric [so we write $(\ell^1)^* = \ell^\infty$].

Let $t = \{t_i\} \in \ell^\infty$. Define $f_t: \ell^1 \rightarrow \mathbb{R}$ by $f_t(\{s_i\}) = \sum_{i=1}^\infty s_i t_i$. Now f_t is obviously linear and $|f_t(\{s_i\})| \leq \|t\|_\infty \|\{s_i\}\|_1$ so $f_t \in (\ell^1)^*$ and $\|f_t\| \leq \|t\|_\infty$. However,

$$\|f_t\| = \sup\left\{|f_t(\{s_i\})|: \|\{s_i\}\|_1 \leq 1\right\}$$

$$\geq \sup\left\{|f_t(e_i)|: i \in \mathbb{N}\right\}$$

$$= \sup\left\{|t_i|: i \in \mathbb{N}\right\} = \|t\|_\infty;$$

thus $\|f_t\| = \|t\|_\infty$.

The map $\Psi: t \rightarrow f_t$ from ℓ^∞ into $(\ell^1)^*$ is obviously linear and therefore an isometry. We claim that Ψ is onto. Let $f \in (\ell^1)^*$. Set $t_i = f(e_i)$. Then $|t_i| \leq \|f\| \|e_i\| = \|f\|$ so that $t = \{t_i\} \in \ell^\infty$ and $f(\{s_i\}) = f(\sum_{i=1}^\infty s_i e_i) = \sum_{i=1}^\infty s_i f(e_i) = \sum_{i=1}^\infty s_i t_i = f_t(\{s_i\}) = \Psi(t)(\{s_i\})$. Hence, $\Psi(t) = f$, and Ψ is onto. $\square$

Products of Metric Spaces

If (S_i, d_i), $i = 1, 2$, are semimetric spaces, then there are several natural candidates for a semimetric on the product space $S_1 \times S_2$; namely,

(1) $d((x_1, x_2), (y_1, y_2)) = d_1(x_1, y_1) + d_2(x_2, y_2)$.

(2) $d'((x_1, x_2), (y_1, y_2)) = \max\{d_1(x_1, y_1), d_2(x_2, y_2)\}$.

(3) $d''((x_1, x_2), (y_1, y_2)) = [d_1(x_1, y_1)^2 + d_2(x_2, y_2)^2]^{1/2}$.

By Proposition 23 it follows that the semimetrics d, d', d'' are equivalent, in the sense that each one will induce the same open and closed sets on $S_1 \times S_2$ (Exercise 23), so it is immaterial which one is used. Because of its simplicity, the semimetric d is the one most often employed.

If $(X_i, \| \cdot \|_i)$, $i = 1, 2$ are seminormed spaces, then the seminorms

(1') $\|(x_1, x_2)\| = \|x_1\|_1 + \|x_2\|_2$,

(2') $\|(x_1, x_2)\|' = \max\{\|x_1\|_1, \|x_2\|_2\}$,

(3') $\|(x_1, x_2)\|'' = (\|x_1\|_1^2 + \|x_2\|_2^2)^{1/2}$,

induce the semimetrics (1), (2), (3) above.

EXERCISES 21

In the following exercises, (S, d) and (S_i, d_i), $i = 1, 2, \ldots$ are metric spaces; X and Y are normed linear spaces.

1. Let d be the usual metric and d' be the distance-1 metric on $[0, 1]$. Define $f: [0, 1] \to [0, 1]$ by $f = C_{\mathbf{Q} \cap [0, 1]}$. (a) Does $\lim_{z \to 0} f(z)$ exist if $d_1 = d$ and $d_2 = d'$? (b) Does $\lim_{z \to 0} f(z)$ exist if $d_1 = d'$ and $d_2 = d$?

2. If S_1 has the distance-1 metric in Definition 4, which functions $f: S_1 \to S_2$ are continuous? If S_2 has the distance-1 metric, which functions are continuous?

3. Show that $f: E \to S_2$ is continuous at $x \in E$ iff $\{x_k\} \subset E$ and $x_k \to x$ implies that $f(x_k) \to f(x)$.

4. If $f: S_1 \to X$ is continuous, show that $\{x: f(x) = 0\}$ is closed.

5. Let $f, g: S_1 \to S_2$ be continuous and let D be a dense subset of S_1. If $f(x) = g(x)$ for all $x \in D$, show that $f = g$.

6. Suppose that $f: S \to \mathbb{R}$ is continuous at x with $f(x) > 0$. Show that there are $m > 0$ and $\delta > 0$ such that $f(z) \geq m$ for all $z \in S(x, \delta)$.

7. Suppose $f: S \to X$ is continuous at x. Show that there are $M > 0$ and $\delta > 0$ such that $\|f(z)\| \leq M$ for all $z \in S(x, \delta)$.

8. Show that the composition of uniformly continuous functions is uniformly continuous.

9. If X is a normed linear space, show that $x \to \|x\|$ is uniformly continuous from X to $\mathbb{R}$.

10. Let A and B be disjoint closed sets in S. (a) Show that there is a continuous function $f: S \to [0, 1]$ such that $f(x) = 0$ for all $x \in A$ and $f(x) = 1$ for all $x \in B$.

Hint: Consider $d(\cdot, A)$ and $d(\cdot, B)$. (b) Show that there are disjoint open sets U, V such that $A \subset U$ and $B \subset V$.
Hint: Use f.

11. If $U \in L(X, Y)$ and $V \in L(Y, Z)$, show that $\|VU\| \leq \|V\|\,\|U\|$.

12. Let $T\colon X \to Y$ be linear. Show that T is continuous iff T carries bounded sets into bounded sets.

13. Show that the operators of Examples 18 and 19 are also continuous with respect to $\|\cdot\|_1$ on $\mathscr{C}[a, b]$.

14. Define $L\colon c \to \mathbb{R}$ by $L(\{t_i\}) = \lim t_i$. Show that L is linear and continuous and compute $\|L\|$.

15. Define $\delta\colon \mathscr{C}[0, 1] \to \mathbb{R}$ by $\delta(f) = f(0)$. Show that δ is continuous with respect to $\|\cdot\|_\infty$ and compute $\|\delta\|$. Show that δ is not continuous with respect to $\|\cdot\|_1$.

16. Define $R, L\colon \ell^p \to \ell^p$, $1 \leq p \leq \infty$ by $R(\{t_1, t_2, \dots\}) = \{0, t_1, t_2, \dots\}$ (right shift), and $L(\{t_1, t_2, \dots\}) = \{t_2, t_3, \dots\}$ (left shift). Show that $R, L \in L(\ell^p, \ell^p)$ and compute $\|R\|$ and $\|L\|$.

17. Let $g \in \mathscr{C}[0, 1]$ and define $T\colon \mathscr{C}[0, 1] \to \mathscr{C}[0, 1]$ by $T(f) = fg$. Show that T is linear and continuous and compute $\|T\|$.

18. Show that if $\|\cdot\|_i$, $i = 1, 2, 3$, are norms on the vector space X such that $\|\cdot\|_1$ is equivalent to $\|\cdot\|_2$ and $\|\cdot\|_2$ is equivalent to $\|\cdot\|_3$, then $\|\cdot\|_1$ is equivalent to $\|\cdot\|_3$.

19. Let I be an interval on $\overline{\mathbb{R}}$ and $f\colon S \times I \to \mathbb{R}$ be such that $f(s, \cdot)$ is integrable over I for each $s \in S$ and $f(\cdot, t)$ is continuous for each $t \in I$. If there is an integrable function $g\colon I \to \mathbb{R}$ such that $|f(s, t)| \leq g(t)$ for $t \in I$, show that the function $s \to \int_I f(s, t)\, dt$ is continuous.

20. Exhibit an isometry between $\mathscr{C}[0, 1]$ and $\mathscr{C}[1, 2]$. Do it also for $\mathscr{L}^1[0, 1]$ and $\mathscr{L}^1[1, 2]$.

21. Show that the operator R of Exercise 16 is an isometry. Is it onto? Is L of Exercise 16 an isometry?

22. Show that $(c_0)^*$ and ℓ^1 are linearly isometric.

23. Show that the semimetrics defined in (1), (2) and (3) induce the same open sets on $S_1 \times S_2$.

24. Show that a sequence $\{(x_k, y_k)\}$ in $S_1 \times S_2$ converges to $(x, y) \in S_1 \times S_2$ if, and only if, $x_k \to x$ and $y_k \to y$ with respect to d_1 and d_2.

25. Show that the function $(x, y) \to d(x, y)$, from $S \times S$ into $\mathbb{R}$, is uniformly continuous.

26. Show that the projection $P_1\colon S_1 \times S_2 \to S_1$ defined by $P_1(x, y) = x$ is uniformly continuous. Show that P_1 maps open sets onto open sets.

27. Let X be a real normed linear space. Show that the maps $(x, y) \to x + y$ and $(t, x) \to tx$, from $X \times X$ into X and $\mathbb{R} \times X$ into X, respectively, are continuous.

28. Let $E = E_1 \times E_2 \subseteq S_1 \times S_2$. Show that $\overline{E} = \overline{E}_1 \times \overline{E}_2$ and $E^0 = E_1^0 \times E_2^0$.

29. Show that each half-open interval $[a, b)$ is homeomorphic to each closed half-line $(-\infty, c]$.

30. If $E \subseteq S$ is open (closed) and $F \subseteq T$ is open (closed), show that $E \times F$ is open (closed) in $S \times T$.

31. If $f: [a, b] \to [a, b]$ is a homeomorphism, show that f carries endpoints to endpoints.

32. If E and F are subsets of (S, d), define the *distance* δ from E to F by $\delta(E, F) = \inf \{ d(x, F) : x \in E \}$. Show that $\delta(E, F) = \delta(F, E)$.

33. Show that $\mathrm{diam}(E \cup F) \le \mathrm{diam}\ E + \mathrm{diam}\ F + \delta(E, F)$. Give an example in which strict inequality holds.

34. Let $T \in L(X, Y)$. The *adjoint* of T is a mapping, $T': Y^* \to X^*$, defined by $T'y'(x) = y'(Tx)$. Show that $T' \in L(Y^*, X^*)$ and $\|T'\| \le \|T\|$.

35. Let (S_i, d_i), $i = 1, 2$ be metric spaces and $f: S_1 \to S_2$. Show that f is continuous if, and only if, $f(\overline{A}) \subseteq \overline{f(A)}$ for every $A \subseteq S_1$.

36. Let (S_i, d_i), $i = 1, 2$ be metric spaces. Show that $f: S_1 \to S_2$ is continuous iff $f^{-1}(E^0) \subseteq (f^{-1}(E))^0$ for every subset $E \subseteq S_2$.

37. If S is a metric space, show that the diagonal $\{(x, x): x \in S\}$ is closed in $S \times S$.

38. Define $f: c_{00} \to \mathbb{R}$ by $f(\{t_i\}) = \Sigma_{i=1}^{\infty} t_i$. Show that f is linear. Is f continuous with respect to $\| \cdot \|_{\infty}$?

39. (This is a continuation of Exercise 20.29.) Show that the quotient map $x \to \hat{x}$ from X into X/L is continuous.

CHAPTER 22

Complete Metric Spaces

In the discussion of sequences in $\mathbb{R}^n$, it was shown that questions on convergence could often be settled by showing whether or not the sequence satisfied the Cauchy condition. This concept has important applications in the study of metric spaces and is easily extended to that setting.

Let (S, d) be a semimetric space. A sequence $\{x_k\}$ in S is a *Cauchy sequence*, with respect to d, if for each $\varepsilon > 0$ there is N such that $d(x_m, x_n) < \varepsilon$ whenever $m, n \geq N$. In our earlier discussion, we gave examples of spaces, for example, the rational numbers $\mathbb{Q}$, in which some Cauchy sequences failed to converge. In fact, we used this particular example to motivate the "completion" of the reals by insisting that there be "enough" elements so that every Cauchy sequence in $\mathbb{R}$ would converge. Since $\mathbb{Q}$ and $\mathbb{R}$, with the usual metric induced by the Euclidean norm, are metric spaces, this situation is also a metric space problem and the procedure for handling it carries over directly to metric spaces.

Definition 1. The semimetric space (S, d) is *complete* if every Cauchy sequence in S converges to a point in S.

273

Thus, $\mathbb{R}^n$ with the Euclidean norm is a complete metric space. If $\| \ \|_1$ and $\| \ \|_2$ are equivalent norms on a vector space X and $(X, \| \ \|_1)$ is complete, then $(X, \| \ \|_2)$ is complete. So, $\mathbb{R}^n$ is complete with respect to any norm. The rational numbers $\mathbb{Q}$ with the usual metric is, of course, an example of a metric space that is not complete.

EXAMPLE 2. If K is a compact subset of $\mathbb{R}^n$, then $\mathscr{C}(K)$ is complete with respect to $\| \cdot \|_\infty$ (**11.4**). $\square$

A normed linear space that is complete with respect to the metric induced by the norm is called a *Banach space* or a *B-space*. The appellation honors a brilliant, young Polish mathematician Stefan Banach (1892–1945). For example, $\mathbb{R}^n$ with the Euclidean norm is a Banach space (**5.9**); moreover it follows from **21.23** that $\mathbb{R}^n$ is a Banach space under every norm. From Example 2, $\mathscr{C}(K)$ is a B-space under the sup-norm.

EXAMPLE 3. $\mathscr{C}^1[a, b]$ is a B-space with respect to $\| \cdot \|_{\infty,1}$. Let $\{ f_k \}$ be a Cauchy sequence with respect to $\| \cdot \|_{\infty,1}$. Then both $\{ f_k \}$ and $\{ f_k' \}$ are Cauchy sequences with respect to $\| \cdot \|_\infty$, and are, therefore, convergent with respect to $\| \cdot \|_\infty$. So, by Proposition **11.6**, $\{ f_k \}$ converges to an element of $\mathscr{C}^1[a, b]$ with respect to $\| \cdot \|_{\infty,1}$.

Note that $\mathscr{C}^1[a, b]$ is not complete with respect to $\| \cdot \|_\infty$ (Exercise 7). $\square$

EXAMPLE 4. For $1 \le p < \infty$, ℓ^p is a B-space. The proof that we give for completeness of ℓ^p is a prototype for most completeness proofs. Let $x_k = \{ t_{kj} \}_{j=1}^\infty$ be a Cauchy sequence in ℓ^p. First we find a candidate $x = \{ t_j \}$ for the limit of the sequence $\{ x_k \}$, and then show that $x \in \ell^p$ and $x_k \to x$ in $\| \cdot \|_p$. For each j,

$$|t_{mj} - t_{nj}| \le \|x_m - x_n\|_p;$$

thus, $\{ t_{nj} \}_{n=1}^\infty$ is a Cauchy sequence in $\mathbb{R}$ and converges. Let $t_j = \lim_n t_{nj}$, and $x = \{ t_j \}$.

We next show that $x \in \ell^p$ and $\|x_k - x\|_p \to 0$. Let $\varepsilon > 0$. There is N such that $m, n \ge N$ implies $\|x_m - x_n\|_p < \varepsilon$. Then, for any M,

$$\sum_{j=1}^M |t_{mj} - t_{nj}|^p \le \|x_m - x_n\|_p^p < \varepsilon^p.$$

Letting $n \to \infty$, with M and m fixed, yields

$$\sum_{j=1}^M |t_{mj} - t_j|^p \le \varepsilon^p$$

for $m \geq N$. Since this holds for every M, we have

$$\sum_{j=1}^{\infty} |t_{mj} - t_j|^p \leq \varepsilon^p.$$

It follows that $\{t_{mj} - t_j\}_{j=1}^{\infty} \in \ell^p$; so $x \in \ell^p$, and $\|x_m - x\|_p \leq \varepsilon$ for $m \geq N$. $\square$

The case when $p = \infty$ is left to Exercise 9. See also Remark 9.

Proposition 5. Let (S, d) be a complete metric space. A subset $E \subseteq S$ is complete with respect to d if, and only if, E is a closed subset of S.

Proof. Suppose E is closed and $\{x_k\}$ is a Cauchy sequence in E. Since S is complete, there is $x \in S$ such that $x_k \to x$. Since E is closed, $x \in E$ (**20.4**).

Suppose E is complete. Let $\{x_k\}$ be a sequence in E such that $x_k \to x \in S$. Then $\{x_k\}$ is a Cauchy sequence in E and, therefore, converges to a point $y \in E$. So $y = x \in E$ and E is closed (**20.4**). $\square$

EXAMPLE 6. The space c is complete with respect to $\| \cdot \|_\infty$. It suffices to show that c is a closed subspace of ℓ^∞ (Exercise 9 or Remark 9). Let $\{x_k\}$, $x_k = \{t_{kj}\}_{j=1}^{\infty}$, be a sequence in c that converges to a point $x = \{t_j\} \in \ell^\infty$. Now $\lim_j t_{kj}$ exists for all k and

$$\lim_k t_{kj} = t_j$$

uniformly for $j \in \mathbb{N}$. So, by **11.5**,

$$\lim_j \lim_k t_{kj} = \lim_j t_j$$

exists and $x \in c$. $\square$

We next consider completeness for the space of gauge integrable functions and establish the important property that the space $\mathscr{L}^1(I)$ of integrable functions is complete with respect to the seminorm induced by the integral (Example **19**.16). The corresponding space of Riemann integrable functions is *not* complete with respect to the seminorm induced by the Riemann integral. This is the primary motivation for a more general integration theory.

Theorem 7 (Riesz–Fischer). Let I be an interval in $\mathbb{R}^n$. Then $\mathscr{L}^1(I)$ is complete with respect to $\| \cdot \|_1$ (Example **19**.16).

Proof. Let $\{f_n\}$ be a Cauchy sequence in $\mathscr{L}^1(I)$. It suffices to produce a subsequence $\{f_{n_k}\}$ that converges to a function $f \in \mathscr{L}^1(I)$ (Exercise 3).

Pick a subsequence $\{f_{n_k}\}$ such that

$$\|f_{n_{k+1}} - f_{n_k}\|_1 < \frac{1}{2^k}.$$

Then

$$\sum_{k=1}^{\infty} \|f_{n_{k+1}} - f_{n_k}\|_1 < \infty.$$

If

$$g_k = \sum_{i=1}^{k} |f_{n_{i+1}} - f_{n_i}|,$$

then $\int_I g_k \le 1$ for all k and Theorem **17.17** implies that

$$\lim g_k(x) = \sum_{k=1}^{\infty} |f_{n_{k+1}} - f_{n_k}|(x) = g(x)$$

is finite for almost all $x \in I$. If

$$g(x) = \begin{cases} \lim g_k(x) & \text{when this limit exists,} \\ 0 & \text{otherwise,} \end{cases}$$

then g is integrable and

$$\int_I g = \lim \int_I g_k = \sum_{i=1}^{\infty} \|f_{n_{i+1}} - f_{n_i}\|_1.$$

Define

$$f(x) = \begin{cases} \sum_{k=1}^{\infty} \{f_{n_{k+1}}(x) - f_{n_k}(x)\}, & \text{when the series converges absolutely,} \\ 0, & \text{otherwise.} \end{cases}$$

Then

$$\left| \sum_{i=1}^{k} (f_{n_{i+1}} - f_{n_i})(x) \right| \le g_k(x) \le g(x)$$

for almost all $x \in I$. So, f is absolutely integrable and the dominated convergence theorem (**17.18** and Exercise **15.31**) implies that

$$\left\| f - \sum_{i=1}^{k} (f_{n_{i+1}} - f_{n_i}) \right\|_1 = \|f + f_{n_1} - f_{n_{k+1}}\|_1 \to 0 \qquad \text{as } k \to \infty.$$

Thus, $f_{n_k} \to f + f_{n_1}$ in $\|\cdot\|_1$. $\square$

To show that the space of Riemann integrable functions is not complete with respect to the seminorm induced by the integral is somewhat technical. The interested reader is encouraged to examine the discussion of this subject given in M. E. Monroe, *Measure and Integration*, 1953, p. 242.

Linear Operators

Much of what has been done to this point has important consequences for spaces of linear operators. We shall give only sufficient conditions for certain spaces of linear operators to be complete and a brief discussion on the existence of inverses of linear operators. The reader who wants to learn more about this fascinating subject is encouraged to consult one of the many excellent books available on linear operator theory (e.g., N. Dunford and J. T. Schwartz, *Linear Operators*, 1958).

Proposition 8. Let X be a normed linear space and Y a B-space. Then $L(X, Y)$ is a B-space under the operator norm.

Proof. Let $\{T_m\}$ be a Cauchy sequence in $L(X, Y)$. Let $x \in X$. Then $\|T_n(x) - T_m(x)\| \leq \|T_n - T_m\| \|x\|$ implies that $\{T_n x\}$ is a Cauchy sequence in Y, and so, converges to a point $T(x) = \lim T_n(x) \in Y$. The map $T: x \to T(x)$ is clearly linear. We claim that $T \in L(X, Y)$ and $T_n \to T$.

Let $\varepsilon > 0$. There is N such that $n, m \geq N$ implies $\|T_n - T_m\| < \varepsilon$. If $\|x\| \leq 1$, $x \in X$, then $\|T_n(x) - T_m(x)\| < \varepsilon$. Letting $m \to \infty$ implies that $\|T_n(x) - T(x)\| \leq \varepsilon$. Thus $T_n - T \in L(X, Y)$ and $T = (T - T_n) + T_n \in L(X, Y)$. Since $\|T_n - T\| \leq \varepsilon$ for $n \geq N$, we have $T_n \to T$. □

Remark 9. In particular, it follows from Proposition 8 that any dual X^* of a normed linear space X is complete with respect to the dual norm. Thus ℓ^∞ is complete with respect to $\|\cdot\|_\infty$ (**21.28**).

We give an application of Proposition 8 to invertible operators. Let X be a B-space. An element $T \in L(X, X)$ is said to be *invertible* if T is 1-1, onto, and has a continuous inverse $T^{-1} \in L(X, X)$.

Proposition 10. Let X be a B-space and $T \in L(X, X)$ with $\|T\| < 1$. Then $I - T$ is invertible and $(I - T)^{-1} = \sum_{n=0}^\infty T^n$, where $T^0 = I$ is the identity operator on X.

Proof. For $m > n$,

$$\left\| \sum_{k=n+1}^m T^k \right\| \leq \sum_{k=n+1}^m \|T^k\| \leq \sum_{k=n+1}^m \|T\|^k \leq \frac{\|T\|^{n+1}}{1 - \|T\|};$$

so the partial sums, $S_n = \sum_{k=0}^n T^k$, form a Cauchy sequence and, from Proposition 8, converge to an element $U \in L(X, X)$.

Since

$$U(I - T) = \sum_{n=0}^{\infty} T^n - \sum_{n=0}^{\infty} T^{n+1} = I = (I - T)U,$$

we have $U = (I - T)^{-1}$. $\qquad\qquad\qquad\qquad\qquad\qquad\qquad\qquad \square$

Corollary 11. Let X be a B-space. If $A \in L(X, X)$ is invertible and if $B \in L(X, X)$ is such that $\|A - B\| < 1/\|A^{-1}\|$, then B is invertible and

(i) $\|B^{-1}\| \leq \|A^{-1}\|/(1 - \|A^{-1}\|\|A - B\|)$.

(ii) $\|B^{-1} - A^{-1}\| \leq \|A^{-1}\|^2\|A - B\|/(1 - \|A^{-1}\|\|A - B\|)$.

Proof. By Proposition 10, $I - A^{-1}(A - B)$ is invertible. Consequently, $B = A - (A - B) = A(I - A^{-1}(A - B))$ is invertible. Since

$$\left(I - A^{-1}(A - B)\right)^{-1} = \sum_{k=0}^{\infty} \left(A^{-1}(A - B)\right)^k,$$

$$B^{-1} = \sum_{k=0}^{\infty} \left(A^{-1}(A - B)\right)^k A^{-1};$$

so, (i) follows immediately.

For (ii),

$$\|B^{-1} - A^{-1}\| = \|(B^{-1}A - I)A^{-1}\| \leq \|A^{-1}\|\|B^{-1}A - I\|$$

$$= \|A^{-1}\| \left\| I - \sum_{k=0}^{\infty} \left(A^{-1}(A - B)\right)^k \right\|$$

$$= \|A^{-1}\| \left\| \sum_{k=1}^{\infty} \left(A^{-1}(A - B)\right)^k \right\|$$

$$\leq \|A^{-1}\|\|A^{-1}\|\|A - B\|/(1 - \|A^{-1}\|\|A - B\|). \qquad \square$$

From Corollary 11 we have

Corollary 12. Let X be a B-space and let $\mathcal{G}$ be the invertible elements in $L(X, X)$. Then $\mathcal{G}$ is open in $L(X, X)$ and the map $A \to A^{-1}$ is continuous from $\mathcal{G}$ into $\mathcal{G}$.

Completion

A metric space (S, d) can be made complete by adding to it, in an appropriate way, its set of accumulation points. This is described more precisely in the following definition.

Definition 13. Let (S, d) be a metric space. A *completion* of (S, d) is a pair, $((S^*, d^*), \varphi)$, where (S^*, d^*) is a complete metric space, $\varphi \colon S \to S^*$ is an isometry, and $\varphi(S)$ is dense in S^*.

If we identify S and $\varphi(S)$ under the isometry φ, then a completion is a complete space S^* that contains S as a dense subspace.

Any completion is unique up to isometry, that is, for any two completions of a metric space, there is an isometry from one onto the other. To establish this, we need Lemma 14.

Lemma 14. Let (S_i, d_i), $i = 1, 2$ be metric spaces with S_2 complete and let E be a dense subset of S_1. If $f \colon E \to S_2$ is uniformly continuous, then f has a unique extension $\bar{f} \colon S_1 \to S_2$, which is uniformly continuous.

Proof. Let $x \in S_1$, and $\{x_k\} \subset E$ such that $x_k \to x$. Then $\{x_k\}$ is a Cauchy sequence and, since f is uniformly continuous, $\{f(x_k)\}$ is a Cauchy sequence in S_2. Since S_2 is complete, $\{f(x_k)\}$ converges, say to $\bar{f}(x)$. The value $\bar{f}(x)$ is independent of the particular sequence $\{x_k\}$ that converges to x; for, if $y_k \to x$, then the interlaced sequence $x_1, y_1, x_2, y_2, \ldots$ also converges to x (Exercise **19**.22). It follows that $\lim f(x_k) = \lim f(y_k) = \bar{f}(x)$. Moreover, if $x \in E$, then $\bar{f}(x) = f(x)$ so that the map $x \to \bar{f}(x)$ is an extension of f.

We claim that $\bar{f}$ is uniformly continuous on S_1. Let $\varepsilon > 0$. There is $\delta > 0$ such that $d_1(x, y) < \delta$ and $x, y \in E$ imply $d_2(f(x), f(y)) < \varepsilon$. Let $x, y \in S_1$ be such that $d_1(x, y) < \delta$. Pick $\{x_k\} \subset E$ and $\{y_k\} \subset E$ such that $x_k \to x$ and $y_k \to y$. Then $d_2(\bar{f}(x), \bar{f}(y)) = \lim d_2(f(x_k), f(y_k))$ and $d_1(x, y) = \lim d_1(x_k, y_k) < \delta$ imply $d_2(\bar{f}(x), \bar{f}(y)) \leq \varepsilon$.

The uniqueness follows from Exercise **21**.5. $\qquad\square$

To show that a completion is unique, let $((S^*, d^*), \varphi)$ and $((S', d'), \psi)$ be completions of (S, d). Then $\theta = \psi \circ \varphi^{-1}$ is an isometry from $\varphi(S)$ into S'. By Lemma 14, θ has a unique extension $\bar{\theta}$ from S^* into S', and $\bar{\theta}$ is an isometry from S^* onto S'.

We next show that a completion always exists.

Theorem 15. Every metric space (S, d) has a completion.

Proof. Let $\mathscr{BC}(S)$ be the space of all bounded continuous real-valued functions defined on S. We show later that $\mathscr{BC}(S)$ is complete under the sup-norm, $\|f\|_\infty = \sup\{|f(s)| \colon s \in S\}$ (**26**.6).

We define an isometry from S into $\mathscr{BC}(S)$. Fix $a \in S$. For $x \in S$, define $f_x \colon S \to \mathbb{R}$ by $f_x(y) = d(x, y) - d(y, a)$. Then f_x is continuous on S and $\|f_x\|_\infty \leq d(x, a)$; so $f_x \in \mathscr{BC}(S)$. Define $\varphi \colon S \to \mathscr{BC}(S)$ by $\varphi(x) = f_x$. Now $|f_x(z) - f_y(z)| = |d(z, x) - d(z, y)| \leq d(x, y)$ and $|f_x(x) - f_y(x)| = d(x, y)$ imply $\|f_x - f_y\|_\infty = d(x, y)$; so, φ is an isometry.

By Proposition 5, $((\overline{\varphi(S)}, \|\cdot\|_\infty), \varphi)$ is a completion. $\qquad\square$

A property is called a *topological property* if it is preserved under homeomorphisms. Example **21.26** shows that completeness is not a topological property.

EXERCISES 22

1. Show that every convergent sequence is Cauchy.

2. Show that every Cauchy sequence is bounded.

3. Show that if $\{x_k\}$ is a Cauchy sequence and has a subsequence $\{x_{n_k}\}$ that converges to a point x, then $x_k \to x$.

4. Show that a uniformly continuous function between metric spaces carries Cauchy sequences into Cauchy sequences. Can "uniformly continuous" be replaced by "continuous"?

5. Characterize the Cauchy sequences in the distance-1 metric. Is a space with the distance-1 metric complete?

6. Let $\mathscr{P}[a, b]$ be the (real) polynomials equipped with $\| \cdot \|_\infty$. Is $\mathscr{P}[a, b]$ complete?
   Hint: Recall the Weierstrass approximation theorem.

7. Show that $\mathscr{C}^1[a, b]$ is not complete with respect to $\| \cdot \|_\infty$.

8. Let $x_k = \{1, \frac{1}{2}, \ldots, \frac{1}{k}, 0, 0, \ldots\} \in c_{00}$. Show that $\{x_k\}$ is a Cauchy sequence in c_{00} with respect to $\| \cdot \|_\infty$, but that $\{x_k\}$ does not converge to an element in c_{00}.

9. Using only the definition, show that ℓ^∞ is complete.

10. Show that c_0 is complete with respect to $\| \cdot \|_\infty$.

11. A (formal) series $\sum_{k=1}^\infty x_k$ in a normed linear space is *absolutely convergent* if $\sum_{k=1}^\infty \|x_k\|$ converges. Show that a normed linear space is complete if, and only if, every absolutely convergent series is convergent.

12. Show that the series $\sum_{i=1}^\infty (1/i)e_i$ converges in ℓ^∞, but does not converge absolutely. How does this compare with the situation in $\mathbb{R}^n$?

13. Give a specific example of a series that is absolutely convergent, but not convergent.

14. In Lemma 14, can "uniform continuity" be replaced by "continuity"?

15. Suppose that Y is a dense subset of a metric space X and that every Cauchy sequence in Y converges to a point in X. Show that X is complete.

16. Let (S, d) be complete. Suppose that $\{x_k\}$ satisfies $\sum_{k=1}^\infty d(x_k, x_{k+1}) < \infty$. Show that $\{x_k\}$ converges.

17. Suppose that $\{x_k\}$ is a Cauchy sequence in a nls and that $\{t_k\}$ is convergent in $\mathbb{R}$. Show that $\{t_k x_k\}$ is Cauchy.

18. Let X be a nls and $B = \{x: \|x\| \leq 1\}$. Show that X is complete if, and only if, B is complete.

19. Let $\mathscr{P}[a, b]$ be the polynomials over $[a, b]$. What is the completion of $\mathscr{P}[a, b]$ with respect to $\| \cdot \|_{\infty}$?

20. Let (S_i, d_i), $i = 1, 2$ be metric spaces. Show that $S_1 \times S_2$ is complete iff both S_1 and S_2 are complete.

21. Let (S, d) be complete and $F_k \subseteq S$ closed $\forall\, k \in \mathbb{N}$. If $F_k \supseteq F_{k+1}$ and diam $F_k \to 0$, show that $\bigcap_{k=1}^{\infty} F_k \neq \varnothing$. Can completeness be dropped?

22. Give an example of a complete space containing a sequence $\{F_k\}$ of nonempty closed sets such that $F_k \supseteq F_{k+1}$ and $\bigcap_{k=1}^{\infty} F_k = \varnothing$.

23. Is $BV[a, b]$ complete under the seminorm of Exercise **19**.41?

24. Show that s is complete under the Frechet metric.

25. Show that a complete subset of a metric space is closed.

26. (This exercise is a continuation of Exercise **20**.29.) Show that X/L is complete if X is complete.

27. Let $\{x_k\}$ be a bounded sequence in a Banach space X. Show that $T(\{t_k\}) = \sum_{k=1}^{\infty} t_k x_k$ defines a continuous linear operator T from ℓ^1 into X. Calculate $\|T\|$.

CHAPTER 23

Contraction Mappings

The number of applications of the completeness of $\mathbb{R}^n$ and of certain metric spaces in previous chapters gives evidence of the importance of this concept. The role of completeness in the convergence of Cauchy sequences is sufficient justification for an extensive study of its role in other situations. In this chapter, we discuss several interesting and important applications of the completeness of certain function spaces.

Let (S, d) be a metric space. A function $f: S \to S$ is a *contraction* if there is α, $0 < \alpha < 1$, such that

$$d(f(x), f(y)) \leq \alpha d(x, y) \,\forall\, x, y \in S.$$

A contraction map "contracts" or "shrinks" the distance between points by the factor α. Clearly any contraction map is uniformly continuous on S. A function $f: S \to S$ has a *fixed point* x if $f(x) = x$. We show that any contraction on a complete metric space always has a unique fixed point.

Theorem 1 (Banach). If $f: S \to S$ is a contraction and S is complete, then f has a unique fixed point.

Proof. Let $x_0 \in S$. Define $f(x_k) = x_{k+1}$ for $k \geq 0$. We claim that $\{x_k\}$ is a Cauchy sequence. Note that $d(x_2, x_1) = d(f(x_1), f(x_0)) \leq \alpha d(x_1, x_0)$, $d(x_3, x_2) = d(f(x_2), f(x_1)) \leq \alpha d(x_2, x_1) \leq \alpha^2 d(x_1, x_0)$ and, in general, $d(x_{k+1}, x_k) \leq \alpha^k d(x_1, x_0)$. If $k > j$,

$$d(x_k, x_j) \leq \sum_{i=j}^{k-1} d(x_{i+1}, x_i) \leq \sum_{i=j}^{k-1} \alpha^i d(x_1, x_0) \leq \frac{\alpha^j}{1-\alpha} d(x_1, x_0), \quad (1)$$

and since $\alpha < 1$, (1) shows that $\{x_k\}$ is Cauchy.

Let $x = \lim x_k$. Since f is (uniformly) continuous, $f(x) = \lim f(x_k) = \lim x_{k+1} = x$; so x is a fixed point.

For the uniqueness, suppose z is also a fixed point of f. Then $d(z, x) = d(f(z), f(x)) \leq \alpha d(z, x)$, and $0 < \alpha < 1$ implies $z = x$. □

Note that the proof above is constructive in the sense that the fixed point is the limit of the iterates defined by $x_{k+1} = f(x_k)$ where the initial point or initial guess x_0 is an arbitrary point in S. Equation (1) also gives an estimate for the rapidity of the convergence of $x_k \to x$:

$$d(x, x_j) \leq \frac{\alpha^j}{1-\alpha} d(x_0, f(x_0)). \quad (2)$$

For an interesting proof of the contraction principle in $\mathbb{R}^n$, see L. D. Drager and R. L. Foote, 1986, 52–54.

The following applications of the fixed point theorem to integral and differential equations are indicative of its importance.

EXAMPLE 2. Let $k: [a, b] \times [a, b] \to \mathbb{R}$ be continuous and $\varphi \in \mathscr{C}[a, b]$. An equation of the type

$$f(x) = \lambda \int_a^b k(x, y) f(y) \, dy + \varphi(x), \qquad \text{where } \lambda \in \mathbb{R} \quad (3)$$

and f is an unknown function, is called a *Fredholm integral equation of the second kind*. We show that this equation has a unique solution $f \in \mathscr{C}[a, b]$ for certain λ. Define $K: \mathscr{C}[a, b] \to \mathscr{C}[a, b]$ by

$$K(f)(x) = \lambda \int_a^b k(x, y) f(y) \, dy + \varphi(x).$$

Solving (3) is equivalent to showing that K has a fixed point. Now

$$\|K(f) - K(g)\|_\infty \leq |\lambda| M(b - a) \|f - g\|_\infty,$$

where $M = \sup\{|k(x, y)|: a \leq x, y \leq b\}$. Thus, if $|\lambda|M(b - a) < 1$, K is a contraction and, therefore, has a unique fixed point. $\qquad\square$

EXAMPLE 3. Let D be an open set in $\mathbb{R}^2$ and $(x_0, y_0) \in D$. Let $f: D \rightarrow \mathbb{R}$ be continuous and satisfy a Lipschitz condition of the form

$$|f(x, y) - f(x, y')| \leq L|y - y'| \; \forall \, (x, y), (x, y') \in D. \qquad (4)$$

Consider the Initial Value Problem (IVP):

$$y'(x) = f(x, y(x))$$
$$y(x_0) = y_0. \qquad (5)$$

We show that $\exists \, \delta > 0$ such that (5) has a unique solution in the interval $[x_0 - \delta, x_0 + \delta]$.

Solving (5) is equivalent to solving the integral equation

$$y(x) = y_0 + \int_{x_0}^x f(t, y(t)) \, dt. \qquad (6)$$

So, we wish to show that (6) has a unique solution.

Let Γ be a closed rectangle centered at (x_0, y_0), which lies inside D. Then $\exists$ $M > 0$ such that $|f(x, y)| \leq M \; \forall (x, y) \in \Gamma$. Now choose $\delta > 0$ such that

$$[x_0 - \delta, x_0 + \delta] \times [y_0 - M\delta, y_0 + M\delta] = I \times J \subseteq \Gamma \qquad \text{and} \qquad L\delta < 1.$$

Let $\mathscr{C}$ be the subset of $\mathscr{C}(I)$, which consists of those functions whose range is in J. Since $\mathscr{C}$ is a closed subset of $\mathscr{C}(I)$, $\mathscr{C}$ is complete.

Define $F: \mathscr{C} \rightarrow \mathscr{C}$ by

$$F(\varphi)(x) = y_0 + \int_{x_0}^x f(t, \varphi(t)) \, dt.$$

Note that $F(\varphi)$ is continuously differentiable and $F(\varphi) \in \mathscr{C}$, since $|F(\varphi)(x) - y_0| \leq M|x - x_0| \leq M\delta$. Showing that (6) has a unique solution in $\mathscr{C}$ is equivalent to showing that F has a unique fixed point. Thus, it suffices to show that F is a contraction. For this, if $x \in I$ and $\varphi, \psi \in \mathscr{C}$, then

$$|F(\varphi)(x) - F(\psi)(x)| \leq \left| \int_{x_0}^x |f(t, \varphi(t)) - f(t, \psi(t))| dt \right|$$

$$\leq \left| \int_{x_0}^x L|\varphi(t) - \psi(t)| dt \right|$$

$$\leq L|x - x_0| \, \|\psi - \varphi\|_\infty$$

$$\leq (L\delta)\|\varphi - \psi\|_\infty,$$

and the result follows. $\qquad\square$

Any existence theorem for the nonlinear IVP (5) must be local in nature. For example, $y' = y^2, y(1) = -1$, has the solution $y(x) = -1/x$, which is not defined at $x = 0$ even though $f(x, y) = y^2$ is continuous there.

If the Lipschitz condition (4) is dropped, the IVP still has a solution, but the solution may fail to be unique (Exercise 6).

When the variable x also appears in the upper limit of the integral in (3), then (3) is called a *Volterra integral equation*. In order to give an application to Volterra integral equations, we need a more general fixed point theorem.

Corollary 4. Let (S, d) be complete and $f: S \to S$. Suppose that there exists a positive integer n such that f^n is a contraction, then f has a unique fixed point. (Here f^n is f composed with itself n times.)

Proof. The uniqueness is immediate for if x is a fixed point of f, it is also a fixed point of f^n.

Since f^n is a contraction, it has a unique fixed point x. We claim that x is also a fixed point of f, for if α is the contraction constant for f^n, we have

$$d(f(x), x) = d(f(f^n(x)), f^n(x))$$

$$= d(f^n(f(x)), f^n(x))$$

$$\leq \alpha d(f(x), x);$$

so, $f(x) = x$.   □

Even a discontinuous function f can be such that its iterate becomes a contraction. For example, let $f(t) = \frac{1}{4}$ for $0 \leq t \leq \frac{1}{2}$ and $f(t) = \frac{1}{2}$ for $\frac{1}{2} < t \leq 1$. Then $f^2(t) = \frac{1}{4}$ for $0 \leq t \leq 1$. See also Exercise 11 for an example of an integral operator that is not a contraction, but whose square is.

EXAMPLE 5. Consider the Volterra integral equation

$$f(x) = \lambda \int_a^x k(x, y) f(y) \, dy + \varphi(x), \tag{7}$$

where the notation is as in Example 2. Define $K: \mathscr{C}[a, b] \to \mathscr{C}[a, b]$ by

$$K(f)(x) = \lambda \int_a^x k(x, y) f(y) \, dy + \varphi(x).$$

We claim that (7) has a unique solution in $[a, b]$; for this, it suffices to show that K has a unique fixed point. If $f_1, f_2 \in \mathscr{C}[a, b]$, then

$$|K(f_1)(x) - K(f_2)(x)| \leq |\lambda||M|\|f_1 - f_2\|_\infty (x - a),$$

$$|K^2(f_1)(x) - K^2(f_2)(x)| \leq |\lambda|^2 M^2 \|f_1 - f_2\|_\infty \int_a^x (t - a)\, dt$$

$$= |\lambda M|^2 \|f_1 - f_2\|_\infty \frac{(x - a)^2}{2},$$

and

$$|K^n(f_1)(x) - K^n(f_2)(x)| \leq |\lambda M|^n \|f_1 - f_2\|_\infty \frac{(x - a)^n}{n!}.$$

Hence,

$$\|K^n(f_1) - K^n(f_2)\|_\infty \leq (|\lambda M|^n (b - a)^n / n!) \|f_1 - f_2\|_\infty.$$

Since $|\lambda M|^n (b - a)^n / n! \to 0$, K^n is a contraction map for large n and, therefore, has a unique fixed point for any value of the parameter λ by Corollary 4. $\square$

Finally, we consider a family of contractions that vary continuously with respect to a parameter and show that the corresponding fixed points also vary continuously.

Corollary 6. Let (S, d) be a complete metric space and (S', d') a metric space. Suppose $f: S' \times S \to S$ is such that $f(\cdot, s)$ is continuous $\forall s \in S$ and $\exists\, 0 < \alpha < 1$ such that $d(f(s', s_1), f(s', s_2)) \leq \alpha d(s_1, s_2)\ \forall\, s' \in S',\ s_1, s_2 \in S$. For each $s' \in S'$, let $x_{s'}$ be the unique fixed point of the contraction $f(s', \cdot)$. Then the map $s' \to x_{s'}$ is continuous from S' to S.

Proof. For $y \in S'$, consider the equation

$$x = f(y, x). \tag{8}$$

Let $y_0 \in S'$. Construct a solution of (8) starting with the initial point $x_0 = x_{y_0}$ so that $x_{k+1} = f(y, x_k) \to x_y$. By (2),

$$d(x_{y_0}, x_y) \leq \frac{1}{1 - \alpha} d(x_{y_0}, f(y, x_{y_0}))$$

$$= \frac{1}{1 - \alpha} d(f(y_0, x_{y_0}), f(y, x_{y_0}))$$

so that $y \to x_y$ is continuous at y_0 since $y \to f(y, x_{y_0})$ is continuous at y_0. $\square$

See C. H. Wagner, 1982, 259–273, for additional applications of the contraction principle.

EXERCISES 23

1. A metric space (S, d) has the *fixed point property* if every continuous function $f: S \rightarrow S$ has a fixed point.
   (a) Show that this property is a topological property.
   (b) Give an example of a space that has more than one point and has the fixed point property.

2. Give an example showing that the completeness in Theorem 1 cannot be dropped.

3. Show that the contraction assumption in Theorem 1 cannot be replaced by the weaker condition

$$d(f(x), f(y)) < d(x, y) \; \forall \, x, y \in S.$$

   Hint: Consider $f(x) = 1/(1 + x^2)$ for $x \geq 0$.

4. Show that y satisfies (5) iff it satisfies (6).

5. Suppose that D is convex and $D_2 f$ is bounded in D. Show that f satisfies a Lipschitz condition of the form (4).

6. Show that the IVP $y' = y^{1/3}$, $y(0) = 0$, has an infinite number of solutions

$$y_c(x) = \begin{cases} 0 & 0 \leq x \leq c, \\ \left(\dfrac{2(x - c)}{3} \right)^{3/2} & c < x \leq 1, \end{cases} \qquad (0 \leq c \leq 1).$$

   Why doesn't this violate Example 3?

7. Solve the IVP, $y' = y$, $y(0) = 1$. Using the initial point $\varphi_0 = 1$, calculate the iterates $\varphi_{k+1} = F(\varphi_k)$ and show that they converge to the solution.

8. Consider the nonlinear integral equation

$$f(x) = \lambda \int_a^b k(x, y, f(y)) \, dy + \varphi(x),$$

   with continuous k and φ, where k satisfies a Lipschitz condition

$$\left| k(x, y, z_1) - k(x, y, z_2) \right| \leq L |z_1 - z_2|.$$

   Show that this equation has a unique solution if $|\lambda| < 1/L(b - a)$.

9. Let a, b be continuous functions on $[x_0 - A, x_0 + A]$ and $y_0 \in \mathbb{R}$. Show that the linear IVP

$$y'(t) = a(t)y(t) + b(t),$$

$$y(x_0) = y_0,$$

has a unique solution y defined on $[x_0 - A, x_0 + A]$. How does this compare with Example 3?

10. Let $f: \mathbb{R} \to \mathbb{R}$ be differentiable with $|f'(t)| \leq \alpha$, where $0 < \alpha < 1$. Show that f is a contraction.

11. Define $T: \mathscr{C}[0,1] \to \mathscr{C}[0,1]$ by $Tf(x) = \int_0^x f$. Show that
   (a) T is not a contraction.
   (b) T has a unique fixed point.
   (c) T^2 is a contraction.

CHAPTER 24

The Baire Category Theorem

It was shown in Chapter 2 that between any two real numbers there is a rational number. An equivalent statement is that in every neighborhood of every real number, there is a rational number. This, of course, says that $\overline{\mathbb{Q}} = \mathbb{R}$. We also say that $\mathbb{Q}$ is *dense* in $\mathbb{R}$. Since $\mathbb{Q}$ is countable, its Lebesgue measure (see Chapter **16**) is zero. In a way, this is an unpalatable fact for, as we have just observed, $\mathbb{Q}$ is large enough and distributed in such a way that its closure is all of $\mathbb{R}$. The situation is even more bizarre than is indicated by this property of the rationals. There are sets of real numbers larger than $\mathbb{Q}$, that is, uncountable, which are bounded, "evenly" distributed, whose measure is zero, and whose interior is empty, that is, they contain no neighborhoods of their points.

A surprising consequence of such sets is the number of their applications and the richness of the mathematical areas arising from their investigation. In order to describe some of these sets, we begin, as usual, with an introduction to the necessary equipment.

292 Introduction to Real Analysis

Let (S, d) be a metric space. A subset $E \subseteq S$ is *nowhere dense* if $\overline{E}$ contains no sphere (i.e., $\overline{E}\,^0 = \varnothing$).

EXAMPLE 1. The Cantor set K, described in Example **16**.7, is nowhere dense. In that example, the "distance" between the adjacent open intervals of the set

$$\bigcup_{n=0}^{N} \bigcup_{k=1}^{2^n} E_n^k$$

is $1/3^{N+1}$ ($N = 1, 2, \ldots$). Thus, if K were to contain an open interval (a, b), then for $1/3^{N+1} < b - a$ the "distance" between two intervals in $\{E_n^k:$ $k = 1, \ldots, 2^n;\ n = 0, \ldots, N\}$ would be greater $1/3^{N+1}$. Thus, $K^0 = \varnothing$.

A subset $E \subseteq S$ is *first category* in S if E is a countable union of nowhere dense sets; E is *second category* in S if E is not first category. For example, $\mathbb{Q}$ is first category in $\mathbb{R}$; $\mathbb{R}$ is second category in itself. In fact, every complete metric space is second category in itself. This latter result is called the Baire category theorem, after the French mathematician René Louis Baire (1874–1932). To establish it, we need a couple of definitions and the characterization of complete spaces given below in Proposition 2.

If $x \in S$ and $r > 0$, the *closed ball* with center at x and radius r is $B(x, r) = \{y: d(x, y) \leq r\}$. A sequence of closed balls $\{B(x_n, r_n)\}$ is *nested* if $B(x_k, r_k) \supseteq B(x_{k+1}, r_{k+1})$ for $k = 1, 2 \ldots$.

Proposition 2. (S, d) is complete iff every nested sequence of closed balls $\{B(x_n, r_n)\}$ with $r_n \to 0$ has nonvoid intersection.

Proof. Suppose (S, d) is complete. Then $\{x_n\}$ is Cauchy in S since $d(x_n, x_m) \leq r_n$ for $m \geq n$, and $r_n \to 0$. Therefore, there is $x \in S$ such that $x_n \to x$. We claim that $x \in \bigcap_{n=1}^{\infty} B(x_n, r_n)$. Each $B(x_n, r_n)$ contains all but possibly a finite number of the $\{x_k: k \in \mathbb{N}\}$; since $B(x_n, r_n)$ is closed, it follows that $x \in B(x_n, r_n)$.

Let $\{x_n\}$ be Cauchy in S. There is n_1 such that $n \geq n_1$ implies $d(x_n, x_{n_1})$ $\leq 1/2$. Suppose $x_{n_1}, \ldots, x_{n_k}$ have been chosen such that $n_{j+1} > n_j$ $(j = 1, \ldots, k - 1)$ and $d(x_n, x_{n_j}) < 1/2^j$ for $n \geq n_j$ Choose $n_{k+1} > n_k$ so that $d(x_n, x_{n_{k+1}}) < 1/2^{k+1}$ for $n \geq n_{k+1}$.

Consider $B(x_{n_k}, 1/2^{k-1})$. We claim that this sequence is nested. If $x \in B(x_{n_{k+1}}, 1/2^k)$, then $d(x, x_{n_{k+1}}) \leq 1/2^k$ and

$$d(x, x_{n_k}) \leq d(x, x_{n_{k+1}}) + d(x_{n_{k+1}}, x_{n_k}) < 1/2^k + 1/2^k = 1/2^{k-1};$$

so $x \in B(x_{n_k}, 1/2^{k-1})$. By hypothesis, there is

$$x \in \bigcap_{k=1}^{\infty} B(x_{n_k}, 1/2^{k-1}).$$

Obviously, $x_{n_k} \to x$ and so $x_n \to x$ (Exercise **22**.3). $\qquad\square$

Theorem 3 (The Bair Category Theorem). A complete metric space (S, d) is second category in itself.

Proof. Let A_n be nowhere dense in S $\forall n \in \mathbb{N}$, and let $E = \bigcup_{n=1}^{\infty} A_n$. We will show that there is $x \in S \setminus E$ and this will establish the result. Let $x_0 \in S$. Since A_1 is nowhere dense, there is a closed ball B_1 of radius less than $1/2$ inside the closed ball $B_0 = B(x_0, 1)$ and such that $B_1 \cap A_1 = \varnothing$ (Exercise 1). Since A_2 is nowhere dense, there is also a closed ball B_2 of radius smaller that $1/3$ such that $B_2 \subseteq B_1$ *and* $B_2 \cap A_2 = \varnothing$. Continuing this construction produces a nested sequence of closed balls, $\{B_n\}$, such that the radius of B_n is less than $1/(n+1)$. By Proposition 2, there is $x \in \bigcap_{n=1}^{\infty} B_n$. But $x \notin A_n$ $\forall n$; so $x \notin E$. $\qquad\square$

Remark 4. The proof shows that the complement of a first category set in a complete metric space is dense.

Some applications will help illustrate the utility of the Baire category theorem. Recall that the real polynomials $\mathscr{P}[a, b]$ on $[a, b]$ with the sup-norm is not a complete normed linear space. We show that there is *no* norm on $\mathscr{P}[a, b]$ under which it is complete. For this we need the following observation.

Proposition 5. Any finite dimensional subspace F of a normed linear space E is closed.

Proof. For convenience, assume that the scalar field is $\mathbb{R}$. Now F is algebraically isomorphic to $\mathbb{R}^n$ for some n. But all norms on $\mathbb{R}^n$ are equivalent; so F is complete and, hence, closed. $\qquad\square$

Let $\mathscr{P}_n = \{ f \in \mathscr{P}[a, b]: \text{degree } f \le n \}$. Then $\mathscr{P}[a, b] = \bigcup_{n=1}^{\infty} \mathscr{P}_n$. The set $\mathscr{P}_n$ is closed with respect to any norm by Proposition 5 so, if there were a norm under which $\mathscr{P}[a, b]$ were complete, some $\mathscr{P}_n$ would contain a sphere (Exercise 5) and, hence, would coincide with $\mathscr{P}[a, b]$ (Exercise 7). $\qquad\square$

We next give an application to the existence of continuous, nowhere differentiable functions. An example of such a function was given in Chapter 11. Here, the Baire category theorem is used to show the existence of such a function and actually show that in a sense "most" continuous functions are nowhere differentiable.

Let E be the subset of $\mathscr{C}[0, 1]$ consisting of those functions that have a finite right-hand derivative at one point at least. We show that E is of first category in $\mathscr{C}[0, 1]$ (with respect to $\| \cdot \|_{\infty}$) so that $\mathscr{C}[0, 1] \setminus E$ is nonvoid. Thus, there exist functions that fail to have a finite right-hand derivative at any point of $[0, 1]$. Intuitively, a second category set is much "larger" than a first category set (in particular, see Remark 4), so "most" functions are in $\mathscr{C}[0, 1] \setminus E$.

Let $E_n = \{f \in \mathscr{C}[0,1]:$ there is a point $x \in (0, 1 - 1/n)$ such that

$$\left| \frac{f(x+h) - f(x)}{h} \right| \leq n \ \forall h \in \left(0, \frac{1}{n}\right)\right\}.$$

Clearly, $\bigcup_{n=1}^{\infty} E_n \supseteq E$. We show that each E_n is closed and nowhere dense, so that $\bigcup_{n=1}^{\infty} E_n$ is first category.

First, E_n is closed. Let f be an accumulation point of E_n and $\{f_k\} \subseteq E_n$ be such that $f_k \to f$. Let $\varepsilon > 0$. For each k, there is $t_k \in (0, 1 - 1/n)$ such that $|(f_k(t_k + h) - f(t_k))/h| \leq n$ for $0 < h < 1/n$. Let t be an accumulation point of $\{t_k\}$, and for convenience of notation assume that $t_k \to t$. Let $0 < h < 1/n$. There is N such that $k \geq N$ implies $\|f_k - f\|_\infty < \varepsilon h/4$. Since f is continuous and $t_k \to t$, there is $M > N$ such that $k \geq M$ implies $|f(t + h) - f(t_k + h)| < \varepsilon h/4$ and $|f(t_k) - f(t)| < \varepsilon h/4$. Hence, for $k \geq M$,

$$|f(t+h) - f(t)|/h \leq |f_k(t_k + h) - f_k(t_k)|/h$$

$$+ \frac{1}{h} \{ |f(t+h) - f(t_k + h)|$$

$$+ |f(t_k + h) - f_k(t_k + h)|$$

$$+ |f_k(t_k) - f(t_k)| + |f(t_k) - f(t)| \}$$

$$\leq n + \varepsilon$$

It follows that $f \in E_n$ and E_n is closed.

Next, we show that $\mathscr{C}[0,1] \setminus E_n$ is dense. Let $f \in \mathscr{C}[0,1]$ and $\varepsilon > 0$. Partition $[0, 1]$ into k equal subintervals by $\{x_0 < x_1 < \cdots < x_k\}$ such that if $x, y \in [x_{j-1}, x_j]$, then $|f(x) - f(y)| < \varepsilon/2$. Consider the rectangle

$$\{(x, y): x_{j-1} \leq x \leq x_j; \ f(x_{j-1}) - \varepsilon/2 \leq y \leq f(x_{j-1}) + \varepsilon/2\}$$

(Figure 24.1). Then $(x_j, f(x_j))$ lies on the right side of the rectangle. We can construct a "sawtooth" function g in this rectangle that has slopes exceeding n in absolute value so that g is continuous on $[0, 1]$. Then $g \in \mathscr{C}[0,1] \setminus E_n$ and $\|g - f\|_\infty \leq \varepsilon$.

Since E_n is closed and has a dense complement, E_n is nowhere dense.

The Baire category theorem can be used in a similar manner to show the existence of functions with "strange" behavior. (See R. Boas, 1960, for several such applications.)

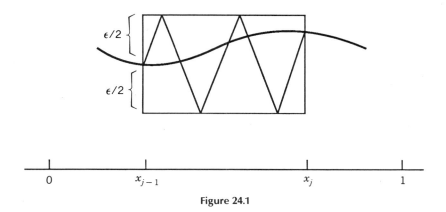

x_{j-1} x_j

0   1

Figure 24.1

EXERCISES 24

1. Show that E is nowhere dense iff every sphere S contains a sphere S' such that $S' \cap E = \varnothing$.

2. If S has the distance-1 metric, which subsets are nowhere dense?

3. Show that the countable union of first category sets is first category.

4. Is $\mathbb{R} \setminus \mathbb{Q}$ first or second category in $\mathbb{R}$?

5. If (S, d) is complete and $S = \bigcup_{n=1}^{\infty} F_n$, where F_n is closed, show that some F_n must contain a sphere.

6. Show that $\mathbb{R}^2$ is not a countable union of lines.

7. Let E be a normed linear space and M a linear subspace of E. Show that if M contains a sphere, then $M = E$.

8. Show that a Banach space cannot have countably infinite algebraic dimension.
   Hint: Let $\{e_i\}$ be linearly independent and consider $E_k = $ span $\{e_1, \ldots, e_k\}$.

9. Give examples of countable bases in $\mathscr{P}[a, b]$ and c_{00}.

10. Let $f_k \colon \mathbb{R} \to \mathbb{R}$ be continuous, nonnegative, and such that $\sum_{k=1}^{\infty} f_k(t)$ converges $\forall\, t \in \mathbb{R}$. Show that there is an interval in $\mathbb{R}$ where the convergence is uniform.

11. Let (S, d) be complete and have no isolated points. Show that S is uncountable.

CHAPTER 25

Compactness

The utility of the concept of compactness has already been well documented in $\mathbb{R}^n$ (see Chapter 8). The characterization of compactness as using coverings by open sets seems to best extend the idea to more general topological spaces and it is this characterization that we use to define compactness in metric spaces.

Let (S, d) be a metric space. An *open cover* of a subset $E \subseteq S$ is a family $\mathscr{G}$ of open subsets of S such that $E \subseteq \bigcup_{G \in \mathscr{G}} G$. We use the characterization of compactness in $\mathbb{R}^n$ given in **8**.7 as our definition of compactness in S. A subset $K \subseteq S$ is *compact* iff every open cover of K has a finite subcover.

Proposition 1. If $K \subseteq S$ is compact, then K is closed.

Proof. Let $x \in S \setminus K$. For each $y \in K$, $\exists$ open neighborhoods N_y of y and U_y of x such that $N_y \cap U_y = \varnothing$. $\{N_y : y \in K\}$ is an open cover of K, and so there is a finite subcover, say, $N_{y_1}, \ldots, N_{y_k}$. Then $V = \bigcap_{i=1}^{k} U_{y_i}$ is a neighborhood of x such that $V \cap K = \varnothing$. Thus $V \subseteq S \setminus K$ and $S \setminus K$ is open. □

Proposition 2. Let $K \subseteq S$ be compact and $F \subseteq K$ closed. Then F is compact.

Proof. Let $\mathcal{G}$ be an open cover of F. Then $\mathcal{G}' = \mathcal{G} \cup \{S \setminus F\}$ is an open cover of K and, therefore, has a finite subcover, $\mathcal{F}$, which also covers F. If $S \setminus F \notin \mathcal{F}$, then $\mathcal{F}$ is a finite subcover that also covers F; if $S \setminus F \in \mathcal{F}$, then we may remove $S \setminus F$ from $\mathcal{F}$ and obtain a finite subcover of $\mathcal{G}$ that covers F. □

Proposition 3. If K is compact, then K is bounded.

Proof. Fix $x \in K$. Consider the open cover consisting of spheres of radius n about x; $\{S(x, n): n = 1, 2, \dots\}$. This cover has a finite subcover of nested spheres that reduces to a single element, say $S(x, N)$, so that $K \subset S(x, N)$. □

It follows from Propositions 1 and 3 that a compact set is always closed and bounded. Recall that in $\mathbb{R}^n$ the converse holds; that is, a closed, bounded set is compact. This is not the case in a general metric space. For example, consider $\{e_i: i \in \mathbb{N}\} = E$ in ℓ^∞. Since $\|e_i - e_j\|_\infty = 1$ for $i \ne j$, $\{S(e_i, 1/2): i \in \mathbb{N}\}$ is an open cover of E that has no finite subcover. Thus, E is closed and bounded; but it is not compact.

The definition that has been given for compactness in a metric space is satisfying both mathematically and as a "geometric" visualization of the concept, but it is often difficult to use in establishing some of the consequences of compactness. It would be convenient to have equivalent statements on compactness that are easier to use in some of the applications. As is often the case, some new concepts and new language are necessary before we can proceed.

Let $E \subseteq S$. A set $A \subseteq S$ is called an *ε-net for E* if $E \subseteq \bigcup_{x \in A} S(x, \varepsilon)$. A subset $E \subseteq S$ is *totally bounded* (*precompact*) iff E has a finite ε-net $\forall \varepsilon > 0$.

It is easily seen that each compact set K is totally bounded. For if $\varepsilon > 0$, then $\{S(x, \varepsilon): x \in K\}$ is an open cover of K and, therefore, has a finite subcover, say $S(x_1, \varepsilon), \dots, S(x_k, \varepsilon)$. Then $\{x_1, \dots, x_k\}$ is a finite ε-net for K. Since subsets of totally bounded sets are clearly totally bounded, it is easy to see that a totally bounded set need not be compact. For example, $(0, 1)$ is totally bounded but not compact.

Proposition 4. A totally bounded set is bounded.

Proof. Let E be totally bounded and let $\{x_1, \dots, x_k\}$ be a 1-net for E. If $R = \max\{d(x_1, x_j) + 1: 1 \le j \le k\}$, then $E \subseteq S(x_1, R)$. □

The converse of Proposition 4 is false. For example, consider $E = \{e_i: i \in \mathbb{N}\}$ in ℓ^∞ with $\varepsilon = 1/2$. In $\mathbb{R}^n$, the converse does hold (Exercise 7).

Another important property of totally bounded sets is that of separability. A metric space S is *separable* if S contains a dense set that is at most countable. For example, $\mathbb{R}^n$ is separable, but ℓ^∞ is not separable. (Let E be the subset of ℓ^∞, which consists of those sequences that take on the values 0 and 1; then E is uncountable and $\|e - f\|_\infty = 1$ if $e, f \in E$ and $e \neq f$. So any dense set must be uncountable.)

Proposition 5. If (S, d) is totally bounded, then S is separable.

Proof. For each n let A_n be a finite $1/n$-net for S. Then $D = \bigcup_{n=1}^\infty A_n$ is a dense set that is at most countable. $\square$

The ideas just discussed, along with some studied earlier, enable us to state conditions equivalent to compactness.

Theorem 6. For $E \subseteq S$, the following are equivalent:

(i) E is compact.
(ii) Every infinite subset of E has an accumulation point in E.
(iii) E is complete and totally bounded.
(iv) Every sequence in E has a convergent subsequence that converges to a point in E.

Proof. We demonstrate equivalence by showing that (i) $\Rightarrow$ (ii), (iv) $\Rightarrow$ (iii), (ii) $\Rightarrow$ (iv), and (iii) $\Rightarrow$ (i).

First, (i) $\Rightarrow$ (ii). We may assume that the infinite subset of E is a sequence $\{x_k\}$. Let $H = \{x_k: k \in \mathbb{N}\}$ and suppose that no point of E is an accumulation point of H. Then $\forall x \in E$, $\exists$ a neighborhood N_x of x such that $N_x \cap H$ contains at most one point of H. $\{N_x: x \in E\}$ is an open cover of E, but clearly no finite number of the $\{N_x: x \in E\}$ covers H. Therefore, no finite number covers E.

Next, (iv) $\Rightarrow$ (iii). We begin by showing that E is complete. If $\{x_k\}$ is a Cauchy sequence in E, then by hypothesis, $\{x_k\}$ has a convergent subsequence that converges to some point $x \in E$. Since $\{x_k\}$ is Cauchy, $x_k \to x$.

Next assume that (iv) holds but that E is not totally bounded. Then $\exists\, \varepsilon > 0$ such that E has no finite ε-net. Define inductively a sequence $\{x_k\}$ in E as follows: Let $x_1 \in E$ and suppose that $x_1, \ldots, x_k \in E$ have been chosen so that

$$d(x_i, x_j) \geq \varepsilon \qquad \text{for } i \neq j, \qquad 1 \leq i, j \leq k. \tag{1}$$

Now $\bigcup_{i=1}^k S(x_i, \varepsilon)$ is not all of E; so $\exists\, x_{k+1} \in E$ such that $d(x_{k+1}, x_i) \geq \varepsilon$ for $i = 1, \ldots, k$. Then (1) holds with k replaced by $k + 1$. The inequalities (1) imply that the sequence $\{x_k\}$ has no convergent subsequences. This contradicts (iv).

Next, (ii) $\Rightarrow$ (iv). Let $\{x_k\}$ be a sequence in E and set $H = \{x_k: k \in \mathbb{N}\}$. If H is finite, then $\{x_k\}$ clearly has a convergent subsequence. If H is infinite, it has an accumulation point $p \in E$. Choose $n_1 \in \mathbb{N}$ so that $d(x_{n_1}, p) < 1$.

Suppose that $n_1, \ldots, n_k$ have been chosen such that $n_{i+1} > n_i$ and $d(x_{n_i}, p) < 1/i$. Choose $n_{k+1} > n_k$ and such that $d(x_{n_{k+1}}, p) < 1/(k+1)$. Then $x_{n_k} \to p$.

Finally, (iii) $\Rightarrow$ (i). First, note that if E is totally bounded, then we may assume that E has a finite ε-net A with $A \subseteq E$. For if B is a finite $\varepsilon/2$-net for E, then for each $b \in B$ there is $x_b \in E$ such that $d(b, x_b) < \varepsilon/2$ (if no such x_b exists, remove b from the $\varepsilon/2$-net). Then $\{x_b : b \in B\} = A$ is an ε-net for E and $A \subseteq E$.

Now suppose that (iii) holds, but (i) fails. Then there is an open cover $\mathscr{G}$ of E such that no finite subset of $\mathscr{G}$ covers E. Let E be covered by a finite number of spheres with centers in E and radius $1/2$. At least one of these spheres, say $S(x_1, 1/2)$, has the property that $S(x_1, 1/2) \cap E$ cannot be covered by a finite number of the elements of $\mathscr{G}$. Let E be covered by a finite number of spheres with centers in E and radius $1/2^2$. At least one of these spheres that has nonvoid intersection with $S(x_1, 1/2)$, say, $S(x_2, 1/2^2)$, has the property that $S(x_2, 1/2^2) \cap E$ cannot be covered by a finite number of the elements of $\mathscr{G}$. Continuing in this way by induction produces a sequence of spheres $\{S(x_i, 1/2^i)\}$ with centers in E having the property that $S(x_i, 1/2^i) \cap E$ cannot be covered by a finite number of the elements of $\mathscr{G}$. The sequence $\{x_k\}$ is Cauchy since

$$d(x_n, x_{n+p}) \le d(x_n, x_{n+1}) + d(x_{n+1}, x_{n+2}) + \cdots + d(x_{n+p-1}, x_{n+p})$$
$$\le [2^{-n} + 2^{-(n+1)}] + [2^{-(n+1)} + 2^{-(n+2)}] + \cdots$$
$$+ [2^{-(n+p-1)} + 2^{-(n+p)}]$$
$$< 2[2^{-n} + 2^{-(n+1)} + \cdots] = 2^{-n+2}.$$

E is complete, and so there is $x \in E$ such that $x_k \to x$. Then $x \in G$ for some $G \in \mathscr{G}$. Hence, $x \in S(x, \varepsilon) \subseteq G$ for some $\varepsilon > 0$. Choose n such that $2^{-n} < \varepsilon/2$ and $d(x_n, x) < \varepsilon/2$. Then

$$S(x_n, 2^{-n}) \subseteq S(x_n, \varepsilon/2) \subseteq S(x, \varepsilon) \subseteq G,$$

in contradiction to our construction. $\qquad \square$

Note that (iv) was our definition of compactness in $\mathbb{R}^n$ in Chapter 8. The following Corollary gives the metric space analogue of the Heine–Borel theorem in $\mathbb{R}^n$ (8.6).

Corollary 7. Let (S, d) be complete. A subset $E \subseteq S$ is compact if, and only if, E is closed and totally bounded.

Proof. Since a subset of a complete space is complete iff it is closed, the result follows from Theorem 6. $\qquad \square$

An application of Theorem 6 in an infinite dimensional space is given by Example 8.

EXAMPLE 8 (The Hilbert Cube). Let $H \subseteq \ell^2$ be the set of all sequences $\{t_j\}$ such that $|t_j| \leq 1/2^{j-1}$ for $j \in \mathbb{N}$. We shall show that H is compact.

First, H is totally bounded. Let $\varepsilon > 0$. Choose n such that $1/2^{n-1} < \varepsilon/2$ and define $Q_n : \ell^2 \to \ell^2$ by

$$Q_n(x) = (t_1, \ldots, t_n, 0, 0, \ldots),$$

where $x = (t_1, t_2, \ldots) \in \ell^2$. Now

$$\|x - Q_n(x)\|_2 = \left[\sum_{j=n+1}^{\infty} |t_j|^2 \right]^{1/2} < 1/2^{n-1} < \varepsilon/2$$

for $x = \{t_j\} \in H$. Also, $Q_n(H)$ is bounded in an n-dimensional subspace of ℓ^2 and, therefore, is totally bounded (Exercise 7). Hence $Q_n(H)$ has a finite $\varepsilon/2$-net, say, A. Thus, A is a finite ε-net for H.

Since H is closed, Corollary 7 implies that H is compact.   □

Relative Metrics

Let (S, d) be a metric space and $X \subseteq S$. For $E \subseteq X$, we may consider E as a subset of the metric space (S, d) or the metric space $(X, d|_{X \times X})$ where $d|_{X \times X}$ is the restriction of the metric d to $X \times X$. We say that E is open relative to X iff $\forall p \in E \; \exists \varepsilon > 0$ such that $x \in X$, $d(x, p) < \varepsilon$ implies $x \in E$.

For example, if $S = \mathbb{R}^2$, $X = \mathbb{R}$ and $E = (0, 1)$, then E is open relative to X, but E is not open in S.

Proposition 9. A subset $E \subseteq X$ is open relative to X if, and only if, $E = G \cap X$ for some open subset G of S.

Proof. Suppose that E is open relative to X. For each $p \in E \; \exists \varepsilon_p > 0$ such that $d(p, x) < \varepsilon_p$, $x \in X$ implies $x \in E$. Let $G = \bigcup_{p \in E} S(p, \varepsilon_p)$ (spheres in S). Then G is open in S and $E \subseteq G \cap X$. For each $p \in E$, $S(p, \varepsilon_p) \cap X \subseteq E$. Thus, $\bigcup_{p \in E} (S(p, \varepsilon_p) \cap X) = X \cap G \subseteq E$ and $E = G \cap X$.

If G is open in S and $E = G \cap X$, then $\forall p \in E \; \exists \varepsilon_p > 0$ such that $S(p, \varepsilon_p) \subseteq G$ and, therefore, $S(p, \varepsilon_p) \cap X \subseteq G \cap X = E$; so E is open relative to X.   □

In the example preceding Proposition 9, note that $E = S((1/2, 0), 1/2) \cap X$.

In a similar fashion, we say that a subset $K \subseteq X$ is *compact relative to* X iff K is a compact subset of $(X, d|_{X \times X})$. In contrast to Proposition 9, the situation for compactness is given in Proposition 10.

Proposition 10. A subset $K \subseteq X$ is compact relative to X iff K is compact relative to S.

Proof. Suppose that K is compact relative to S. Let $\{U_\alpha: \alpha \in A\}$ be a cover of K by sets that are open relative to X. By Proposition 9, $\forall \alpha \ \exists \ G_\alpha$ open in S such that $U_\alpha = G_\alpha \cap X$. Then $\{G_\alpha: \alpha \in A\}$ is a cover of K by sets that are open in S; so $\exists \ G_{\alpha_1}, \ldots, G_{\alpha_k}$, which cover K. But then $U_{\alpha_1}, \ldots, U_{\alpha_k}$ cover K.

For the converse, let $\{G_\alpha: \alpha \in A\}$ be a cover of K by sets that are open relative to S. Then $\{U_\alpha = G_\alpha \cap X: \alpha \in A\}$ is a cover of K by sets that are open relative to X and, therefore, $\exists \ U_{\alpha_1}, \ldots, U_{\alpha_k}$ covering K. Then $G_{\alpha_1}, \ldots, G_{\alpha_k}$ cover K. $\qquad\square$

Compactness and Continuity

We show in this section that many of the results concerning continuous functions on compact subsets of $\mathbb{R}^n$ generalize to metric spaces. Let (S_1, d_1) and (S_2, d_2) be metric spaces.

Proposition 11. If $f: S_1 \to S_2$ is continuous and S_1 is compact, then $f(S_1)$ is compact.

Proof. Let $\mathscr{V}$ be an open cover of $f(S_1)$. Then $\mathscr{U} = \{f^{-1}(V): V \in \mathscr{V}\}$ is an open cover of S_1; therefore, $\exists \ V_1, \ldots, V_k$ such that $S_1 \subseteq \bigcup_{i=1}^k f^{-1}(V_i)$. Then $f(S_1) \subseteq \bigcup_{i=1}^k V_i$, and it follows that $f(S_1)$ is compact. $\qquad\square$

Corollary 12. Let $f: S_1 \to \mathbb{R}$ be continuous and S_1 compact. Then the range of f is closed and bounded and $\exists \ x, y \in S_1$ such that

$$f(x) = \sup\{f(s): s \in S_1\}, \ f(y) = \inf\{f(s): s \in S_1\}.$$

Corollary 13. Let $f: S_1 \to S_2$ be continuous and S_1 compact. If f is one-to-one and onto S_2, then f^{-1} is continuous.

Proof. It suffices to show that $(f^{-1})^{-1}(F) = f(F)$ is closed whenever F is closed. But F closed in S_1 implies that F is compact. So $f(F)$ is compact, and therefore, closed. $\qquad\square$

Recall that the condition that S_1 be compact cannot be dropped.

Proposition 14. Let $f: S_1 \to S_2$ be continuous and S_1 compact. Then f is uniformly continuous on S_1.

Proof. Let $\varepsilon > 0$. Since f is continuous, $\forall x \in S_1 \ \exists \delta_x > 0$ such that $d_1(x, y) < \delta_x$ implies $d_2(f(x), f(y)) < \varepsilon/2$. $\{S(x, \delta_x/2): x \in S_1\}$ is an open cover of S_1 and, therefore, has a finite subcover, say

$$S(x_1, \delta_{x_1}/2), \ldots, S(x_k, \delta_{x_k}/2).$$

Let $\delta = \min\{\delta_{x_i}/2\colon 1 \le i \le k\}$. Suppose $p, q \in S_1$ are such that $d_1(p, q) < \delta$. Then, there is i such that $p \in S(x_i, \delta_{x_i}/2)$; thus

$$d_1(q, x_i) \le d_1(p, q) + d_1(p, x_i) < \delta + \delta_{x_i}/2 \le \delta_{x_i}$$

and

$$q \in S(x_i, \delta_{x_i}).$$

Thus,

$$d_2(f(p), f(q)) \le d_2(f(p), f(x_i)) + d_2(f(x_i), f(q)) < \varepsilon/2 + \varepsilon/2 = \varepsilon.$$

$\square$

The proofs of Propositions 11 and 14 and Corollary 13 above should be contrasted with the corresponding results in $\mathbb{R}^n$ given in Chapter 8.

EXERCISES 25

1. If $x_k \to x$ in S, show that $\{x_k\colon k \in \mathbb{N}\} \cup \{x\}$ is compact.

2. If S has the distance-1 metric, what are the compact subsets of S?

3. Is the converse of Proposition 1 true?

4. If K is compact and F is closed, show that $K \cap F$ is compact.

5. If K_1 and K_2 are compact, show that $K_1 \cup K_2$ is compact. What about infinite unions? What about intersections?

6. Give an example of a closed, bounded subset of ℓ^p ($1 \le p < \infty$) which is not compact.

7. Show that in $\mathbb{R}^n$ a set is bounded iff it is totally bounded.

8. Show that $E \subseteq S$ is totally bounded iff $\overline{E}$ is totally bounded.

9. Show that ℓ^p, $1 \le p < \infty$, is separable.

10. Show that $C[a, b]$ is separable.
    Hint: Use the Weierstrass approximation theorem.

11. When is a space with the distance-1 metric separable? When is it totally bounded?

12. Give an example that shows that the completeness condition in Corollary 7 cannot be dropped.

13. Show that a subset $E \subseteq S$ is totally bounded iff every sequence in E has a Cauchy subsequence.
    Hint: Consider the proof of Theorem 6.

14. Let K_i be compact for $i \in \mathbb{N}$ and let $K_{i+1} \subseteq K_i$. Show that $\bigcap_{i=1}^{\infty} K_i \neq \varnothing$. Can compactness be replaced by "closed and bounded"?

15. Let (S, d) be complete. Show that $E \subseteq S$ is totally bounded iff $\bar{E}$ is compact.

16. Let $(S_1, d_1), (S_2, d_2)$ be compact (totally bounded). Show that $S_1 \times S_2$ is compact (totally bounded). What about the converse(s)?

17. If X is open relative to S and $E \subseteq X$ is open relative to X, show that E is open relative to S.

18. The *distance* between two subsets $E, F \subseteq S$ is defined to be $d(E, F) = \inf \{ d(x, F): x \in E \}$ (**21**.11). Show that $d(E, F) = d(F, E)$.

19. Let $F \subseteq S$ be closed, $K \subseteq S$ compact, and $K \cap F = \varnothing$. Show that $d(K, F) > 0$. Can compactness be replaced by closed?

20. Show that a uniformly continuous function between metric spaces carries totally bounded sets into totally bounded sets. Can uniform continuity be replaced by continuity?

21. Let S be compact and suppose that $f: S \to S'$ is continuous. Show that f carries closed sets into closed sets. Can compactness be dropped?

22. Suppose that X is a metric space such that closed, bounded subsets of X are compact. Show that X is complete. Is the converse true?

23. Show that $K \subseteq S$ is compact if, and only if, every continuous real-valued function on K is uniformly continuous.

24. Show that a metric space is separable if, and only if, for each $\varepsilon > 0$, there is a countable covering of subsets of diameter smaller than ε.

25. If E and F are compact subsets of a metric space, show that there are $x \in E$ and $y \in F$ such that $d(x, y) = d(E, F)$.

26. Show that a nls is separable if, and only if, its unit sphere is separable.

27. Let S be compact and $\{K_n\}$ a decreasing sequence of closed subsets of S. If $f: S \to S$ is continuous, show that

$$f\left(\bigcap_{n=1}^{\infty} K_n \right) = \bigcap_{n=1}^{\infty} f(K_n).$$

28. Let (S_i, d_i), $i = 1, 2$ be metric spaces. Show that $f: S_1 \to S_2$ is continuous iff f is continuous on every compact subset of S_1.
Hint: Exercise 1.

29. Let (S, d) be a metric space and let $f: S \to S$ satisfy $d(f(x), f(y)) < d(x, y) \, \forall x, y \in S$. If S is compact, show that f has a unique fixed point. [*Hint:* Consider the minimum of the function $\varphi(x) = d(x, f(x))$.] Can compactness be dropped?
[*Hint:* Exercise **23**.3.]

CHAPTER 26

Compactness in $\mathscr{C}(S)$: Arzela–Ascoli

The Bolzano–Weierstrass theorem (**8**.4) states that every bounded sequence in $\mathbb{R}^n$ has a convergent subsequence. This leads directly to the result (**8**.6) of Heine and Borel that a subset of $\mathbb{R}^n$ is compact if, and only if, it is closed and bounded. These results have analogs in the space $\mathscr{C}(S)$ of continuous functions on a metric space (S, d). The extension leads to the concept of an equicontinuous subset of functions in $\mathscr{C}(S)$, an idea closely tied to uniform convergence by the Arzela–Ascoli theorem (Theorem 11). Our statement of this theorem gives necessary and sufficient conditions for a subset of $\mathscr{C}(S)$ to be compact. The results in this chapter have significant applications in the development of differential and integral equations.

Let (S, d) be a metric space. The vector space of all bounded real-valued, continuous functions on S will be denoted by $\mathscr{BC}(S)$. This agrees with our previous notation in Definition **19**.12, where E was a subset of $\mathbb{R}^n$. As before, we can equip $\mathscr{BC}(S)$ with the sup-norm, $\|f\|_\infty = \sup\{|f(t)|: t \in S\}$. We shall show that $\mathscr{BC}(S)$ is complete with respect to the sup-norm. For this, some of our previous results on uniform convergence must be recast in the setting of metric spaces.

Definition 1. Let $E \neq \varnothing$ and $f_n, f \colon E \to S$. The sequence $\{f_n\}$ *converges pointwise* to f if $\lim f_n(x) = f(x) \; \forall \, x \in E$. The sequence $\{f_n\}$ *converges uniformly* to f on E iff $\forall \, \varepsilon > 0 \; \exists \, N$ such that $n \geq N$ implies $d(f_n(x), f(x)) < \varepsilon$ $\forall \, x \in E$. We write $f_n \to f$ pointwise on E and $f_n \to f$ uniformly on E, respectively.

Note that convergence in $\mathscr{BC}(S)$ with respect to $\| \cdot \|_\infty$ is just uniform convergence on S.

Proposition 2. Let (S_1, d_1), (S_2, d_2) be metric spaces. If $f_n \colon S_1 \to S_2$ is continuous at $x_0 \in S_1$ $\forall \, n$ and $f_n \to f$ uniformly on S_1, then f is continuous at x_0.

Proof. Let $\varepsilon > 0$. $\exists \, N$ such that $n \geq N$ implies $d_2(f_n(x), f(x)) < \varepsilon/3$ $\forall \, x \in S_1$. Since f_N is continuous at x_0, there is $\delta > 0$ such that $d_1(x, x_0) < \delta$ implies $d_2(f_N(x), f_N(x_0)) < \varepsilon/3$. Thus, $d_1(x, x_0) < \delta$ implies

$$d_2(f(x), f(x_0)) \leq d_2(f(x), f_N(x)) + d_2(f_N(x), f_N(x_0)) +$$
$$d_2(f_N(x_0), f(x_0))$$
$$< \varepsilon. \qquad \square$$

Proposition 3. If $f_n \to f$ uniformly on E, then $\forall \, \varepsilon > 0 \; \exists \, N$ such that $n, m \geq N$ implies

$$d(f_n(x), f_m(x)) < \varepsilon \; \forall \, x \in E \; \text{(uniform Cauchy condition)}. \qquad (1)$$

Proof. Equation (1) follows immediately from the triangle inequality

$$d(f_n(x), f_m(x)) \leq d(f_n(x), f(x)) + d(f(x), f_m(x)). \qquad \square$$

Proposition 4. If $f_n \to f$ pointwise on E and $\{f_n\}$ satisfies the uniform Cauchy condition (1), then $f_n \to f$ uniformly on E.

Proof. Let $\varepsilon > 0$ and let N be as in (1). Letting $m \to \infty$ with n fixed in (1) gives $d(f_n(x), f(x)) \leq \varepsilon \; \forall \, x \in E, \, n \geq N$. $\qquad \square$

Corollary 5. Let (S, d) be a complete metric space. If the sequence $f_n \colon E \to S$ satisfies the uniform Cauchy condition (1), then there is $f \colon E \to S$ such that $f_n \to f$ uniformly on E.

Proof. By (1), $\forall \, x \in E$, $\{f_n(x)\}$ is a Cauchy sequence in S, and therefore converges to some $f(x) \in S$. Hence, $f_n \to f$ pointwise on E and Proposition 4 applies. $\qquad \square$

Corollary 6. $\mathscr{BC}(S)$ is complete.

Proof. If $\{f_k\}$ is a Cauchy sequence in $\mathscr{BC}(S)$ with respect to $\|\cdot\|_\infty$, then $\{f_k\}$ satisfies (1). Hence, $\exists f\colon S \to \mathbb{R}$ such that $f_k \to f$ uniformly on S. If $|f_k(x) - f(x)| < 1$ $\forall x \in S$, then $\|f\|_\infty \le 1 + \|f_k\|_\infty$. It follows that $f \in \mathscr{BC}(S)$ and $\|f_k - f\|_\infty \to 0$. $\qquad\square$

For the remainder of the chapter, we assume that S is compact. Since every continuous real-valued function on S is bounded, we abbreviate $\mathscr{BC}(S) = \mathscr{C}(S)$. This agrees with the notation used in **18**.13. We seek a characterization of the compact subsets of $\mathscr{C}(S)$. Every compact subset of $\mathscr{C}(S)$ is closed and bounded; however, the following example shows that, in general, closed and bounded subsets of $\mathscr{C}(S)$ are not compact.

EXAMPLE 7. In $\mathscr{C}[0,1]$, let

$$f_n(t) = \begin{cases} 0 & 0 \le t \le 2^{-n-1} \\ 2^{n+2}(t - 2^{-n-1}) & 2^{-n-1} < t < 3 \cdot 2^{-n-2} \\ -2^{n+2}(t - 2^{-n}) & 3 \cdot 2^{-n-2} \le t < 2^{-n} \\ 0 & 2^{-n} \le t \le 1 \end{cases}$$

The graph of f_n is shown in Figure 26.1. Then $\|f_n - f_m\|_\infty = 1$ $(n \ne m)$ so $E = \{f_n\colon n \in \mathbb{N}\}$ is closed and bounded, but not compact. $\qquad\square$

Consequently, an additional condition is necessary to guarantee the compactness of the closed and bounded subsets of $\mathscr{C}(S)$.

Definition 8. Let (S_1, d_1), (S_2, d_2) be metric spaces. A family $\mathscr{F}$ of functions from S_1 into S_2 is *equicontinuous* if $\forall \varepsilon > 0$, $\exists \delta > 0$ such that $d_1(x, y) < \delta$ implies $d_2(f(x), f(y)) < \varepsilon$ $\forall f \in \mathscr{F}$.

Any member of an equicontinuous family is uniformly continuous, and a family $\mathscr{F}$ of uniformly continuous functions is equicontinuous if $\forall \varepsilon > 0$, $\exists \delta > 0$ that satisfies the condition for uniform continuity *simultaneously* for all the functions $f \in \mathscr{F}$.

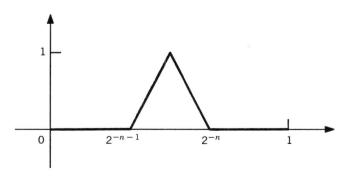

Figure 26.1

Proposition 9. If $\mathscr{F} \subseteq \mathscr{C}(S)$ is totally bounded, then $\mathscr{F}$ is equicontinuous.

Proof. Let $\varepsilon > 0$. $\exists f_1, \ldots, f_k \in \mathscr{F}$ such that $\forall f \in \mathscr{F}$ $\exists i$ with $\|f_i - f\|_\infty < \varepsilon/3$. Each f_i is uniformly continuous; so $\exists \delta > 0$ such that $d(x, y) < \delta$ implies $|f_i(x) - f_i(y)| < \varepsilon/3$ for $1 \le i \le k$.

Suppose that $f \in \mathscr{F}$ and $d(x, y) < \delta$. Choose i so that $\|f - f_i\|_\infty < \varepsilon/3$. Then

$$|f(x) - f(y)| \le |f(x) - f_i(x)| + |f_i(x) - f_i(y)| + |f_i(y) - f(y)| < \varepsilon. \quad \square$$

It follows that if a subset of $\mathscr{C}(S)$ is compact, it must be closed, bounded, and equicontinuous. We next show that the converse of this statement holds. We first establish a lemma. The method of proof, called the diagonal method, is an important technique of proof in analysis.

Lemma 10. Suppose that $\{f_n\}$ is a sequence of real-valued functions, defined on a countable set $E = \{x_k : k \in \mathbb{N}\}$, which is pointwise bounded on E. Then there is a subsequence $\{g_n\}$ of $\{f_n\}$ that converges pointwise on E.

Proof. (Diagonalization Procedure). The sequence $\{f_n(x_1)\}_{n=1}^\infty$ is bounded in $\mathbb{R}$ and, therefore, has a convergent subsequence that we denote by $\{f_{n,1}(x_1)\}_{n=1}^\infty$ (the reason for the slightly unorthodox notation will become apparent). The sequence $\{f_{n,1}(x_2)\}_{n=1}^\infty$ is bounded and, therefore, also has a convergent subsequence, $\{f_{n,2}(x_2)\}_{n=1}^\infty$. The sequence $\{f_{n,2}\}$ is a subsequence of $\{f_{n,1}\}$; so $\{f_{n,2}(x_1)\}_{n=1}^\infty$ also converges. Proceeding by induction produces a sequence $\{S_n\}$ as follows:

$$S_1: f_{1,1}, f_{2,1}, f_{3,1}, \ldots$$
$$S_2: f_{1,2}, f_{2,2}, f_{3,2}, \ldots$$
$$S_3: f_{1,3}, f_{2,3}, f_{3,3}, \ldots$$
$$\vdots \quad \vdots$$

such that

(i) S_{n+1} is a subsequence of S_n ($n \ge 1$) and

(ii) $\{f_{n,k}(x_j)\}_{n=1}^\infty$ converges for $1 \le j \le k$.

Finally, choose $\{g_n\}$ to be the "diagonal sequence" $g_n = f_{n,n}$ and note that $\forall k$, $\{g_n(x_k)\}_{n=1}^\infty$ converges. $\quad \square$

The necessary tools and definitions are now in place to achieve the goal stated at the beginning of the chapter, that is, to establish a necessary and sufficient condition for compactness.

Theorem 11 (Arzela–Ascoli). A subset $\mathscr{F} \subseteq \mathscr{C}(S)$ is compact if, and only if, $\mathscr{F}$ is closed, bounded and equicontinuous.

Proof. The necessity has been established.

Suppose that $\mathscr{F}$ is closed, bounded, and equicontinuous. To show that $\mathscr{F}$ is compact, it suffices to show that any sequence $\{f_n\} \subseteq \mathscr{F}$ has a subsequence that converges uniformly on S.

Since S is compact, it is separable. Let D be a countable dense subset of S. By Lemma 10, $\{f_n\}$ has a subsequence, $\{g_n\}$, such that $\{g_n\}$ converges pointwise on D. [Note that $\{f_n\}$ bounded in $\mathscr{C}(S)$ implies $\{f_n\}$ is pointwise bounded.] Let $\varepsilon > 0$. Since $\mathscr{F}$ is equicontinuous, $\exists\, \delta > 0$ such that $d(x, y) < \delta$ implies

$$|f(x) - f(y)| < \varepsilon/3 \qquad \forall f \in \mathscr{F}.$$

Consider $\mathscr{G} = \{S(x, \delta): x \in D\}$. Since D is dense in S, $\mathscr{G}$ is an open cover of S and, therefore, has a finite subcover, say, $S(x_1, \delta), \ldots, S(x_k, \delta)$, where $x_i \in D$. Choose N such that $n, m \geq N$ imply

$$|g_n(x_i) - g_m(x_i)| < \varepsilon/3 \qquad \text{for } i = 1, \ldots, k.$$

Let $x \in S$ and $n, m \geq N$. Then $\exists\, i \in \{1, \ldots, k\}$ such that $d(x, x_i) < \delta$, and so

$$|g_n(x) - g_m(x)| \leq |g_n(x) - g_n(x_i)| + |g_n(x_i) - g_m(x_i)|$$
$$+ |g_m(x_i) - g_m(x)| < \varepsilon.$$

Hence, $\{g_n\}$ converges uniformly on S by Corollary 5. $\qquad \square$

Note that the proof actually shows that if $\{f_n\}$ is pointwise bounded and equicontinuous, then $\{f_n\}$ has a uniformly convergent subsequence.

EXAMPLE 12. A subset E of a metric space (S, d) is *relatively compact* if $\overline{E}$ is compact. If X and Y are nls and $T \in L(X, Y)$, then T is called a *compact operator* (*completely continuous operator*) if T carries bounded subsets of X into relatively compact subsets of Y. We show that the integral operator of Example 21.18 is a compact operator. Let the notation be as in Example 21.18.

Let E be a bounded subset of $\mathscr{C}[a, b]$. To show that $K(E)$ is relatively compact it suffices to show that $K(E)$ is equicontinuous (Exercise 2 and Theorem 11). Let $\varepsilon > 0$ and let M be such that $\|f\|_\infty \leq M \,\,\forall f \in E$. Since k is uniformly continuous, there is $\delta > 0$ such that $|s - s'| < \delta$ implies $|k(s, t) - k(s', t)| < \varepsilon \,\,\forall t \in [a, b]$. If $f \in E$ and $|s - s'| < \delta$, then

$$|K(f)(s) - K(f)(s')| = \left| \int_a^b (k(s, t) - k(s', t))f(t)\, dt \right| \leq \varepsilon M(b - a).$$

It follows that $K(E)$ is equicontinuous. $\qquad \square$

EXERCISES 26

1. Let $f_k: S \to \mathbb{R}$ be continuous $\forall k \in \mathbb{N}$ and $f_k \to f$ uniformly on S. Show that if $x_k \to x$ in S, then $\lim f_k(x_k) = f(x)$. Does the converse hold?

2. Show that if $\mathscr{F} \subseteq \mathscr{C}(S)$ is equicontinuous, then $\overline{\mathscr{F}}$ is equicontinuous.

3. Suppose that $\{f_n\} \subseteq \mathscr{C}(S)$ is equicontinuous, and $\{f_n\}$ converges point-wise to f on S. Show that $\{f_n\}$ converges uniformly to f on S.

4. Let $\mathscr{F} \subseteq \mathscr{C}[a, b]$ be such that each element $f \in \mathscr{F}$ is continuously differentiable and suppose $\mathscr{F}' = \{f': f \in \mathscr{F}\}$ is bounded in $\mathscr{C}[a, b]$ with respect to $\|\cdot\|_\infty$. Show that $\mathscr{F}$ is equicontinuous.

5. Let $\mathscr{F} \subseteq \mathscr{C}[a, b]$ be bounded. For $f \in \mathscr{F}$, let $f^{(1)}$ be the indefinite integral of f. Show that $\{f^{(1)}: f \in \mathscr{F}\}$ is relative compact in $\mathscr{C}[a, b]$.

6. Show that the operator $T: \mathscr{C}[a, b] \to \mathscr{C}[a, b]$ defined by $T(f)(x) = \int_a^x f$ is compact.

7. Let S_1, S_2 be compact metric spaces that are homeomorphic. Show that $\mathscr{C}(S_1)$ and $\mathscr{C}(S_2)$ are linearly isometric.

8. Show that the identity operator on ℓ^p $(1 \le p \le \infty)$ is not a compact operator. What about the identity operator on $\mathscr{C}[a, b]$?

9. Give a characterization of the relatively compact subsets of $\mathbb{R}^n$.

10. Let $T \in L(X, Y)$. Show that if either X or Y is finite dimensional, then T is a compact operator.

11. Let $1 \le p < \infty$. Show that a subset $E \subseteq \ell^p$ is relatively compact iff E is bounded and $\lim_n \sum_{i=n}^\infty |t_i|^p = 0$ uniformly for $\{t_i\} \in E$.

12. Let D be a dense subset of the metric space X and let $f_k: X \to \mathbb{R}$ be continuous. Show that if $\sum_{k=1}^\infty f_k$ converges uniformly on D, then $\sum_{k=1}^\infty f_k$ converges uniformly on X.

13. Let $f_k(t) = \sin kt$, $0 \le t \le 2\pi$. Show that $\{f_k: k \in \mathbb{N}\}$ is not equicontinuous.
    Hint: If it is, there is a subsequence $\{f_{k_j}\}$ such that $\{f_{k_{j+1}} - f_{k_j}\} = \{g_j\}$ converges uniformly to 0. Now apply the dominated convergence theorem to $\{g_j^2\}$.

14. Let $f: \mathbb{R} \to \mathbb{R}$ be uniformly continuous. For $a \in \mathbb{R}$, let $f_a(x) = f(x - a)$. Show that $\{f_a: a \in \mathbb{R}\}$ is equicontinuous.

15. Let $\mathscr{F} \subseteq \mathscr{C}(S)$ be equicontinuous and pointwise bounded on S. Show that $\mathscr{F}$ is uniformly bounded, that is, norm bounded in $\mathscr{C}(S)$.

16. Let $\{f_n\} \subseteq \mathscr{C}[0, 1]$ be uniformly convergent on $[0, 1]$. Show that $\{f_n\}$ is equicontinuous.

17. Does the converse of Proposition 9 hold?

18. Let (S, d) be a metric space and $\{f_n\}$ an equicontinuous sequence of real-valued functions on S. If $\{f_n\}$ converges pointwise on a dense subset of S, show that $\{f_n\}$ converges pointwise on S.

CHAPTER 27

Connectedness

The elementary form of the intermediate value theorem states that a real-valued continuous function defined on an interval in $\mathbb{R}$ takes on all intermediate values in its range. This is, of course, a statement about continuous functions; however, in this case, it is also a statement about the character of the domain of definition of the function, that is, an interval. The significant property of an interval that enables this theorem is the proximity of its points and, in particular, the nonexistence of "gaps" in the interval. The lack of gaps can be well-described by saying that the points of an interval are "connected" in the sense that any two distinct points of an interval I are endpoints of an interval $J \subseteq I$. Another way of viewing this is that any pair of distinct points of an interval cannot be "separated" by disjoint nonempty open intervals. By this, we mean that there do not exist two disjoint nonempty open intervals, A and B, such that $A \cap I \neq \varnothing$, $B \cap I \neq \varnothing$, and $I \subseteq A \cup B$. This characterization of connectedness of intervals generalizes nicely to a definition for connected sets in metric spaces. We proceed to make this definition and then to study the basic properties of connected sets.

Let (S, d) be a metric space. A subset $E \subseteq S$ is *connected* if there fail to exist two disjoint open sets $A, B \subseteq S$ such that $A \cap E \neq \varnothing$, $B \cap E \neq \varnothing$ and $E \subseteq A \cup B$. A pair of open sets, (A, B), with this property is called a *disconnection* of E. Thus, E is connected if it has no disconnection. If the set E is not connected, it is said to be *disconnected*.

Lemma 1. Let $E \subseteq S$, $E = G \cup H$, where $G \neq \varnothing$, $H \neq \varnothing$, $G \cap H = \varnothing$, and G and H are open relative to E. Then there are sets, A, B, open in S, such that $A \cap B = \varnothing$, $G = A \cap E$, and $H = B \cap E$.

Proof. Since G is open relative to E, $\forall p \in G \; \exists \varepsilon_p > 0$ such that $x \in E$ and $d(p, x) < \varepsilon_p$ imply $x \in G$. Similarly, $\forall q \in H \; \exists \delta_q > 0$ such that $x \in E$ and $d(q, x) < \delta_q$ imply $x \in H$. Let $A = \bigcup_{p \in G} S(p, \varepsilon_p/2)$, $B = \bigcup_{q \in H} S(q, \delta_q/2)$ be the union of spheres in S about p and q, respectively. Then A and B are open in S and $G = A \cap E$, $H = B \cap E$.

We must still show that $A \cap B = \varnothing$. Suppose that $x \in A \cap B$. Then $\exists p \in G$ such that $x \in S(p, \varepsilon_p/2)$ and $\exists q \in H$ such that $x \in S(q, \delta_q/2)$. If $\varepsilon_p \leq \delta_q$, then $d(p, q) \leq d(p, x) + d(x, q) < \delta_q$; so $p \in H$ and $G \cap H \neq \varnothing$. A similar conclusion holds if $\delta_q < \varepsilon_p$. $\square$

Corollary 2. A subset $E \subset S$ is connected in S if, and only if, E is connected relative to E.

From our introductory paragraph, one would expect that the connected subsets of $\mathbb{R}$ are particularly easy to characterize; they are the intervals (possibly "degenerate").

Lemma 3. A subset $E \subseteq \mathbb{R}$ is connected if, and only if, E has the property that if $x, y \in E$ and $x < y$, then $x < z < y$ implies $z \in E$.

Proof. Suppose that E does not have the property above. Then $\exists x, y \in E$, $x < y$ and $z \notin E$ with $x < z < y$. Put $A = (-\infty, z)$, $B = (z, \infty)$. Then (A, B) is a disconnection of E; so, E is disconnected.

Suppose that E is not connected and let (A, B) be a disconnection of E. Suppose $x \in A \cap E$, $y \in B \cap E$ with $x < y$. Let $Z = A \cap [x, y]$ and set $z = \sup Z$.

Since $y \in B$ and B is open in $\mathbb{R}$ with $B \cap A = \varnothing$, $z < y$. Also, $x \in A$ and A open in $\mathbb{R}$ implies $x < z$. That is, $x < z < y$.

If $z \in A$, then A open implies z is not an upper bound of Z; so $z \notin A$. If $z \in B$, B open implies z is not $\sup Z$; so $z \notin B$. Since $E \subseteq A \cup B$, $z \notin E$. Thus E fails to have the property in the hypothesis. $\square$

Corollary 4. A subset $E \subseteq \mathbb{R}$ is connected if, and only if, E is a singleton or an interval (perhaps infinite).

The image of a connected set under a continuous mapping is again connected.

Proposition 5. Let $(S_1, d_1), (S_2, d_2)$ be metric spaces and $f\colon S_1 \to S_2$ be continuous. If S_1 is connected, then $f(S_1)$ is connected.

Proof. Suppose $f(S_1)$ is disconnected and let (U, V) be a disconnection of $f(S_1)$. But then $(f^{-1}(U), f^{-1}(V))$ is a disconnection of S_1, which implies that S_1 is not connected. $\qquad\square$

From Proposition 5 and Corollary 4, we obtain the intermediate value theorem as a corollary. More generally, we have the following corollary.

Corollary 6. Let (S, d) be connected and $f\colon S \to \mathbb{R}$ continuous. If $x, y \in S$ and $f(x) < \zeta < f(y)$, then $\exists z \in S$ such that $f(z) = \zeta$.

Proof. $f(S)$ is an interval in $\mathbb{R}$. $\qquad\square$

Note again that the essential ingredient in establishing the intermediate value theorem is the connectedness of the domain.

Definition 7. A *path* in a metric space (S, d) is a continuous function φ from a compact interval $[a, b]$ into S. The points $\varphi(a)$ and $\varphi(b)$ are called the *initial* and *terminal points* of the path; the path is said to join $\varphi(a)$ to $\varphi(b)$.

A metric space (S, d) is *path-connected* if any two points of S can be joined by a path.

Proposition 8. A path-connected metric space is connected.

Proof. Suppose (S, d) is not connected and let (A, B) be a disconnection. Pick $x \in A$, $y \in B$ and let $\varphi\colon [a, b] \to S$ be a path in S joining x to y. Then $(\varphi^{-1}(A), \varphi^{-1}(B))$ is a disconnection of $[a, b]$, and this contradicts Corollary 4. $\qquad\square$

The converse of Proposition 8 is false (Exercise 4). There is, however, a partial converse in $\mathbb{R}^n$. To establish it, we need Proposition 9.

Proposition 9. S is connected if, and only if, the only subsets of S that are both open and closed in S are S and $\varnothing$.

Proof. If S is disconnected and (A, B) is a disconnection of S, then A and B are both open and closed in S.

If $A \subseteq S$ is both open and closed in S and $A \neq \varnothing$, $A \neq S$, then $(A, S \setminus A)$ is a disconnection of S. $\qquad\square$

Using Proposition 9, we obtain the following partial converse of Proposition 8.

Proposition 10. Let $D \subseteq \mathbb{R}^n$ be open. Then D is connected if, and only if, every two points of D can be joined by a polygonal line that lies entirely in D.

Proof. Sufficiency follows from Proposition 8.

Suppose D is connected. Fix $p_0 \in D$. Let $E = \{ p \in D: p$ can be joined to p_0 by a polygonal line lying entirely in $D \}$. Since D is open, $E \neq \varnothing$. If $q \in E$, $\exists$ an ε-neighborhood N_q of q such that $N_q \subseteq D$. Since p_0 and q can be joined by a polygonal line lying entirely in D, any point in N_q can be joined to p_0 by such a polygonal line. Hence, E is open in D. If $D \setminus E \neq \varnothing$, the same argument shows that $D \setminus E$ is open in D. Since D is connected, $E = D$, and the result follows. $\qquad\square$

Proposition 11. Let $\{ E_\alpha: \alpha \in A \}$ be a family of connected subsets of S such that $\bigcap_{\alpha \in A} E_\alpha \neq \varnothing$. Then $E = \bigcup_{\alpha \in A} E_\alpha$ is connected.

Proof. Suppose E is not connected and let (G, H) be a disconnection of E. Let $a \in \bigcap_{\alpha \in A} E_\alpha$. Then either $a \in G$ or $a \in H$; assume $a \in G$. Then $H \cap E \neq \varnothing$ implies that $\exists \alpha_0 \in A$ such that $H \cap E_{\alpha_0} \neq \varnothing$. But $G \cap E_{\alpha_0} = \varnothing$; so E_{α_0} is then disconnected. $\qquad\square$

Proposition 12. Let (S_1, d_1) and (S_2, d_2) be connected. Then $S_1 \times S_2$ is connected.

Proof. We show that any two points of $S_1 \times S_2$ are contained in a connected subset of $S_1 \times S_2$, and from this it follows that $S_1 \times S_2$ is connected. Let $(x_1, x_2), (y_1, y_2)$ be distinct points in $S_1 \times S_2$. $\{x_1\} \times S_2$ is connected, since it is the image of S_2 under the continuous map $\rho: S_2 \to S_1 \times S_2$ defined by $\rho(y) = (x_1, y)$. Similarly, $S_1 \times \{y_2\}$ is connected. Since the point

$$(x_1, y_2) \in (\{x_1\} \times S_2) \cap (S_1 \times \{y_2\}), \qquad (\{x_1\} \times S_2) \cup (S_1 \cup \{y_2\})$$

is connected by Proposition 11. $\qquad\square$

See R. L. Wilder, 1978, 720–726, for a historical sketch of the evolution of connectedness.

EXERCISES 27

1. Which subsets are connected with respect to the distance-1 metric?

2. If $E \subseteq S$ is connected, show that $\overline{E}$ is connected. Is the converse true?

3. Let I be an interval in $\mathbb{R}$ and $f: I \to \mathbb{R}$ continuous. Show that the graph of f in $\mathbb{R}^2$, $G(f) = \{(x, f(x)): x \in I\}$, is connected.

4. Show that $E = \{(x, \sin(1/x)): 0 < x \leq 1\} \cup \{(0, y): -1 \leq y \leq 1\}$ is connected in $\mathbb{R}^2$. Is E path-connected?

5. A subset K of a vector space is *convex* if $x, y \in K$ and $t \in [0, 1]$ imply that $tx + (1 - t)y \in K$. Show that a convex subset K of a nls is connected.

6. Show that $S(x, \varepsilon)$ and $\{y: \|y - x\| \leq \varepsilon\}$ are convex in any nls.

7. Let $D \subseteq \mathbb{R}^n$ be open and $f: D \to \mathbb{R}$ differentiable with $\nabla f(\mathbf{x}) = 0$, $\forall \mathbf{x} \in D$. If D is connected, show that $f = $ constant. (Recall Exercise 10.13.)

   Hint: Pick $x_0 \in D$ and consider $E = \{x \in D: f(x) = f(x_0)\}$.

8. Is the union of connected sets connected? The intersection?

9. Prove the converse of Proposition 12.

10. Show that $[0, 1]$ and $\{(x, y) \in \mathbb{R}^2: x^2 + y^2 = 1\}$ are not homeomorphic.

11. Is Corollary 4 valid in $\mathbb{Q}$?

12. Let $T = \{0, 1\}$ be equipped with the distance-1 metric. Show that a metric space (S, d) is connected iff the only continuous mappings $f: S \to T$ are the constant functions.

13. Let (S, d) be connected. Which functions $f: S \to \mathbb{Q}$ are continuous?

14. Show that if S is connected and has at least two points, then S is uncountable.

15. Show that S is connected iff every continuous real-valued function on S has the intermediate value property.

16. Show that if A is a countable subset of $\mathbb{R}^2$, then $\mathbb{R}^2 \setminus A$ is connected.

17. If E_k, $k \in \mathbb{N}$, is connected and $E_k \cap E_{k+1} \neq \emptyset$, show that $\bigcup_{k=1}^{\infty} E_k$ is connected.

18. Let E be the subset of $\mathbb{R}^2$ such that $(x, y) \in E$ iff either x or y is rational. Let $F = \mathbb{Q} \times \mathbb{R}$, $G = \mathbb{Q} \times \mathbb{Q}$. Show that E and the complement of G are connected, but that F and G are not connected.

19. Show that $\mathbb{R}$ and $\mathbb{R}^2$ are not homeomorphic.

CHAPTER 28

The Stone–Weierstrass Theorem

The Weierstrass approximation theorem, discussed in Chapter 18, asserts that the class of polynomials is dense in $\mathscr{C}[a, b]$, the class of all real-valued functions that are continuous on $[a, b]$. The American mathematician, Marshall H. Stone (1903–), generalized this theorem to the setting $\mathscr{C}(S)$, where (S, d) is a compact metric space. This abstract result, called the Stone–Weierstrass theorem, contains the Weierstrass approximation theorem as a very special case and is surely one of the most beautiful results in all of mathematics.

We seek algebraic conditions, related to the algebraic properties of the polynomials, that will guarantee that certain subsets of $\mathscr{C}(S)$ are dense in $\mathscr{C}(S)$. Let (S, d) be a compact metric space. A subset $\mathscr{A} \subseteq \mathscr{C}(S)$ is called an *algebra* if

(i) $f, g \in \mathscr{A}$ imply $f + g$ and fg belong to $\mathscr{A}$.
(ii) $f \in \mathscr{A}$ implies $tf \in \mathscr{A} \ \forall t \in \mathbb{R}$.

317

EXAMPLE 1. $\mathscr{C}(S)$ is an algebra. The set of polynomials, $\mathscr{P}$, is an algebra in $\mathscr{C}[a, b]$. The set of even polynomials, $\mathscr{E}$, is an algebra in $\mathscr{C}[a, b]$; the set of odd polynomials, $\mathscr{O}$, is not an algebra in $\mathscr{C}[a, b]$. The polynomials with 0 constant term is an algebra in $\mathscr{C}[a, b]$. If $K \subseteq \mathbb{R}^n$ is compact, the set of polynomials in n real variables forms an algebra in $\mathscr{C}(K)$.   □

A subset $\mathscr{B} \subseteq \mathscr{C}(S)$ *separates the points of* S iff for $t, s \in S, t \neq s, \exists f \in \mathscr{B}$ such that $f(t) \neq f(s)$.

EXAMPLE 2. $\mathscr{P}$ separates the points of $[a, b]$; $\mathscr{E}$ does not separate the points of $[-1, 1]$; $\mathscr{O}$ separates the points of $[a, b]$.   □

Lemma 3. Let $\mathscr{L}$ be a vector subspace of $\mathscr{C}(S)$ that separates the points of S and is such that $1 \in \mathscr{L}$. Then given $t, s \in S, t \neq s$, and $a, b \in \mathbb{R} \ \exists f \in \mathscr{L}$ such that $f(s) = a$ and $f(t) = b$.

Proof. There is $g \in \mathscr{L}$ such that $g(s) \neq g(t)$. Put $c = g(s) - g(t)$. Then the function

$$f = \frac{(a - b)g + (bg(s) - ag(t)) \cdot 1}{c} \in \mathscr{L}$$

and satisfies $f(s) = a$, $f(t) = b$.   □

Definition 4. A subset $\mathscr{L}$ of $\mathscr{C}(S)$ is called a *function space* if $\mathscr{L}$ is a vector space and if $f, g \in \mathscr{L}$ implies $f \vee g \in \mathscr{L}$ and $f \wedge g \in \mathscr{L}$.

Note that if $\mathscr{L}$ is a function space, then whenever $f \in \mathscr{L}$, $f^+ = f \vee 0$, $f^- = (-f) \vee 0$ and $|f|$ also belong to $\mathscr{L}$.

EXAMPLE 5. Let $\mathscr{PL}$ be the collection of all piecewise linear, continuous functions on $[0, 1]$ (i.e., $f \in \mathscr{PL}$ iff $f \in \mathscr{C}[0, 1]$ and $\exists$ a partition $x_0 < x_1 < \cdots < x_n$ of $[0, 1]$ such that f is linear on each subinterval $[x_{i-1}, x_i]$.) Then $\mathscr{PL}$ is a function space but is not an algebra.   □

Lemma 6. Let $\mathscr{L}$ be a function space in $\mathscr{C}(S)$ that contains the constant function 1 and separates the points of S. Then given $g \in \mathscr{C}(S)$ and $t_0 \in S$ and $\varepsilon > 0$, $\exists f \in \mathscr{L}$ such that $f(t_0) = g(t_0)$ and $f(t) > g(t) - \varepsilon \ \forall t \in S$.

Proof. By Lemma 3, $\forall t \in S \ \exists f_t \in \mathscr{L}$ such that $f_t(t_0) = g(t_0)$ and $f_t(t) = g(t)$. Since f_t and g are continuous, there is a δ-neighborhood V_t of t such that $f_t(s) > g(s) - \varepsilon \ \forall s \in V_t$.

Then $\{V_t : t \in S\}$ is an open cover of S and, therefore, there are $t_1, \ldots, t_n \in S$ such that $\bigcup_{i=1}^n V_{t_i} = S$. Let $f = f_{t_1} \vee \cdots \vee f_{t_n}$. Then $f \in \mathscr{L}$ and $f(t_0) =$

$g(t_0)$. Also, if $s \in S$, then $\exists \, k$ such that $s \in V_{t_k}$. Then $f(s) \geq f_{t_k}(s) > g(s) - \varepsilon$; so f is the desired function. $\square$

We can now give the lattice version of the Stone–Weierstrass theorem.

Theorem 7. Let $\mathscr{L}$ be a function space in $\mathscr{C}(S)$ that contains the constant function 1 and separates the points of S. Then $\mathscr{L}$ is dense in $\mathscr{C}(S)$.

Proof. Let $g \in \mathscr{C}(S)$ and $\varepsilon > 0$. By Lemma 6, $\forall \, t \in S \; \exists \, f_t \in \mathscr{L}$ such that $f_t(t) = g(t)$ and $f_t(s) > g(s) - \varepsilon \; \forall \, s \in S$. By the continuity of f_t and g, $\exists$ a neighborhood U_t of t such that $f_t(s) < g(s) + \varepsilon \; \forall \, s \in U_t$. Since S is compact, $\exists \, t_1, \ldots, t_n \in S$ such that $\bigcup_{i=1}^{n} U_{t_i} = S$. Put $f = f_{t_1} \wedge \cdots \wedge f_{t_n}$. Then $f \in \mathscr{L}$.

Since $f_{t_i} > g - \varepsilon$, $f > g - \varepsilon$. If $s \in S$, then $s \in U_{t_k}$ for some k, so that $f(s) \leq f_{t_k}(s) < g(s) + \varepsilon$. Thus

$$g(s) - \varepsilon \leq f(s) \leq g(s) + \varepsilon \quad \forall \, s \in S \qquad \text{or} \qquad \|f - g\|_\infty \leq \varepsilon. \qquad \square$$

In order to state the algebraic version of the Stone–Weierstrass theorem, a lemma is needed.

Lemma 8. There is a sequence of polynomials $\{p_n\}$ such that $p_n(t) \to \sqrt{t}$ uniformly for $t \in [0, 1]$.

Proof. Set $p_1 = 0$ and $p_{n+1}(t) = p_n(t) + \frac{1}{2}(t - p_n(t)^2)$ for $n \geq 1$. Clearly, each p_n is a polynomial.

We first claim that: $0 \leq p_n(t) \leq \sqrt{t}$, $0 \leq t \leq 1$. This certainly holds for $n = 1$; assume that it holds for $n \leq k$. $p_{k+1}(t) \geq 0$ and $\sqrt{t} - p_{k+1}(t) = \sqrt{t} - p_k(t) - \frac{1}{2}[t - p_k(t)^2] = [\sqrt{t} - p_k(t)]\{1 - \frac{1}{2}[\sqrt{t} + p_k(t)]\} \geq 0$. Thus, the claim is established by induction.

Since $p_n(t)^2 \leq t \; \forall \, t \in [0, 1]$, it follows that $\{p_n(t)\} \uparrow \; \forall \, t \in [0, 1]$. Put $p(t) = \lim p_n(t)$. Since $p(t) \geq 0$ and $p(t)^2 = t$, we have $p(t) = \sqrt{t}$, that is, $p_n(t) \to \sqrt{t} \; \forall \, t \in [0, 1]$. The convergence is uniform on $[0, 1]$ by Dini's theorem **(11.18)**. $\square$

Of course, the conclusion of Lemma 8 follows directly from the Weierstrass approximation theorem. We gave an independent proof in order to show that the Weierstrass approximation theorem truly is a corollary of the Stone–Weierstrass theorem.

Theorem 9 (Stone–Weierstrass). Let $\mathscr{A}$ be an algebra in $\mathscr{C}(S)$ such that $\mathscr{A}$ contains the constant function 1 and separates the points of S. Then $\mathscr{A}$ is dense in $\mathscr{C}(S)$.

Proof. By Theorem 7, it suffices to show that $\overline{\mathscr{A}}$ is a function space.

We first claim that $f \in \overline{\mathscr{A}}$ implies $|f| \in \overline{\mathscr{A}}$. Let $\{p_n\}$ be the polynomials in Lemma 8, let $f \in \overline{\mathscr{A}}$ be such that $f \neq 0$, and put $a = \|f\|_\infty > 0$. Then

$g_n = p_n \circ (f^2/a^2) \in \mathscr{A}$ $\forall$ n (Exercise 1) and, since $p_n(t) \to \sqrt{t}$ uniformly for $0 \le t \le 1$, $g_n \to \sqrt{f^2/a^2} = |f|/a$ in $\| \ \|_\infty$. Hence, $|f| = a(|f|/a) \in \mathscr{A}$.

But $f \vee g = \frac{1}{2}(f + g + |f - g|)$, and $f \wedge g = \frac{1}{2}(f + g - |f - g|)$; thus, $\mathscr{A}$ is indeed a function space.   □

Corollary 10. (Weierstrass Approximation Theorem) The polynomials are dense in $\mathscr{C}[a, b]$.

A more general statement is given by Corollary 11.

Corollary 11. Let $K \subseteq \mathbb{R}^n$ be compact. The polynomials in n-variables are dense in $\mathscr{C}(K)$.

Finally, we should check the necessity of the various hypotheses in Theorem 9. The algebra $\mathscr{E}$ in $\mathscr{C}[-1, 1]$ does not separate the points of $[-1, 1]$ and is not dense in $\mathscr{C}[-1, 1]$, so this condition cannot be dropped. Because the algebra of polynomials in $\mathscr{C}[0, 1]$ that vanish at 0 is not dense in $\mathscr{C}[0, 1]$, the condition that the algebra contains the constant function 1 cannot be dropped.

For further remarks on the Stone–Weierstrass theorem, see the article by M. Stone, in *Studies in Modern Analysis*, Volume 1 in *Mathematical Association of America Studies in Mathematics*, edited by R. C. Buck.

EXERCISES 28

1. If $\mathscr{A}$ is an algebra in $\mathscr{C}(S)$, show that $\bar{\mathscr{A}}$ is an algebra.

2. Is $\mathscr{P}$ a function space in $\mathscr{C}[a, b]$?

3. If $\mathscr{L}$ is a function space in $\mathscr{C}(S)$, show that $\bar{\mathscr{L}}$ is a function space.

4. Show that $\mathscr{PL}$ (see Example 15) is dense in $\mathscr{C}[0, 1]$.

5. Let $\mathscr{A}$ be the vector space in $\mathscr{C}[0, 1]$ generated by the functions $1, \sin^1 t, \sin^2 t, \ldots$. [$f \in \mathscr{A}$ iff $f(t) = \sum_{k=0}^n a_k \sin^k t$ for some $a_i \in \mathbb{R}$.] Show that $\mathscr{A}$ is dense in $\mathscr{C}[0, 1]$.

6. Show that the algebra generated by the functions $\{1, t^2\}$ is dense in $\mathscr{C}[0, 1]$ but is not dense in $\mathscr{C}[-1, 1]$.

7. Let S, T be compact metric spaces. If $f \in \mathscr{C}(S)$, $g \in \mathscr{C}(T)$, write $f \otimes g$ for the function $(s, t) \to f(s)g(t)$. Show that the functions of the form $\Sigma f_k \otimes g_k$ (finite sum) are dense in $\mathscr{C}(S \times T)$.

8. Give an example of a situation where Theorem 7 is applicable but Theorem 9 is not.

9. If $f_k \in \mathscr{BV}[a, b]$ and $f_k \to f$ uniformly on $[a, b]$, is it necessarily true that $f \in \mathscr{BV}[a, b]$?

10. If $g \in \mathscr{C}[0,1]$ and $\varepsilon > 0$, show that $\exists \, \alpha_0, \alpha_1, \ldots, \alpha_k \in \mathbb{R}$ such that

$$\left| g(t) - \sum_{j=0}^{k} \alpha_j e^{jt} \right| < \varepsilon, \qquad \forall \, t \in [0,1].$$

11. Show that the polynomials with rational coefficients are dense in $\mathscr{C}[a, b]$.

12. Let $\mathscr{A}$ be the set of all functions of the form $\sum_{k=0}^{n} c_k e^{kt}$, $n \in \mathbb{N}$, $c_k \in \mathbb{R}$. Show that $\mathscr{A}$ is dense in $\mathscr{C}[a, b]$.

13. Let $\mathscr{A}$ be the set of all functions of the form $\sum_{k=0}^{n} c_k \cos kt$, $n \in \mathbb{N}$, $c_k \in \mathbb{R}$. Show that $\mathscr{A}$ is dense in $\mathscr{C}[0, \pi]$. Is $\mathscr{A}$ dense in $\mathscr{C}[-\pi, \pi]$?

CHAPTER 29

Differentiation of Vector-Valued Functions

Many of the results in Chapter 10 on differentiation of real-valued functions can be generalized to functions with vector values, that is, functions with values in a normed linear space.

Let X and Y be normed linear spaces, D an open subset of X, $x \in D$ and $f: D \to Y$. We want to define differentiability of f at x. Recall that for the case when $X = \mathbb{R}^n$ and $Y = \mathbb{R}$, the differential of f at x is a (continuous) linear functional $df(x)$: $X \to Y$ which, in a certain sense, approximates increments in f at x. The notion of a continuous linear operator enables us to directly extend the definition.

Definition 1. The function f is (*Frechet*) *differentiable* at x if there is $T \in L(X, Y)$ such that

$$\lim_{h \to 0} \| f(x + h) - f(x) - T(h) \| / \|h\| = 0. \qquad (1)$$

It is shown below in Proposition 2 that the operator T, when it exists, is unique; T is called the *differential* of f at x and is denoted by $df(x)$. If f is differentiable at every point of D, then f is said to be *differentiable on D*.

Proposition 2. Suppose T_1 and T_2 are in $L(X, Y)$ and satisfy

$$\lim_{h \to 0} \|f(x + h) - f(x) - T_i(h)\| / \|h\| = 0.$$

Then $T_1 = T_2$.

Proof. Let $A = T_1 - T_2$. Then

$$\|A(h)\| = \|f(x + h) - f(x) - T_1(h) - [f(x + h) - f(x) - T_2(h)]\|$$

$$\leq \|f(x + h) - f(x) - T_1(h)\| + \|f(x + h) - f(x) - T_2(h)\|.$$

It follows that $\|A(h)\| / \|h\| \to 0$ as $h \to 0$. Thus, if $h \neq 0$ is fixed in X, then $\|A(th)\| / \|th\| \to 0$ as $t \to 0$. But $\|A(th)\| / \|th\|$ is independent of t; so $\|A(th)\| = 0$ and $Ah = 0$. ☐

As in the scalar case, differentiability implies continuity.

Proposition 3. If f is differentiable at x, then f is continuous at x.

Proof. Equation (1) implies $\|f(x + h) - f(x) - df(x)(h)\| / \|h\| \leq 1$ for h sufficiently small. Consequently,

$$\left[\|f(x + h) - f(x)\| - \|df(x)\| \|h\|\right] / \|h\| \leq 1,$$

and

$$\|f(x + h) - f(x)\| \leq \|h\| + \|df(x)\| \|h\|.$$

Let $h \to 0$. ☐

As in the scalar case, it is routine to check that the differential is linear (Exercise 1).

The utility of the concept of differentiation can best be illustrated by several examples.

EXAMPLE 4. Let $D \subseteq \mathbb{R}^n$ be open and $\mathbf{f} = (f_1, \ldots, f_m): D \to \mathbb{R}^m$. Since convergence in $\mathbb{R}^m$ is equivalent to coordinatewise convergence, it is easy to see that $\mathbf{f}$ is differentiable at $\mathbf{x} \in D$ iff each f_i $(i = 1, \ldots, m)$ is differentiable at $\mathbf{x}$, and in this case, $d\mathbf{f}(\mathbf{x})(\mathbf{e}_j) = (D_j f_1(\mathbf{x}), \ldots, D_j f_m(\mathbf{x}))$, $j = 1, \ldots, n$ (Exercise 2). ☐

Recall that if $T: \mathbb{R}^n \to \mathbb{R}^m$ is linear, then T can be represented by a matrix, which is unique with respect to the canonical bases in $\mathbb{R}^n$ and $\mathbb{R}^m$, as follows. For each $j = 1, \ldots, n$, let $T(\mathbf{e}_j) = \sum_{i=1}^m t_{ij}\mathbf{e}_i$ and let $[T]$ be the matrix

$$\left[t_{ij} \right]_{\substack{i=1,\ldots,m \\ j=1,\ldots,n}}.$$

Note that the coordinates of $T\mathbf{e}_j$ appear on the jth column of the matrix $[T]$. If $\mathbf{x} = \sum_{j=1}^n x_j\mathbf{e}_j \in \mathbb{R}^n$, then

$$T(\mathbf{x}) = \sum_{j=1}^n x_j T(\mathbf{e}_j) = \sum_{i=1}^m \left(\sum_{j=1}^n x_j t_{ij} \right) \mathbf{e}_j;$$

that is, $T(\mathbf{x})$ is the transpose of the matrix product

$$[T]\mathbf{x}^t = \left[t_{ij} \right] \begin{bmatrix} x_1 \\ \vdots \\ x_n \end{bmatrix} = (T(\mathbf{x}))^t.$$

If $\mathbf{f}$ is differentiable at $\mathbf{x}$, then the matrix representation of $d\mathbf{f}$ at $\mathbf{x}$ is

$$d\mathbf{f}(\mathbf{x}) = \begin{bmatrix} D_1 f_1(\mathbf{x}) & \cdots & D_n f_1(\mathbf{x}) \\ \vdots & & \\ D_1 f_m(\mathbf{x}) & \cdots & D_n f_m(\mathbf{x}) \end{bmatrix} = \left[\frac{\partial f_i}{\partial x_j}(\mathbf{x}) \right];$$

this matrix is called the *Jacobian matrix* of $\mathbf{f}$ at $\mathbf{x}$.

Suppose $m = n$. If $\mathbf{f}$ is written in component form as

$$y_1 = f_1(x_1, \ldots, x_n),$$
$$\vdots$$
$$y_n = f_n(x_1, \ldots, x_n),$$

then the determinant of the Jacobian matrix is often denoted by

$$\frac{\partial(y_1, \ldots, y_n)}{\partial(x_1, \ldots, x_n)} = \frac{\partial(f_1, \ldots, f_n)}{\partial(x_1, \ldots, x_n)}$$

and is called the *Jacobian* of $\mathbf{f}$ at $\mathbf{x}$. $\quad\square$

EXAMPLE 5. If $f: (a, b) \to Y$, then, as in the scalar case, f is differentiable at $t \in (a, b)$ iff $\lim_{h \to 0} (f(t + h) - f(t))/h = f'(t)$ exists in Y. In this case, the linear transformation $df(t)$ from $\mathbb{R}$ to Y is just $df(t)(h) = hf'(t)$. We usually write $df(t) = f'(t)$ and call $f'(t)$ the *derivative* of f at t. $\quad\square$

If f is differentiable at x and $h \neq 0$, then

$$df(x)(h) = \lim_{t \to 0} \frac{f(x + th) - f(x)}{t}.$$

This formula is the analog of the directional derivative of the function f at the point x in the direction h discussed for scalar-valued functions in Chapter 10. It is useful for discovering the Frechet differential of certain functions. To use the formula, we merely compute the derivative of the function $t \to f(x + th)$ at $t = 0$ for fixed x and h. We illustrate these observations in the following example.

EXAMPLE 6. Let $f: [a, b] \times \mathbb{R} \to \mathbb{R}$ be continuous. Define $F: \mathscr{C}[a, b] \to \mathbb{R}$ by $F(\varphi) = \int_a^b f(s, \varphi(s))\, ds$. Assume that $D_2 f$ is continuous. Then, if $\varphi, \xi \in \mathscr{C}[a, b]$, the derivative of the function $t \to F(\varphi + t\xi)$ at $t = 0$ is found by Leibniz' rule to be $\int_a^b D_2 f(s, \varphi(s))\xi(s)\, ds$. Thus, if F has a differential at φ, it must be given by $dF(\varphi)(\xi) = \int_a^b D_2 f(s, \varphi(s))\xi(s)\, ds$. The linear map $dF(\varphi)$ is continuous by Example 21.18, and we now show that F is actually differentiable. By the mean value theorem,

$$F(\xi) - F(\varphi) - dF(\varphi)(\xi - \varphi)$$

$$= \int_a^b \{f(t, \xi(t)) - f(t, \varphi(t)) - D_2 f(t, \varphi(t))(\xi(t) - \varphi(t))\}\, dt \quad (2)$$

$$= \int_a^b \{ D_2 f[t, \varphi(t) + \theta(t)(\xi(t) - \varphi(t))] - D_2 f(t, \varphi(t))\}.$$

$$(\xi(t) - \varphi(t))\, dt,$$

where $0 \leq \theta(t) \leq 1$. By (2) and the continuity of $D_2 f$, it follows that F is differentiable. □

An interesting application of differentiability in infinite dimensional spaces is provided by an optimization problem in the calculus of variations. A function $f: D \to \mathbb{R}$ has a *local minimum* at $x \in D$ if there is an ε-neighborhood N of x such that $f(x) \leq f(y)$ for all $y \in N \cap D$. The following is the analog of 10.16.

Proposition 7. If f has a local minimum at $x \in D$, then $df(x) = 0$.

Proof. If $h \in X$, then the function $t \to f(x + th)$ has a local minimum at $t = 0$. It follows that $df(x)(h) = 0$, that is, $df(x) = 0$. □

The calculus of variations is concerned with the problem of minimizing certain functionals defined over some family of curves. One of the early

investigations in the calculus of variations involved the brachistochrone problem introduced by Johann Bernoulli: Given two points $P_0 = (x_0, y_0)$ and
$P_1 = (x_1, y_1)$ with $x_0 < x_1$ and $y_0 > y_1$, find the curve joining P_0 to P_1 with
the property that a point mass sliding without friction along this curve starting
from P_0 will reach P_1 in minimal time. If the curve is the graph of $y = \varphi(x)$,
the transit time, up to a constant factor, is given by the expression

$$F(\varphi) = \int_{x_0}^{x_1} \sqrt{\frac{1 + \varphi'(x)^2}{y_0 - \varphi(x)}} \, dx.$$

Note that the functional F has the general form

$$F(\varphi) = \int_{x_0}^{x_1} f(t, \varphi(t), \varphi'(t)) \, dt.$$

A necessary condition (called the *Euler-Lagrange equation*) for such a function
to have a minimum can be derived as follows.

Let $f: [a, b] \times \mathbb{R} \times \mathbb{R} \to \mathbb{R}$ be continuous and such that $D_2 f$ and $D_3 f$ are
continuous. Let X be the subset of $\mathscr{C}^1[a, b]$ that consists of those functions φ
that satisfy $\varphi(a) = L$, $\varphi(b) = R$, where L and R are fixed. Define $F: X \to \mathbb{R}$
by

$$F(\varphi) = \int_a^b f(t, \varphi(t), \varphi'(t)) \, dt.$$

Suppose that F has a local minimum in X at φ^*.

Let X_0 be the linear subspace of $\mathscr{C}^1[a, b]$ that consists of those functions ξ
satisfying $\xi(a) = \xi(b) = 0$. Now define $I: X_0 \to \mathbb{R}$ by $I(\xi) = F(\varphi^* + \xi)$ and
note that I has a local minimum at 0. By Leibniz' rule and the chain rule
(**10**.9), the derivative of the function $t \to I(0 + t\xi)$ at $t = 0$ is

$$\int_a^b \{ D_2 f(t, \varphi^*(t), \varphi^{*\prime}(t)) \xi(t) + D_3 f(t, \varphi^*(t), \varphi^{*\prime}(t)) \xi'(t) \} \, dt, \; \xi \in X_0.$$

$$(3)$$

So, if I has a differential at 0, its value at ξ must be given by (3). By using the
methods of Example 6, it can be shown that I is actually differentiable at 0
with its differential given by the expression in (3). By Proposition 7, the
expression in (3) must vanish for ξ in X_0. To translate this into a condition on
φ^*, we employ the fundamental lemma of the calculus of variations.

Lemma 8 (DuBois–Reymond). Let $h \in \mathscr{C}[a, b]$ and suppose that

$$\int_a^b h\xi' = 0 \; \forall \, \xi \in X_0.$$

Then $h = $ constant.

Proof. Define c by $c(b - a) = \int_a^b h$. Then $\forall \, \xi \in X_0$, we have by hypothesis that

$$\int_a^b (h - c)\xi' = \int_a^b h\xi' - c\int_a^b \xi' = 0. \tag{4}$$

Define ξ by $\xi(t) = \int_a^t (h - c)$ for $a \le t \le b$. Since $\xi(a) = 0$ and $\xi(b) = \int_a^b h - c(b - a) = 0$, $\xi \in X_0$, and (4) implies that

$$0 = \int_a^b (h - c)\xi' = \int_a^b (h - c)^2.$$

By Exercise **13.**18, $h = c$.   □

Corollary 9. If $g, h \in \mathscr{C}[a, b]$ and $\int_a^b (g\xi + h\xi') = 0$ $\forall \, \xi \in X_0$, then h is differentiable on $[a, b]$ with $h' = g$.

Proof. Set $A(t) = \int_a^t g$ for $a \le t \le b$. Integration by parts implies $\int_a^b (A - h)\xi' = 0$ for $\xi \in X_0$. Lemma 8 implies there is a constant c such that $h = A + c$. By the fundamental theorem of calculus, $h' = g$.   □

If we apply Corollary 9 to (3), we have that φ^* must satisfy

$$\frac{d}{dt}\{D_3 f(t, \varphi^*(t), \varphi^{*\prime}(t))\} - D_2 f(t, \varphi^*(t), \varphi^{*\prime}(t)) = 0;$$

this is the *Euler–Lagrange equation* from the calculus of variations.

To illustrate the use of the Euler–Lagrange equation, consider the problem of finding the function whose graph has minimal arc length and joins two fixed points (a, L) and (b, R). For $\varphi \in \mathscr{C}^1[a, b]$, its arclength is given by

$$\ell(\varphi) = \int_a^b \sqrt{1 + (\varphi')^2}$$

(Exercise **13.**30). If φ minimizes ℓ, the Euler–Lagrange equation for this problem becomes

$$\frac{d}{dt}\left(\varphi'/\sqrt{1 + (\varphi')^2}\right) = 0.$$

Thus, $\varphi' = $ constant and φ must be a straight line. Note that the Euler–Lagrange equation is only a necessary condition for a minimum. What has really been shown is that if this problem has a solution in the class of $\mathscr{C}^1$ functions, it must be a straight line joining the two points. For the solution to the brachistochrone problem mentioned earlier, see G. A. Bliss, 1925.

Theorem 10 (Chain Rule). Let X, Y and Z be normed linear spaces, D an open subset of X, and $f: D \to Y$ differentiable at $x_0 \in D$. Suppose that g maps an open set $G \subseteq Y$ containing $f(D)$ into Z and g is differentiable at $f(x_0)$. Then $F = g \circ f$ is differentiable at x_0 with $dF(x_0) = dg(f(x_0)) \, df(x_0)$.

Proof. Let $y_0 = f(x_0)$, $A = df(x_0)$, and $B = dg(y_0)$. Define

$$u(x) = f(x) - f(x_0) - A(x - x_0), \qquad x \in D$$

$$v(y) = g(y) - g(y_0) - B(y - y_0), \qquad y \in G$$

$$r(x) = F(x) - F(x_0) - BA(x - x_0), \qquad x \in D.$$

To show that $dF(x_0) = BA$ it suffices to show that $\|r(x)\|/\|x - x_0\| \to 0$ as $x \to x_0$. Note that

$$
\begin{aligned}
r(x) &= g(f(x)) - g(y_0) - B(f(x) - y_0) \\
&\quad + B[f(x) - f(x_0) - A(x - x_0)] \\
&= v(f(x)) + B(u(x)).
\end{aligned}
\tag{5}
$$

Let $\varepsilon > 0$. $\exists \, \eta > 0$, $\delta > 0$ such that $\|v(y)\| \le \varepsilon \|y - y_0\|$ for $\|y - y_0\| < \eta$, and $\|f(x) - y_0\| < \eta$ and $\|u(x)\| < \varepsilon \|x - x_0\|$ for $\|x - x_0\| < \delta$. Thus

$$\|v(f(x))\| \le \varepsilon \|f(x) - y_0\| = \varepsilon \|u(x) + A(x - x_0)\|$$

$$\le \varepsilon^2 \|x - x_0\| + \varepsilon \|A\| \, \|x - x_0\|,$$

and

$$\|B(u(x))\| \le \|B\| \, \|u(x)\| \le \varepsilon \|B\| \, \|x - x_0\|,$$

for $\|x - x_0\| < \delta$. So, from (5), $\|r(x)\|/\|x - x_0\| \to 0$ as $x \to x_0$. $\square$

EXAMPLE 11. Let

$$\mathbf{f} = (f_1, \ldots, f_m): D \subseteq \mathbb{R}^n \to \mathbb{R}^m \qquad \text{and} \qquad \mathbf{g} = (g_1, \ldots, g_k): \mathbb{R}^m \to \mathbb{R}^k.$$

If $\mathbf{f}$ is differentiable at $\mathbf{x}$ and $\mathbf{g}$ is differentiable at $\mathbf{y} = \mathbf{f}(\mathbf{x})$, then the Jacobian matrices of $\mathbf{f}$ and $\mathbf{g}$ satisfy the following special chain rule:

$$\left[D_j F_i(\mathbf{x}) \right] = \left[D_p g_i(\mathbf{y}) \right] \left[D_j f_p(\mathbf{x}) \right],$$

where $\mathbf{F} = \mathbf{g} \circ \mathbf{f}$. In particular,

$$D_j F_i(\mathbf{x}) = \sum_{p=1}^m D_p g_i(\mathbf{y}) D_j f_p(\mathbf{x}) \qquad \text{for } j = 1, \ldots, n; \qquad i = 1, \ldots, k.$$

This is a generalization of the chain rule of Theorem **10**.9. $\square$

The Mean Value Theorem

There is a mean value theorem (MVP) for vector-valued functions; however, a straightforward analog of the mean value theorems given for scalar-valued functions in **9.5** and **10.11** will not generally hold for such functions. For example, consider $\mathbf{f} \colon \mathbb{R} \to \mathbb{R}^2$ defined by $\mathbf{f}(t) = (t - t^2, t - t^3)$. Then $\mathbf{f}'(t) = d\mathbf{f}(t) = (1 - 2t, 1 - 3t^2)$ and $\mathbf{f}(1) - \mathbf{f}(0) = (0,0)$; so there is no point ζ between 0 and 1 such that $\mathbf{f}(1) - \mathbf{f}(0) = (1)\, d\mathbf{f}(\zeta) = \mathbf{f}'(\zeta)$.

The following vector version of the mean value theorem is sufficient for most applications.

Theorem 12 (Mean Value Theorem). Let $a, b \in \mathbb{R}$, Y be a normed linear space, and $\varphi \colon [a, b] \to Y$ be continuous on $[a, b]$ and differentiable on (a, b). Then there is $\zeta \in (a, b)$ such that

$$\|\varphi(b) - \varphi(a)\| \leq \|\varphi'(\zeta)\|(b - a).$$

Proof. Set $L = (b - a)/3$, $M = \|\varphi(b) - \varphi(a)\|$. Define $g \colon [a, a + 2L] \to Y$ by $g(s) = \varphi(s + L) - \varphi(s)$. Since

$$\varphi(b) - \varphi(a) = g(a) + g(a + L) + g(a + 2L),$$

we have

$$M \leq \|g(a)\| + \|g(a + L)\| + \|g(a + 2L)\|. \tag{5}$$

We claim that $\|g(s_1)\| \geq M/3$ for some $a < s_1 < a + 2L$. If not, then $\|g(s)\| < M/3$ for all $a < s < a + 2L$ and the continuity of $s \to \|g(s)\|$ would imply that $\|g(a)\| \leq M/3$ and $\|g(a + 2L)\| \leq M/3$. This contradicts (5). Let $t_1 = s_1 + L$. Note that $a < s_1 < t_1 < b$, $t_1 - s_1 = (b - a)/3$ and $\|g(s_1)\| = \|\varphi(t_1) - \varphi(s_1)\| \geq M/3$. The above construction can be repeated in the subinterval $[s_1, t_1]$. Induction gives two sequences $\{s_k\}$ and $\{t_k\}$ in (a, b) such that $t_k - s_k = (b - a)/3^k$, $[s_k, t_k] \subseteq (s_{k-1}, t_{k-1})$ and $\|\varphi(t_k) - \varphi(s_k)\| \geq M/3^k$. Thus,

$$\|\varphi(t_k) - \varphi(s_k)\| / (t_k - s_k) \geq \|\varphi(b) - \varphi(a)\| / (b - a).$$

Let $\{\zeta\} = \bigcap_{k=1}^{\infty}[s_k, t_k]$. Since the proof of Lemma **11.19** is also valid for vector-valued functions, it can be applied here to yield

$$\|\varphi'(\zeta)\| \geq \|\varphi(b) - \varphi(a)\| / (b - a). \qquad \square$$

Corollary 13. If φ in Theorem 12 is such that

$$M = \sup\{\|\varphi'(t)\| \colon a < t < b\},$$

then

$$\|\varphi(b) - \varphi(a)\| \le M(b - a).$$

Corollary 14. Let D be convex and open and let $f: D \to Y$ be differentiable on D. If x, y and $x_0 \in D$, then

  (i) $\|f(y) - f(x)\| \le \|y - x\| \sup \{\|df(\zeta)\|: \zeta \in [x, y]\}$
  (ii) $\|f(y) - f(x) - df(x_0)(y - x)\| \le \|y - x\| \sup \{\|df(\zeta) - df(x_0)\|:$
    $\zeta \in [x, y]\}$.
    [Here $[x, y] = \{ty + (1 - t)x: 0 \le t \le 1\}$.]

Proof. For (i), define the function $\varphi: [0, 1] \to Y$ by

$$\varphi(t) = f(ty + (1 - t)x).$$

By the chain rule,

$$\varphi'(t) = df(ty + (1 - t)x)(y - x).$$

Now apply Corollary 13.
  For (ii), apply (i) to the function $x \to f(x) - df(x_0)(x)$ and employ Exercise 3. □

$\mathscr{C}^1$-mappings

A function $f: D \to Y$ is *continuously differentiable* or of *class $\mathscr{C}^1$* on D if f is differentiable on D and the mapping $x \to df(x)$ from D into $L(X, Y)$ is continuous (with respect to the operator norm on $L(X, Y)$). The function f is of *class $\mathscr{C}^1$* at $x_0 \in D$ if f is of class $\mathscr{C}^1$ in some open neighborhood of x_0. We often abbreviate these statements to "f is $\mathscr{C}^1$ on D" and "f is $\mathscr{C}^1$ at x_0."
  The mean value theorem has the following important corollary for $\mathscr{C}^1$ functions.

Corollary 15. Let $f: D \to Y$ be $\mathscr{C}^1$ at $x_0 \in D$. For each $\varepsilon > 0$, there is $\delta > 0$ such that if $\|x - x_0\| < \delta$ and $\|y - x_0\| < \delta$, then

$$\|f(y) - f(x) - df(x_0)(y - x)\| \le \varepsilon \|y - x\|. \tag{6}$$

Proof. By the continuity of df at x_0, $\exists \delta > 0$, such that $\|df(z) - df(x_0)\| \le \varepsilon$ for $\|z - x_0\| < \delta$. Equation (6) now follows by applying Corollary 14 (ii) to the sphere $S(x_0, \delta)$. □

The conclusion in (6) is actually used to define a notion of differentiability that is stronger than the Frechet derivative and has proven to be useful in many applications. See A. Nijenhuis, 1974 for an interesting example.

Partial Differentials

Let X, Y, and Z be normed linear spaces, $D \subseteq X \times Y$ open and $(x, y) \in D$. Let U and V be open sets in X and Y, respectively, with $x \in U$, $y \in V$ and $U \times V \subseteq D$. A function $f: D \to Z$, has a *differential at* (x, y) *with respect to the 1st coordinate* (*2nd coordinate*) if the function $f(\cdot, y): U \to Z$ ($f(x, \cdot): V \to Z$) has a differential at $x(y)$. We write

$$d(f(\cdot, y))(x) = d_1 f(x, y) \qquad (d(f(x, \cdot))(y) = d_2 f(x, y)).$$

If $X = \mathbb{R}^k$, $Y = \mathbb{R}^m$, $Z = \mathbb{R}^n$, $D \subseteq \mathbb{R}^{k+m}$ and $\mathbf{f}: D \to \mathbb{R}^n$, then $d_1 \mathbf{f}(\mathbf{x}, \mathbf{y})$: $\mathbb{R}^k \to \mathbb{R}^n$ is represented by the matrix

$$d_1 \mathbf{f}(\mathbf{x}, \mathbf{y}) = \begin{bmatrix} D_1 f_1(\mathbf{x}, \mathbf{y}) & \cdots & D_k f_1(\mathbf{x}, \mathbf{y}) \\ \vdots & & \\ D_1 f_n(\mathbf{x}, \mathbf{y}) & \cdots & D_k f_n(\mathbf{x}, \mathbf{y}) \end{bmatrix},$$

where $\mathbf{f} = (f_1, \dots, f_n)$. This matrix is called the *Jacobian matrix* of $\mathbf{f} = (f_1, \dots, f_n)$ with respect to the variables $x_1, \dots, x_k$; when $m = k$ the determinant of this matrix is called the *Jacobian* of $\mathbf{f}$ with respect to $x_1, \dots, x_n$ and is often denoted by

$$\frac{\partial(f_1, \dots, f_n)}{\partial(x_1, \dots, x_n)}.$$

Similar remarks apply to $d_2 \mathbf{f}$.

Proposition 16. If f is differentiable at (x, y), then $d_1 f(x, y)$ and $d_2 f(x, y)$ exist with $df(x, y)(h, k) = d_1 f(x, y)(h) + d_2 f(x, y)(k)$.

Proof. Since $\|f(x + h, y) - f(x, y) - df(x, y)(h, 0)\| / \|h\| \to 0$, $d_1 f(x, y)(h) = df(x, y)(h, 0)$; similarly, $d_2 f(x, y)(k) = df(x, y)(0, k)$. This gives the result.   □

Proposition 17. If $d_1 f$ and $d_2 f$ exist and are continuous at every $(x, y) \in D$, then f is $\mathscr{C}^1$ on D.

Proof. It must be shown that f is differentiable and that the mapping $x \to df(x)$ is continuous on D.

We shall show that f is differentiable at $(x, y) \in D$ by showing that $T(h, k) = d_1 f(x, y)(h) + d_2 f(x, y)(k)$ defines a linear operator T satisfying Definition 1. To see that

$$\| f(x + h, y + k) - f(x, y) - d_1 f(x, y)(h) - d_2 f(x, y)(k) \| / (\|h\| + \|k\|)$$
$$\to 0,$$

let $\varepsilon > 0$ and write

$$f(x + h, y + k) - f(x, y)$$
$$= f(x + h, y + k) - f(x + h, y) + f(x + h, y) - f(x, y).$$

Then, $\exists\, \delta_1 > 0$ such that

$$\| f(x + h, y) - f(x, y) - d_1 f(x, y)(h) \| \le \varepsilon \|h\| \qquad \text{for } \|h\| \le \delta_1.$$

Next, the function $k \to f(x + h, y + k) - f(x + h, y) - d_2 f(x, y)(k)$ has for its differential at k the function

$$d_2 f(x + h, y + k) - d_2 f(x, y).$$

So, by Corollary 14(i),

$$\| f(x + h, y + k) - f(x + h, y) - d_2 f(x, y)(k) \|$$
$$\le \|k\| \sup_{\|z\| \le \|k\|} \| d_2 f(x + h, y + z) - d_2 f(x, y) \|.$$

Since $d_2 f$ is continuous, $\exists\, \delta_2 > 0$ such that

$$\| d_2 f(x + h, y + k) - d_2 f(x, y) \| < \varepsilon$$

for $\|h\| < \delta_2$ and $\|k\| < \delta_2$. If $\delta = \min\{\delta_1, \delta_2\}$ and $\|h\| < \delta$, $\|k\| < \delta$, then

$$\| f(x + h, y + k) - f(x, y) - d_1 f(x, y)(h) - d_2 f(x, y)(k) \| \le \varepsilon \|h\| + \varepsilon \|k\|$$

and it follows that f is differentiable at (x, y).

Finally, define $I_1 : X \to X \times Y$, $I_2 : Y \to X \times Y$ by $I_1(x) = (x, 0)$, $I_2(y) = (0, y)$. Then I_1 and I_2 are continuous and the continuity of $d_1 f$ and $d_2 f$ imply that $df = d_1 f \circ I_1 + d_2 f \circ I_2$ is continuous.   □

EXERCISES 29

1. If $f, g \colon D \to Y$ are differentiable at $x \in D$, show that $af + bg$ is differentiable at x with

$$d(af + bg)(x) = a\, df(x) + b\, dg(x)$$

2. Let $D \subseteq \mathbb{R}^n$ be open and $\mathbf{f} = (f_1, \ldots, f_m)$: $D \to \mathbb{R}^m$. Show that $\mathbf{f}$ is differentiable at $\mathbf{x} \in D$ iff each f_j is differentiable at $\mathbf{x}$.

3. If $T \in L(X, Y)$, show that T is differentiable at every $x \in X$ and compute $dT(x)$.

4. Let $f: \mathbb{R} \to \mathbb{R}$ be continuously differentiable. Define $F: \mathscr{C}[a, b] \to \mathbb{R}$ by $F(\varphi) = f(\varphi(a))$. Show that F is differentiable at every $\varphi \in \mathscr{C}[a, b]$.

5. Compute the differentials for the following functions at the indicated points:
   (a) $\mathbf{f}(x_1, x_2) = (x_1 x_2^2 - 3x_1, -5x_2^2)$ at $(1, -1)$, $(1, 3)$.
   (b) $\mathbf{f}(x_1, x_2, x_3) = (x_1 + 6x_2, 3x_3 x_1, x_1^2 - 3x_2^2)$ at $(1, 1, 0)$.

6. Let $\mathbf{f}: \mathbb{R}^2 \to \mathbb{R}^3$ be defined by $\mathbf{f}(x, y) = (x^2 - 2y, x^2 - 2xy, 3x^2 y - 2y)$. Show that $\mathbf{f}$ is differentiable at every point and calculate $d\mathbf{f}(x, y)$.

7. Suppose that the function F of Example 6 has a local minimum at φ. Show that the function $s \to D_2 f(s, \varphi(s))$ must be zero.

8. If $F: \mathscr{C}^1[0, 1] \to \mathbb{R}$ is given by $F(\varphi) = \int_a^b \{\varphi^2 + (\varphi')^2\}$, find $dF(\varphi)$ for any $\varphi \in \mathscr{C}^1[a, b]$.

9. Solve the Euler–Lagrange equation for the problem:

$$\min F(\varphi) = \int_0^1 \left\{ a\varphi'(t)^2 + b\varphi(t) \right\} dt, \qquad \varphi(0) = 0, \qquad \varphi(1) = 1.$$

10. What points $\zeta \in (0, 1)$ will satisfy the conclusion of the MVT, Theorem 12, for the function $\boldsymbol{\varphi}(t) = (t - t^2, t - t^3)$, $0 \le t \le 1$?

11. Let D be a connected open set and $f: D \to Y$ differentiable on D with $df(x) = 0 \; \forall x \in D$. Show that f is a constant.

12. If $\varphi: (a, b) \to Y$ is differentiable on (a, b) and if $\exists M > 0$ such that $\|\varphi'(t)\| \le M \; \forall t \in (a, b)$, show that $\lim_{t \to a} \varphi(t)$ exists provided Y is a Banach space.

13. Give a proof of Corollary 13 for the case when $Y = \mathbb{R}^n$ by using the scalar MVT.

14. Does the converse of Proposition 16 hold?
    Hint: Consider $Y = \mathbb{R}$.

15. Prove the converse of Proposition 17, that is, if $f: D \subseteq X \times Y \to Z$ is $\mathscr{C}^1$ on D, show that both

$$d_1 f: D \to L(X, Z) \qquad \text{and} \qquad d_2 f: D \to L(Y, Z)$$

are continuous.

16. Let $P: \mathscr{C}^2[0, 1] \to \mathscr{C}[0, 1]$ be given by $P\varphi = \varphi'' - e^\varphi$. Find $dP(\varphi)$.

17. The function $f: D \to Y$ has a *Gateaux variation* at $x \in D$ if $\lim_{t \to 0} (f(x + th) - f(x))/t = f'(x; h)$ exists for each $h \in X$. (Note that $f'(x; h)$ is the analog of the directional derivative in **10.1**.)

(a) If f has a Gateaux variation at x, show that $f'(x; th) = tf'(x; h)$, that is, $f'(x; \cdot)$ is homogeneous.

(b) Show that $f'(x; \cdot)$ is not necessarily additive, that is, $f'(x; h_1 + h_2) \neq f'(x; h_1) + f'(x; h_2)$.

Hint: Consider $f: \mathbb{R}^2 \to \mathbb{R}$ defined by

$$f(x_1, x_2) = \begin{cases} x_1 x_2^2 / (x_1^2 + x_2^4), & \text{if } x_1 \neq 0, \\ 0, & \text{if } x_1 = 0. \end{cases}$$

(c) Show that $f'(x; \cdot)$ is not necessarily continuous on X.

(d) State and prove a result analogous to Proposition 7 for the Gateaux variation.

18. The function $f: D \to Y$ is *Gateaux differentiable* at $x \in D$ if f has a Gateaux variation at x and there is $\delta f(x) \in L(X, Y)$ such that $f'(x; h) = \delta f(x)(h)$ for every $h \in X$. [$\delta f(x)$ is called the *Gateaux derivative* of f at x.]

(a) Show that a function with a Gateaux variation need not be Gateaux differentiable.

(b) Let $f: \mathscr{C}[0, 1] \to \mathbb{R}$ be defined by $f(\varphi) = \varphi(0)$. Show that f is Gateaux differentiable if $\mathscr{C}[0, 1]$ has the sup-norm, but is not Gateaux differentiable if $\mathscr{C}[0, 1]$ has the L^1-norm, $\|\varphi\| = \int_0^1 |\varphi|$. (Note that the Gateaux variation is independent of the norm in X.)

(c) Show that if f is (Frechet) differentiable at x, then f is Gateaux differentiable at x with $df(x) = \delta f(x)$.

(d) Give an example of a function that is Gateaux differentiable, but not Frechet differentiable.

Hint: Consider $f(x, y) = \begin{cases} 1, & \text{if } x = y^2, \quad y \neq 0, \\ 0, & \text{otherwise} \end{cases}$.

(e) Does the analogue of Proposition 3 hold for the Gateaux derivative?

(f) State and prove an analogue to Proposition 7 for the Gateaux derivative.

19. Let A be an $n \times n$ symmetric matrix. Let $q: \mathbb{R}^n \to \mathbb{R}$ be defined by $q(\mathbf{x}) = (A\mathbf{x}^t)^t \cdot \mathbf{x}$. Is q Frechet differentiable? If so, what is the differential of q in terms of A?

20. Let $F: \mathscr{C}[0, 1] \to \mathbb{R}$ be defined by $F(\varphi) = \int_0^1 (t^2 + \varphi^2(t)) \, dt$. Compute the Frechet differential of F.

21. Let X be a normed linear space and K a convex subset of X. Assume that $f: K \to \mathbb{R}$ is Gateaux differentiable on K. The function f is *convex* if $f(tx + (1 - t)y) \le tf(x) + (1 - t)f(y)$ for all $x, y \in K$ and $0 \le t \le 1$. Show that f is convex if, and only if, $f(y) - f(x) \ge f'(x; y - x)$ for all $x, y \in K$.

22. Let $T: X \to Y$ be linear. Show that T has a Gateaux variation at every $x \in X$. When is T Gateaux differentiable?

23. Let $f: D \to Y$ have a Gateaux variation on D. If $x, x + h \in D$ and D is convex, show that there exists $t \in (0, 1)$ such that $f(x + h) - f(x) = f'(x + th; h)$. (This is a mean value theorem for the Gateaux derivative.)

24. Let $f: D \to Y$ be Gateaux differentiable in D. If $x, y \in D$ and D is convex, is there necessarily $t \in (0, 1)$ such that $f(y) - f(x) = \delta f(tx + (1 - t)y)(y - x)$?

CHAPTER 30

Mapping Theorems

Finally, we consider mapping properties of differentiable functions. If f has a differential $df(x)$ at a point x, then this differential is a linear approximation to f at x; so it is not unreasonable to expect that the mapping properties of $df(x)$ are inherited at least locally by f. This is essentially the content of our first result, called the inverse function theorem, which states that if $df(x)$ is invertible, then the function f is locally invertible near x.

Throughout this chapter, X and Y denote normed linear spaces.

Theorem 1 (Inverse Function Theorem). Let D be an open subset of the Banach space X and let $f: D \to Y$ be differentiable on D and such that df is continuous at $x_0 \in D$. If $f(x_0) = y_0$ and $df(x_0)$ is invertible, then

    (i) There exists open sets $U \subseteq X, V \subseteq Y$ such that $x_0 \in U$, $y_0 \in V$, f is 1-1 on U, and $f(U) = V$.

    (ii) If g is the inverse of f on V, then g is differentiable on V with $dg(y) = df(x)^{-1}$ for $y \in V$, where $x = g(y)$.

Moreover, if f is $\mathscr{C}^1$ on D, then g is $\mathscr{C}^1$ on V.

Proof. The proof is made easier by the introduction of some reductions. Let $D_1 = D - x_0$ and define $f_1: D_1 \to Y$ by $f_1(x) = f(x + x_0) - y_0$. Then $f_1(0) = 0$. If we can prove the theorem for f_1, then the theorem follows for f. So we may assume that $f(0) = 0$.

Next, let $T = df(0)$. Consider the map $f_2 = T^{-1} \circ f$ on D into X. Note that f_2 is differentiable with $df_2(x) = T^{-1} df(x)$; so, $df_2(0) = I$. If we can establish the theorem for f_2, then it follows for f. So we may assume that $f: D \to X$ and $df(0) = I$.

Since $df(0) = I$, it seems reasonable to regard f as a small pertubation of the identity, $f = I + p$. If g is a local inverse of f near 0, then $I = f \circ g = g + p \circ g$, or $g = I - p \circ g$. That is, g is a fixed point of the transformation $\mathcal{T}: \varphi \to I - p \circ \varphi$. To make this formalism work, we need to set up an appropriate function space on which to define the operator $\mathcal{T}$ and then show that $\mathcal{T}$ is a contraction on this space.

For convenience, let $B(0, r) = \{x: \|x\| \le r\}$. Pick $r > 0$ such that $B(0, 2r) \subseteq D$ and

$$\|I - df(x)\| \le \tfrac{1}{3} \qquad \text{for } x \in B(0, 2r) \tag{1}$$

(the continuity of df at 0 is being used). From (1) it follows by **29**.14(ii) that

$$\|f(x) - x - f(x_1) + x_1\| \le \tfrac{1}{3}\|x - x_1\| \qquad \text{for } x, x_1 \in B(0, 2r). \tag{2}$$

Let $\mathcal{BC}(B(0, r), X) = \mathcal{BC}$ be the space of all bounded, continuous X-valued functions defined on $B(0, r)$ and equip $\mathcal{BC}$ with the sup-norm, $\|\varphi\|_\infty = \sup\{\|\varphi(x)\|: x \in B(0, r)\}$. Then $\mathcal{BC}$ is complete (Proposition **26**.5). Let S be the subset of $\mathcal{BC}$ defined by $\{\varphi: \varphi(0) = 0, \|\varphi\|_\infty \le 2r\}$. Since S is a closed subspace of the complete space $\mathcal{BC}$, S is complete. Now define the transformation $\mathcal{T}$ on S by $\mathcal{T}(\varphi) = I - p \circ \varphi$, where $p = f - I$. Thus, pointwise we have

$$\mathcal{T}(\varphi)(x) = x + \varphi(x) - f(\varphi(x)).$$

We claim that $\mathcal{T}: S \to S$. First, $\mathcal{T}(\varphi)(0) = \varphi(0) - f(\varphi(0)) = 0$. Next, if $\|x\| \le r$, then by setting $x_1 = 0$ in (2), we obtain

$$\|\mathcal{T}(\varphi)(x)\| = \|x + \varphi(x) - f(\varphi(x))\|$$

$$\le \|x\| + \tfrac{1}{3}\|\varphi(x)\| \le 5r/3 < 2r; \tag{3}$$

so that $\|\mathcal{T}(\varphi)\|_\infty \le 2r$. Thus, $\mathcal{T}(\varphi) \in S$ when $\varphi \in S$.

Next, we show that $\mathscr{T}$ is a contraction on S. For $\|x\| \leq r$, and $\varphi, \psi \in S$, we have

$$\|\mathscr{T}(\varphi)(x) - \mathscr{T}(\psi)(x)\| = \|\varphi(x) - f(\varphi(x)) - \psi(x) + f(\psi(x))\|$$

$$\leq \tfrac{1}{3}\|\varphi(x) - \psi(x)\|$$

by (2). Thus $\|\mathscr{T}(\varphi) - \mathscr{T}(\psi)\|_\infty \leq \tfrac{1}{3}\|\varphi - \psi\|_\infty$, and $\mathscr{T}$ is a contraction.

If $g \in S$ is the unique fixed point of T on S, then $f \circ g = I$. Since (2) implies that

$$\|f(x) - f(x_1)\| \geq \tfrac{2}{3}\|x - x_1\| \qquad \text{for } x, x_1 \in B(0, 2r), \qquad (4)$$

f is 1-1 on $B(0, 2r)$. Therefore, $g\colon B(0, r) \to B(0, 2r)$ is the inverse of f on $B(0, 2r)$, and since $g \in S$, it is also continuous.

Put $V = S(0, r)$ and set $U = g(V)$. Then $f(U) = S(0, r)$ and $U = f^{-1}(S(0, r)) \cap S(0, 2r)$ is open (Figure 30.1).

We next show that g is differentiable on V. Let $y, y_1 \in V$ and $x = g(y)$, $x_1 = g(y_1)$. Note that, $x_1, x \in S(0, 2r)$ by (3) and $df(x_1)$ is invertible by (1) and **22.10**. If $A = df(x_1)^{-1}$, then

$$g(y) - g(y_1) - A(y - y_1) = -A[f(x) - f(x_1) - df(x_1)(x - x_1)]. \qquad (5)$$

Now (4) and (5) imply that g is differentiable at y_1 with $dg(y_1) = df(x_1)^{-1}$.

If f is $\mathscr{C}^1$ on D, then since g is continuous and the inverse map $A \to A^{-1}$ is continuous (**22.12**), it follows that $y \to dg(y)$ is continuous on V. $\square$

The proof of Theorem 1 shows that f is an open mapping ; that is, f carries open subsets of U onto open subsets of V.

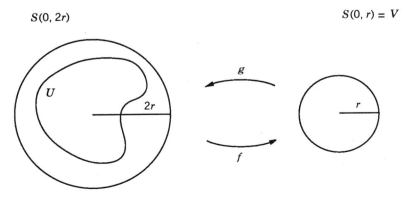

$S(0, 2r)$ ... $S(0, r) = V$

U

$2r$

g

f

r

Figure 30.1

There are several observations that should be made about Theorem 1. First, the invertibility of $df(x_0)$ is a *necessary* condition, for the chain rule implies

$$df(x_0)\, dg(y_0) = I = dg(y_0)\, df(x_0).$$

Next, Theorem 1 cannot be improved to a global result. That is, the invertibility of $df(x_0)$ does not, in general, imply that f is 1-1 on D. For example, consider $f: \mathbb{R}^2 \to \mathbb{R}^2$ defined by $\mathbf{f}(x, y) = (e^x \cos y, e^x \sin y)$. The differential $d\mathbf{f}(x, y)$ is invertible for every (x, y) but $\mathbf{f}$ is not 1-1 on $\mathbb{R}^2$ (Exercise 1).

In the finite dimensional case, the invertibility of $df(x_0)$ implies that both X and Y must be of the same dimension. In the case $X = Y = \mathbb{R}^n$, the inverse function theorem guarantees that a system of n equations in n unknowns

$$y_1 = f_1(x_1, \ldots, x_n),$$
$$\vdots$$
$$y_n = f_n(x_1, \ldots, x_n),$$

can be solved for the x_i in terms of the y_i if we restrict $\mathbf{x}$ and $\mathbf{y}$ to small enough neighborhoods of $\mathbf{x}_0$ and $\mathbf{y}_0$. For example, if $\mathbf{f}(x_1, x_2) = (x_1^3 + x_2^3, x_1^2 + x_2^2)$, then $d\mathbf{f}(1, 2)$ is invertible; so the system of equations,

$$y_1 = x_1^3 + x_2^3,$$
$$y_2 = x_1^2 + x_2^2,$$

has a solution for $\mathbf{x}$ in terms of $\mathbf{y}$ near $(1, 2)$. (We might, however, be hard-pressed to write out such a solution explicitly.)

EXAMPLE 2 (Polar Coordinates). Let D be the subset of $\mathbb{R}^2$ that consists of all pairs (r, θ) with $r > 0$. Define $\mathbf{f} = (x, y): D \to \mathbb{R}^2$ by $\mathbf{f}(r, \theta) = (r \cos \theta, r \sin \theta)$. Then

$$d\mathbf{f}(r, \theta) = \begin{bmatrix} \cos \theta & -r \sin \theta \\ \sin \theta & r \cos \theta \end{bmatrix}.$$

Thus $d\mathbf{f}(r, \theta)$ is invertible at every point and $\mathbf{f}$ is locally invertible at every point. For example, if $V = \{(x, y): x > 0, y > 0\}$, the inverse of $\mathbf{f}$ on V is given by

$$\mathbf{g}(x, y) = (r, \theta) = \left(\sqrt{x^2 + y^2}, \arctan \frac{y}{x} \right). \qquad \square$$

A close relative of the inverse function theorem is the implicit function theorem. In its simplest form, the implicit function theorem gives conditions under which the scalar equation $f(x, y) = 0$ can be solved for the variable y in terms of the variable x. The simple example $f(x, y) = x^2 + y^2 - 1 = 0$ shows that, as in the case of the inverse function theorem, we can at best hope for a local solution. It also indicates the need for an assumption of the form $D_2 f(x, y) \neq 0$.

We shall need the following lemma.

Lemma 3. Let $Z = X \times Y$ and $T_1 \in L(X, Y)$, $T_2 \in L(Y, Y)$. Define $T: Z \to Z$ by $T(x, y) = (x, T_1 x + T_2 y)$. Then $T \in L(Z, Z)$ and T is invertible if, and only if, T_2 is invertible.

Proof. Clearly T is linear and continuous.

First, suppose that T is invertible. The map $\Psi: Y \to Z$ defined by $\Psi(y) = (0, y)$ is a linear isometry from Y onto the subspace $\{0\} \times Y$ of Z, and since $\Psi^{-1} T \Psi = T_2$, T_2 is invertible.

Suppose that T_2 is invertible. Then, since the equation $T(x, y) = (h, k)$ is equivalent to the pair of equations $x = h$, $T_1 x + T_2 y = k$, and these equations have the solution $x = h$, $y = T_2^{-1}(k - T_1 h)$, it follows that T is 1-1 and onto. Since both T_1 and T_2^{-1} are continuous, this formula for $T^{-1}(h, k)$ shows that T^{-1} is continuous. $\square$

Theorem 4 (Implicit Function Theorem). Let X, Y be Banach spaces with $D \subseteq X \times Y$ open and $(x_0, y_0) \in D$. If $f: D \to Y$ is $\mathscr{C}^1$ on D, $f(x_0, y_0) = 0$, and $d_2 f(x_0, y_0)$ is invertible, then there exist open sets $W \subseteq X \times Y$ and $U \subseteq X$ with $(x_0, y_0) \in W$, $x_0 \in U$ and a $\mathscr{C}^1$ function $\varphi: U \to Y$ such that $\varphi(x_0) = y_0$ and $\forall x \in U$, $(x, \varphi(x)) \in W$ with $f(x, \varphi(x)) = 0$. Moreover,

$$d\varphi(x_0) = -[d_2 f(x_0, y_0)]^{-1} d_1 f(x_0, y_0) \tag{6}$$

Proof. Define $F: D \to X \times Y$ by $F(x, y) = (x, f(x, y))$. Then $f(x, y) = 0$ is equivalent to $F(x, y) = (x, 0)$.

If $dF(x_0, y_0)$ is invertible, then by the inverse function theorem, F has a local inverse. That is, there exist open sets W and V in $X \times Y$ such that $F(W) = V$, $(x_0, y_0) \in W$, $(x_0, 0) \in V$, F is 1-1 on W, and there exists a $\mathscr{C}^1$ map $\Psi: V \to W$ that is a local inverse for F. Then $F(\Psi(x, 0)) = (x, 0)$ or $f(\Psi(x, 0)) = 0$ for $x \in U = \{x: (x, 0) \in V\}$. Thus, if $J: X \to X \times Y$, $P: X \times Y \to Y$ are defined by $J(x) = (x, 0)$ and $P(x, y) = y$, then the function $\varphi = P \circ \Psi \circ J$ defined on the open subset U of X satisfies $(x, \varphi(x)) \in W$ and $f(x, \varphi(x)) = 0$ for $x \in U$; since Ψ is $\mathscr{C}^1$, the function φ is also $\mathscr{C}^1$.

It remains to show that $dF(x_0, y_0)$ is invertible. Since

$$dF(x_0, y_0)(h, k) = (h, df(x_0, y_0)(h, k))$$
$$= (h, d_1f(x_0, y_0)(h) + d_2f(x_0, y_0)(k)),$$

Lemma 3 implies that $dF(x_0, y_0)$ is invertible.

Finally, from the chain rule and Proposition 29.16,

$$d_1F(x, \varphi(x)) + d_2f(x, \varphi(x)) \, d\varphi(x) = 0.$$

Since $d_2f(x_0, y_0)$ is invertible, this gives formula (6).   □

If $X = \mathbb{R}^n$ and $Y = \mathbb{R}^m$ and the function $\mathbf{f} = (f_1, \ldots, f_m)$ is written out in its components

$$f_1(x_1, \ldots, x_n, y_1, \ldots, y_m) = 0,$$
$$\vdots$$
$$f_m(x_1, \ldots, x_n, y_1, \ldots, y_m) = 0,$$

then the conclusion of Theorem 4 can be viewed as giving conditions for solving for the variables $\mathbf{y}$ in terms of variables $\mathbf{x}$. Theorem 4 guarantees that such a solution exists for $\mathbf{x}$ near $\mathbf{x}_0$, provided the Jacobian matrix

$$J = \begin{bmatrix} D_{n+1}f_1 \cdots D_{n+m}f_1 \\ D_{n+1}f_m \cdots D_{n+m}f_m \end{bmatrix}$$

is nonsingular at $(\mathbf{x}_0, \mathbf{y}_0)$, that is, if the Jacobian

$$\frac{\partial(f_1, \ldots, f_m)}{\partial(y_1, \ldots, y_m)}(\mathbf{x}_0, \mathbf{y}_0) = \det J(\mathbf{x}_0, \mathbf{y}_0)$$

is nonzero. For example, if

$$f_1(x_1, x_2, x_3, y_1, y_2) = 2e^{y_1} + y_2 x_1 - 4x_2 + 3,$$
$$f_2(x_1, x_2, x_3, y_1, y_2) = y_2 \cos y_1 - 6y_1 + 2x_1 - x_3,$$

then $\mathbf{f}(\mathbf{x}_0, \mathbf{y}_0) = 0$ for $\mathbf{x}_0 = (3, 2, 7)$, $\mathbf{y}_0 = (0, 1)$. Since

$$d_2\mathbf{f}(x, y) = \begin{bmatrix} D_4f_1(x, y) & D_5f_1(x, y) \\ D_4f_2(x, y) & D_5f_2(x, y) \end{bmatrix} = \begin{bmatrix} 2e^{y_1} & x_1 \\ -y_2 \sin y_1 - 6 & \cos y_1 \end{bmatrix}$$

we obtain

$$d_2 \mathbf{f}(\mathbf{x}_0, \mathbf{y}_0) = \begin{bmatrix} 2 & 3 \\ -6 & 1 \end{bmatrix}.$$

Since $\det d_2 \mathbf{f}(\mathbf{x}_0, \mathbf{y}_0) = 20 \neq 0$, $d_2 \mathbf{f}(\mathbf{x}_0, \mathbf{y}_0)$ is invertible and there is a neighborhood about $(3, 2, 7)$ in which the y_i can be solved for the x_i.

We give an application of the implicit function theorem to show the continuous dependence of solutions to Initial Value Problems (IVP) in ordinary differential equations on the initial data of the problem. Let $f: [-\delta, \delta] \times \mathbb{R} \to \mathbb{R}$ be such that f and $D_2 f$ are continuous and bounded. Consider the IVP

$$\varphi'(t) = f(t, \varphi(t)), \qquad -\delta \leq t \leq \delta,$$
$$\varphi(0) = y \tag{7}$$

or, equivalently, the integral equation $\varphi(t) = y + \int_0^t f(s, \varphi(s)) \, ds$. Let $X = \mathscr{C}^1[-\delta, \delta]$ and equip $\mathscr{C}^1$ with the norm, $\| \ \|_{\infty,1}$, so that X is a Banach space. Define $F: \mathbb{R} \times X \to X$ by

$$F(y, \varphi)(t) = \varphi(t) - y - \int_0^t f(s, \varphi(s)) \, ds.$$

Now $d_2 F$ exists at any (y, φ) and is given by

$$d_2 F(y, \varphi)(z, \psi)(t) = \psi(t) - \int_0^t D_2 f(s, \varphi(s)) \psi(s) \, ds \tag{8}$$

(Example **29**.6 and Exercise **29**.3). Formula (8) shows that $d_2 F$ is continuous. Also, $d_1 F$ clearly exists with $d_1 F(y, \varphi)(z, \psi) = -z$ so that $d_1 F$ is $\mathscr{C}^1$.

We next claim that $d_2 F(0, \varphi)$ is invertible. For convenience, put $G = d_2 F(0, \varphi)$. Let $h \in X$ and consider the equation $G(\psi) = h$. This is equivalent to the linear first order differential equation

$$\psi'(t) - D_2 f(t, \varphi(t)) \psi(t) = h'(t), \qquad \psi(0) = h(0).$$

This equation has the solution

$$G^{-1}(h)(t) = \exp\left[-\int_0^t D_1 f(s, \varphi(s)) \, ds \right]$$
$$\times \left(\int_0^t h'(s) \exp\left[\int_0^s D_1 f(x, \varphi(x)) \, dx \right] ds + h(0) \right) \tag{9}$$

and (9) implies that G has a continuous inverse, G^{-1}.

If we fix y_0 and let φ_{y_0} be the unique solution of the IVP (7) with initial condition $\varphi(0) = y_0$ that is valid on the interval $-\delta \leq t \leq \delta$ (Example **23**.3), then $F(y_0, \varphi_{y_0}) = 0$, and the hypotheses of the implicit function theorem are satisfied. Thus, $\exists$ a $\mathscr{C}^1$ map Ψ defined on an open neighborhood I of y_0 such that $\Psi: I \to X$ and $F(y, \Psi(y)) = 0$. The function $\Psi(y)$ is a solution to the IVP (7) and the continuity of the map $y \to \Psi(y)$ means that the solution of the IVP depends continuously on the initial data y.

EXERCISES 30

1. Let $f(x_1, x_2) = (e^{x_1} \cos x_2, \, e^{x_1} \sin x_2)$. Show that $df(x_1, x_2)$ is invertible $\forall (x_1, x_2)$, but f is not 1-1 on $\mathbb{R}^2$. If $\mathbf{x}_0 = (0, \pi/4)$ and $\mathbf{y}_0 = f(\mathbf{x}_0)$, compute a formula for the inverse of f in a neighborhood of y_0.

2. Discuss the invertibility of the transformations

$$f(x_1, x_2) = \left(x_1^2 - x_2^2, \, x_1 x_2\right) \qquad \text{and} \qquad g(x_1, x_2) = \left(x_1^2 + x_2^2, \, x_1 x_2\right).$$

3. Let $f: \mathbb{R} \to \mathbb{R}$ be defined by

$$f(t) = \begin{cases} t + 2t^2 \sin \dfrac{1}{t}, & t \neq 0, \\[2mm] 0, & t = 0. \end{cases}$$

Show that $df(0)$ is 1-1 and onto (invertible), and that f has no inverse near $t = 0$. Why doesn't this contradict Theorem 1?

4. Let D be the subset of $\mathbb{R}^3$ consisting of those points with spherical coordinates (ρ, θ, φ) with $\rho > 0$. Let $f = (x, y, z): D \to \mathbb{R}^3$ be given by $f(\rho, \theta, \varphi) = (\rho \cos \varphi, \rho \sin \varphi \sin \theta, \rho \sin \varphi \cos \theta)$. Show that f is locally invertible when φ is not an integer multiple of π. If

$$V = \{(x, y, z): x, y, z > 0\},$$

find the inverse of f in V.

5. Show that the map $f(x_1, x_2) = (x_1^3, x_2^3)$ is such that $df(0,0)$ is not invertible, but that f has an inverse φ. What can you conclude about differentiability of φ at $(0,0)$?

6. Can the system

$$y_1 = x_1 + x_1 x_2 x_3,$$

$$y_2 = x_2 + x_1 x_2,$$

$$y_3 = x_3 + 2x_1 + 3x_3^2,$$

be solved for $\mathbf{x}$ in terms of $\mathbf{y}$ near $(0, 0, 0)$?

7. Let $\mathbf{f}(x_1, x_2, y_1, y_2) = (x_1 y_2 + x_2 y_1 - 1, x_1 x_2 - y_1 y_2)$. Then $\mathbf{f}(\mathbf{x}_0, \mathbf{y}_0) = 0$ if $\mathbf{x}_0 = (1, 0)$, $\mathbf{y}_0 = (0, 1)$. Show that $\mathbf{y}$ can be solved for $\mathbf{x}$ near $\mathbf{x}_0$, $\mathbf{y} = \boldsymbol{\varphi}(\mathbf{x})$. Use the formula in Theorem 3 to evaluate $d\boldsymbol{\varphi}(\mathbf{x}_0)$.

8. Can the equations

$$\begin{cases} x^2 + \tfrac{1}{2}y^2 + z^3 - z^2 - \tfrac{3}{2} = 0, \\ x^3 + y^3 - 3y + z + 3 = 0, \end{cases}$$

be solved for y and z in terms of x in a neighborhood of $(-1, 1, 0)$?

9. Can the equation $xy - z \ln y + e^{xz} - 1 = 0$ be solved for z in terms of x and y near $(0, 1, 1)$? How about for y in terms of x and z?

Bibliography

Apostol, T. M., *Mathematical Analysis*, Second Edition, Addison-Wesley, Reading, Mass., 1974.

Avriel, M., *Nonlinear Programming*, Prentice-Hall, Englewood Cliffs, N. J., 1976.

Bartle, R. G., *The Elements of Real Analysis*, Wiley, New York, 1976.

Bliss, G. A., *Calculus of Variations*, Mathematical Association of America, LaSalle, Ill. 1925.

Boas, R. P., "Counterexamples to L'Hospital's Rule," *American Mathematical Monthly*, *93* (1986), 644–645.

Boas, R. P., *Primer of Real Functions*, Math. Association of America, Rahway, N. J., 1960.

Buck, R. C. (Ed.), *Studies in Modern Analysis*, Vol. 1, Math. Association of America, Englewood Cliffs, N. J., 1962.

Drager, L. D. and R. L. Foote, "The Contraction Mapping Lemma and the Inverse Function Theorem in Advance Calculus," *American Mathematical Monthly*, *93*, (1986), 52–54.

Dunford, N. and J. T. Schwartz, *Linear Operators*, Interscience, New York, 1958.

Fulks, W., *Advanced Calculus*, Wiley, New York, 1969.

Goldberg, R. R., *Methods of Real Analysis*, Wiley, New York, 1976.

Henstock, R., "A Riemann Type Integral of Lebesgue Power," *Canadian Journal of Math.*, *20* (1968), 79–87.

Henstock, R., "Definitions of Riemann Type of Variational Integral," *Proc. London Math. Soc.*, *11* (1961), 402–418.

Kurzweil, J., *Nichtabsolut Konvergente Integrale*, Tuebner, Leipzig, 1980.

Kurzweil, J., "Generalized Ordinary Differential Equations and Continuous Dependence on a Parameter," *Czech. Math. Journal*, *82* (1957), 418–449.

Landau, E., *Foundations of Analysis*, Chelsea, New York, 1957.

Lewis, J. T. and O. Shisha, "The Generalized Riemann, Simple, Dominated and Improper Integrals," *Journal of Approximation Theory*, *38* (1983), 192–199.

Lord, N. J., "The Irrationality of *e* and Others," *Math Gazette*, *69* (1985), 213–214.

Marsden, J. E., *Elementary Classical Analysis*, Freeman, San Francisco, 1974.

Mawhin, J., *Introduction à L'Analyse*, Université de Louvain, Louvain, 1984.

McLeod, R. M., *The Generalized Riemann Integral*, Mathematical Association of American, Providence, R. I., 1980.

McShane, E. J., *Unified Integration*, Academic Press, New York, 1983.

McShane, E. J., "The Lagrange Multiplier Rule," *American Mathematical Monthly*, *80* (1973), 922–925.

McShane, E. J., "A Unified Theory of Integration," *American Mathematical Monthly*, *80* (1973), 349–359.

McShane, E. J. and T. Botts, *Real Analysis*, Van Nostrand, Princeton. N. J., 1959.

Meyerson, M. D., "Every Power Series Is a Taylor Series," *American Mathematical Monthly*, *88* (1981), 51–52.

Munroe, M. E., *Measure and Integration*, Addison-Wesley, Reading, Mass., 1953.

Nijenhuis, A., "Strong Derivatives and Inverse Mappings," *American Mathematical Monthly*, *81* (1974), 969–980.

Parks, A. E., "π, *e* and Other Irrational Numbers," *American Mathematical Monthly*, *93* (1986), 722–723.

Protter, M. H. and C. B. Morrey, *A First Course in Real Analysis*, Springer-Verlag, New York, 1977.

Rickert, N. W., "A Calculus Counterexample," *American Mathematical Monthly*, *75* (1968), 166.

Rudin, W., *Principles of Mathematical Analysis*, Third Edition, McGraw-Hill. New York, 1976.

Schaefer, P., Sum-Preserving Rearrangements of Infinite Series, *American Mathematical Monthly*, *88* (1981), 33–40.

Spring, O., "On the Second Derivative Test for Constrained Local Extrema," *American Mathematical Monthly*, *92* (1985), 641–643.

Taylor, A. E. and W. R. Mann, *Advanced Calculus*, Wiley, New York, 1983.

Wagner, C. H., "A Generic Approach to Iterative Methods," *Mathematics Magazine*, *55* (1982), 259–273.

Wen, L., "A Space Filling Curve," *American Mathematical Monthly*, *90* (1983), 283.

Wilder, R. L., "Evolution of the Topological Concept of Connected," *American Mathematical Monthly*, *85* (1978), 720–726.

Notation Index

Subject Index